Lecture Notes in Computer Science 7240

Commenced Publication in 1973
Founding and Former Series Editors:
Gerhard Goos, Juris Hartmanis, and Jan van Leeuwen

W0246086

Hwanjo Yu   Ge Yu   Wynne Hsu
Yang-Sae Moon   Rainer Unland
Jaesoo Yoo (Eds.)

Database Systems
for Advanced Applications

17th International Conference, DASFAA 2012
International Workshops:
FlashDB, ITEMS, SNSM, SIM^3, DQDI
Busan, South Korea, April 15-18, 2012
Proceedings

 Springer

Volume Editors

Hwanjo Yu
Pohang University of Science and Technology (POSTECH)
Pohang 790-784, Gyeongbuk, South Korea
E-mail: hwanjoyu@postech.ac.kr

Ge Yu
Northeastern University, Shenyang 110004, Liaoning Province, China
E-mail: yuge@mail.neu.edu.cn

Wynne Hsu
National University of Singapore, Singapore 119077, Singapore
E-mail: whsu@comp.nus.edu.sg

Yang-Sae Moon
Kangwon National University, Chuncheon 200-701, Kangwon, South Korea
E-mail: ysmoon@kangwon.ac.kr

Rainer Unland
University of Duisburg-Essen, 45117 Essen, Germany
E-mail: rainer.unland@icb.uni-due.de

Jaesoo Yoo
Chungbuk National University, Cheongju 361-763, Chungbuk, South Korea
E-mail: yjs@chungbuk.ac.kr

ISSN 0302-9743   e-ISSN 1611-3349
ISBN 978-3-642-29022-0   e-ISBN 978-3-642-29023-7
DOI 10.1007/978-3-642-29023-7
Springer Heidelberg Dordrecht London New York

Library of Congress Control Number: 2012933762

CR Subject Classification (1998): H.3, H.4, I.2, C.2, H.2, H.5, D.2

LNCS Sublibrary: SL 3 – Information Systems and Application, incl. Internet/Web
and HCI

Typesetting: Camera-ready by author, data conversion by Scientific Publishing Services, Chennai, India

Printed on acid-free paper

Springer is part of Springer Science+Business Media (www.springer.com)

Preface

Database Systems for Advanced Applications (DASFAA) is an annual international database conference, located in the Asia-Pacific region, which showcases state-of-the-art R&D activities in database systems and their applications. It provides a forum for technical presentations and discussions among database researchers, developers, and users from academia, business, and industry. DASFAA 2012, the 17th in the series, was held in Busan during April 15–18, 2012. Among the proposals submitted in response to the call-for-workshops, we carefully selected five workshops, each focusing on a specific area that contributes to the main themes of the DASFAA conference. This volume contains the papers accepted for these five workshops that were held in conjunction with DASFAA 2012. They are:

- The Second International Workshop on Flash-Based Database Systems (FlashDB 2012)
- The First International Workshop on Information Technologies for Maritime and Logistics (ITEMS 2012)
- The Third International Workshop on Social Networks and Social Web Mining (SNSM 2012)
- The Second International Workshop on Spatial Information Modeling, Management and Mining (SIM3 2012)
- The 5th International Workshop on Data Quality in Data Integration Systems (DQDI 2012)

We are very grateful to the workshop organizers for their tremendous effort in soliciting papers, selecting papers by peer review, and preparing attractive programs. We asked all workshops to follow a rigid paper-selection process, including the procedure to ensure that any Program Committee members were excluded from the review process of any paper they were involved in. A requirement about the overall paper acceptance rate was also imposed on all the workshops. We would like to express our appreciation to Sang-goo Lee, Bonghee Hong, Won Suk Lee, Wook-Shin Han, Kyuseok Shim, and many other people for their support in the workshop organization. Our thanks also go to Rainer Unland, Jaesoo Yoo, and Yang-Sae Moon for their hard work in compiling this proceedings volume.

April 2012

Hwanjo Yu
Ge Yu
Wynne Hsu

FlashDB 2012 Workshop Organizers' Message

Recently, new storage media such as flash memory and phase change memory have been developed very quickly, bringing big challenges to the architecture of computer systems as well as the design of system software. In particular, NAND flash (either SLC- or MLC-based) in the form of solid state disks (SSDs) has been an alternative to traditional magnetic disks, both in the home-user environment and in the enterprise computing environment, owing to its shock-resistance, low power consumption, non-volatility, and high I/O speed. The special features of flash memory and other new storage media impose new challenges to traditional data management technologies. As a result, traditional database architectures and algorithms designed for magnetic-disk-based storage fail to utilize new storage media efficiently. Meanwhile, the new characteristics of modern storage media, such as not-in-place update and asymmetric read/write/erase latencies of flash memory, also bring great challenges in optimizing database performance, by using new query processing algorithms, indexes, buffer management schemes, and new transaction processing protocols. Consequently, how to exploit the characteristics of flash memory and other new storage media has become an important topic of database systems research. In order to make use of the characteristics of flash memory and other new storage media, the data management community needs to rethink the algorithms and technical issues in magnetic-disk-oriented database systems and make them be adapted to the advances in the underlying storage infrastructure.

The Second International Workshop on Flash-Based Database Systems (FlashDB 2012) was held on April 15, 2012, in Busan, South Korea, in conjunction with DASFAA 2012. The overall goal of the workshop was to bring together researchers who are interested in optimizing database performance on flash memory or other new storage media by designing new data management techniques and tools.

The workshop attracted submissions from Germany, Poland, France, Iceland, Korea, and China. All submissions were peer reviewed by at least three Program Committee members to ensure that high-quality papers were selected. On the basis of the reviews, the Program Committee selected five submissions as full papers and two submissions as short papers for inclusion in the workshop proceedings. The final program of the workshop also consisted of one invited talk from Indilinx Inc.

The Program Committee of the workshop comprised 17 experienced researchers and experts from both industry and academia. We would like to thank the valuable contribution of all the Program Committee members during the peer-review process. Also, we would like to acknowledge the DASFAA 2012 Workshop Co-chairs for their great support in ensuring the success of FlashDB 2012, and the support from the Natural Science Foundation of China (No. 60833005) and Singapore A-Star DSI.

April 2012   Bingsheng He
                                                             Jianliang Xu
                                                           Xiaofeng Meng
                                                              Lihua Yue

ITEMS 2012 Workshop Organizers' Message

According to the International Maritime Organization, more than 90% of the global trade volume, whether it is oil and gas, bulk or containerized cargo, is carried by sea. To this volume, leisure, passenger, and military shipping must also be added to account for the traffic. The stakeholders in the maritime and logistics industry are numerous. Information technologies (IT) for maritime and logistics is a specialized area that deals with important aspects of IT, including but not limited to simulation of maritime and logistics systems; data analytics; acquisition, processing, and management of maritime data; robotics, Web technologies, artificial intelligence, and decision support systems for safety and security.

The First International Workshop on Information Technologies for Maritime and Logistics (ITEMS 2012) was held on April 15, 2012, in Busan, South Korea, in conjunction with DASFAA 2012. The overall goal of the workshop was to bring together researchers, developers, practitioners, and users from academia, business, and industry who are interested in all the important aspects of information technologies dedicated to the maritime and logistics industry.

The workshop attracted submissions from China, France, India, and South Korea. All submissions were peer reviewed by at least three Program Committee members to ensure that high-quality papers were selected. For this first workshop the Program Committee selected two papers for inclusion in the workshop proceedings. The final program of the workshop also consisted of an invited talk on "IT for Sustainable Networks and Logistics" from SAP Singapore.

The Program Committee of the workshop comprised 15 experienced researchers. We would like to thank the valuable contribution of all the Program Committee members during the peer-review process. Also, we would like to acknowledge the DASFAA 2012 Workshop Co-chairs for their great support in ensuring the success of ITEMS 2012, and the support that we received from the Centre for Maritime Studies at the National University of Singapore.

April 2012

Stephane Bressan
Bonghee Hong
Baljeet Malhotra

SNSM 2012 Workshop Organizers' Message

Today the emergence of Web-based communities and hosted services, such as social networking sites, wikis and folksonomies, brings in tremendous freedom of Web autonomy and facilitates collaboration and knowledge sharing between users. Along with the interaction between users and computers, social media are rapidly becoming an important part of our digital experience, ranging from digital textual information to diverse multimedia forms. These aspects and characteristics constitute the core of second generation of the Web.

A prominent challenge lies in modeling and mining this vast pool of data to extract, represent, and exploit meaningful knowledge and to leverage structures and dynamics of emerging social networks residing in the social media. Social networks and social media mining combines data mining with social computing as a promising new direction of research and offers unique opportunities for developing novel algorithms and tools ranging from text and content mining to link mining.

The Third International Workshop on Social Networks and Social Web Mining (SNSM 2012) was held on April 15, 2012, in Busan, Korea, in conjunction with DASFAA 2012. The overall goal of the workshop was to bring together academic researchers and industrial practitioners from computer science, information systems, statistics, sociology, behavioral science, and organization science disciplines, and provide a forum for recent advances in the field of social networks and social media, from the perspectives of data management and mining.

The workshop attracted 16 submissions from Canada, France, Finland, China, Japan, Korea, India, Bangladesh, and Hong Kong. All submissions were peer reviewed by at least three Program Committee members to ensure that high-quality papers were selected. On the basis of the reviews, the Program Committee selected ten papers for inclusion in the workshop proceedings plus an invited paper.

The Program Committee of the workshop consisted of 45 experienced researchers and experts. We would like to thank the valuable contribution of all the Program Committee members during the peer-review process. Also, we would like to acknowledge the DASFAA 2012 Workshop Co-chairs for their great support in ensuring the success of SNSM 2012. Last but not least, we would like to acknowledge all the authors who submitted very interesting and impressive papers from their recent work.

April 2012

Guandong Xu
Lin Li
Wookey Lee

SIM3 2012 Workshop Organizers' Message

Spatial data exists pervasively in various information systems and applications. The unprecedented amount of spatial data that has been amassed and that is being produced in an increasing speed via various facilities, such as sensors, GPS receivers, smart phones, and remote sensing, calls for extensive, deep, and sustaining research on spatial information modeling, management, and mining. In the past decade, we witnessed increasing research interests in these areas from the database, data mining, and geographic information systems (GIS) communities.

Following the success of the First International Workshop on Spatial Information Modeling, Management and Mining (SIM3 2011) held in conjunction with DASFAA 2011, the Second International Workshop on Spatial Information Modeling, Management and Mining (SIM3 2012) held in conjunction with DASFAA 2012, stuck to the tradition that brings together researchers, developers, users, and practitioners carrying out research and development in spatial information modeling, management, and mining, thereby fostering discussions in all aspects of these research areas and providing a forum for original research contributions and practical experiences of spatial information modeling, management, and mining to highlight future trends in these topics.

The workshop received ten submissions. Through a careful review round by the Program Committee, four full papers and one short paper were selected for presentation at the workshop and inclusion in the proceedings. These accepted papers cover various topics of spatial indexing, query processing, and mining. Concretely, He Ma et al. proposed a novel multi-level grid-index and a number of related query types that facilitate application access to augmented, large-scale Geo-tagged video repositories; Ying Fang et al. developed a novel indexing method, History TPR*-tree (HTPR*-tree in short), which can support not only predictive queries but also partial history queries involved from the most recent update instant of each object to the last update time of all objects; Soo Kang et al. introduced a linear bi-directional broadcast indexing scheme for sensor networks in road environments; Wengen Li et al. presented an algorithm for evaluating spatial keyword queries under the MapReduce framework; and Biying Tan et al. proposed an effective method to detect high-risk geographical zones and potentially infected neighbors by capturing the significant changes in the infectious disease monitoring data.

The workshop had an invited talk, given by Gao Cong from the Nanyang Technological University, Singapore. In this talk, Dr. Cong introduced the recent advances and his research group's achievements in spatial keyword query processing.

A successful workshop requires a lot of effort from different people. First, we would like to thank the authors for their contributions and the Program Committee members for their reviewing. We also appreciate the DASFAA 2012 Workshop Co-chairs for their excellent coordination. Finally, we would like to thank the local Organizing Committee for its wonderful arrangements.

April 2012   Jihong Guan
                                                                 Xin Wang

DQDI 2012 Workshop Organizers' Message

Data integration has been a subject of intense research and development for over three decades. Basically, the goal of a data integration system is to provide a uniform interface to a multitude of data sources. Difficulties in overcoming the schematic, syntactic, and semantic differences of data from multiple autonomous and heterogeneous sources are well recognized, and have resulted in a data integration market valued at US$1.34 billion and growing. With the phenomenal increase in the scale and disparity of data, the problems associated with data integration have increased dramatically. A fundamental aspect of user satisfaction from an integration system is the data quality. Industry reports indicate that expensive data integration initiatives stemming from migrations, mergers, legacy upgrades etc., succeed in achieving a common technology platform, but are rejected by the user communities due to the presence (or exposure) of poor data quality. Poor data quality is known to compromise the credibility and efficiency of commercial as well as public endeavors. Several developments from industry as well as academia have contributed significantly toward addressing the problem.

The DQDI workshop (previously titled MCIS) provides a forum to bring together diverse researchers and make a consolidated contribution to new and extended methods to address the challenges of data quality in a collaborative setting. Topics covered by the workshop include data integration, linkage; consistency checking, data profiling, and measurement; methods for data transformation, reconciliation, consolidation; among others. Following the success of this workshop in 2008 in New Delhi, India, 2009 in Brisbane, Australia, 2010 in Tsukuba, Japan, and 2011 in Hong Kong, China, the 5th DQDI was held on April, 2012, in Busan, South Korea, in conjunction with the 17th International Conference on Database Systems for Advanced Applications (DASFAA 2012). This year, the DQDI workshop attracted submissions from Australia, China, Russia, South Korea, and Italy. All submissions were peer reviewed by at least three international reviewers to ensure that high-quality papers were selected. On the basis of technical merit, originality, significance, and relevance to the workshop, the Program Committee decided on five papers to be included in the workshop proceedings (acceptance rate 50%).

The workshop Program Committee consisted of experienced researchers and experts in the area of data analysis and management. We would like to acknowledge the valuable contribution of all the Program Committee members during the peer-review process. Also, we would like to show our gratitude to the DASFAA 2012 Workshop Co-chairs for their great support in ensuring the success of DQDI 2012.

April 2012

Ke Deng
Xiaochun Yang
Shazia Sadiq
Xiaofang Zhou

DASFAA 2012 Workshop Organization

Workshop Co-chairs

Hwanjo Yu   POSTECH, South Korea
Yu Ge   Northeastern University, China
Wynne Hsu   National University of Singapore, Singapore

Publication Co-chairs

Rainer Unland   University of Duisburg-Essen, Germany
Jaesoo Yoo   Chungbuk National University, South Korea
Yang-Sae Moon   Kangwon National University, South Korea

Second International Workshop on Flash-Based Database Systems (FlashDB 2012)

Workshop Co-organizers

Bingsheng He   Nanyang Technological University, China
Jianliang Xu   Hong Kong Baptist University, China

Program Co-chairs

Xiaofeng Meng   Renmin University of China, China
Lihua Yue   University of Science and Technology of China,
                        China

Program Committee

Phillipe Bonnet   IT University of Copenhagen, Denmark
Shimin Chen   HP Labs, China
Tae-Sun Chung   Ajou University, South Korea
Bin Cui   Peking University, China
Theo Härder   University of Kaiserslautern, Germany
Bin He   IBM Almaden Research, USA
Peiquan Jin   University of Science and Technology of China,
                        China
Sang-Wook Kim   Hanyang University, South Korea
Ioannis Koltsidas   IBM Zurich Research, Switzerland
Sang-Won Lee   Sungkyunkwan University, South Korea
Qiong Luo   HKUST, China

Suman Nath	Microsoft Research, USA
Dong-Joo Park	Soongsil University, South Korea
Vijayan Prabhakaran	Microsoft Research, USA
Ha-Joo Song	Pukyoung University, South Korea
Sivan Toledo	Tel Aviv University, Italy
Ming Wu	Microsoft Research Asia, China

First International Workshop on Information Technologies for Maritime and Logistics (ITEMS 2012)

Workshop Co-organizers

Stephane Bressan	National University of Singapore, Singapore
Bonghee Hong	Pusan National University, South Korea
Baljeet Malhotra	National University of Singapore, Singapore

Program Committee

Alex Aravind	University of Northern British Columbia, Canada
Omar Boucelma	Aix-Marseille University, France
Elena Camossi	JRC, ISPRA, Italy
Che Sau Chang	National University of Singapore, Singapore
Li Cheng	A*STAR, Singapore
Christophe Claramunt	Naval Academy Research Institute, France
Isabel F. Cruz	University of Illinois at Chicago, USA
Mark Goh	National University of Singapore, Singapore
Dino Ienco	UMR TETIS, France
Ryszard J. Katulski	Gdansk University of Technology, Poland
Panos Kalnis	KAUST, Saudi Arabia
Hari Krishna Garg	National University of Singapore, Singapore
Joonho Kwon	Pusan National University, South Korea
Robert Laurini	INSA-Lyon, France
Yuxi Li	UESTC, China
Zhanhuai Li	NWPU, China
Anirban Mondal	IIITD, India
Ioanis Nikolaidis	University of Alberta, Canada
Sherif Sakr	NICTA, Australia
Reza Sherkat	University of Hong Kong, China
Ha-Joo Song	Pukyong National University, South Korea
Jacek Stefanski	Gdansk University of Technology, Poland
Maguelonne Teisseire	UMR TETIS, France
Woei Wan Tan	National University of Singapore, Singapore
Osamu Yoshie	Waseda University, Japan
Tingshao Zhu	GUCAS, China

Third International Workshop on Social Networks and Social Web Mining (SNSM 2012)

Workshop Co-organizers

Guandong Xu	Victoria University, Australia
Lin Li	Wuhan University of Technology, China
Wookey Lee	Inha University, South Korea

Program Committee

Nitin Agarwal	University of Arkansas, USA
Toshiyuki Amagasa	University of Tsukuba, Japan
James Bailey	University of Melbourne, Australia
Kevin Chai	IBM Research, China
Ling Chen	UTS, Australia
Wonik Choi	Inha University, South Korea
Soon Ae Chun	City University of New York, USA
Daling Wang	Northeastern University, China
Peter Dolog	Aalborg University, Denmark
Flavius Frasincar	Erasmus University Rotterdam, The Netherlands
Irene Garrigos	University of Alicante, Spain
James Geller	New Jersey Institute of Technology, USA
Yanhui Gu	University of Tokyo, Japan
Hyoil Han	Drexel University, USA
Wonchang Hur	Inha University, South Korea
Seung-won Hwang	POSTECH, South Korea
Deok-Hwan Kim	Inha University, South Korea
Dongsoo Kim	Soongsil University, South Korea
Jinho Kim	Kangwon National University, South Korea
Sang-Wook Kim	Hanyang University, South Korea
Hiroyuki Kitagawa	Tsukuba University, Japan
SangKeun Lee	Korea University, South Korea
Wookey Lee	Inha University, South Korea
Young-Koo Lee	Kyung Hee University, South Korea
Carson Kai-Sang Leung	University of Manitoba, Canada
Wenxin Liang	Dalian University of Technology, China
Mukesh Mohania	IBM Research, India
Yang-Sae Moon	Kangwon National University, South Korea
Satoshi Nakamura	Kyoto University, Japan
Jonghun Park	Seoul National University, South Korea
Tieyun Qian	Wuhan University, China
Sherif Sakr	NICTA, Australia

Munehiko Sasajima	Osaka University, Japan
Myong Keun Shin	SK C&C, South Korea
Kazutoshi Sumiya	University of Hyogo, Japan
Xiaohui Tao	University of Southern Queensland, Australia
Chaokun Wang	Tsinghua University, China
Daling Wang	Northeastern University, China
Jianmin Wang	Tsinghua University, China
Zongda Wu	Wenzhou University, China
Zhenglu Yang	University of Tokyo, Japan
Junjie Yao	Peking University, China
Hwanjo Yu	POSTECH, South Korea
Jianwei Zhang	Kyoto Sangyo University, Japan
Xiuzhen Zhang	RMIT, Australia
Bin Zhao	Shaanxi Normal University, China
Yu Zong	West Anhui University, China
Lei Zou	Peking University, China

Second International Workshop on Spatial Information Modeling, Management and Mining (SIM3 2012)

Workshop Co-organizers

Jihong Guan	Tongji University, China
Xin Wang	University of Calgary, Canada

Program Committee

Michela Bertolotto	University College Dublin, Ireland
Elena Camossi	JRC, ISPRA, Italy
Christophe Claramunt	Naval Academy Research Institute, France
Georg Gartner	Vienna University of Technology, Austria
Liqiang Geng	National Research Council, Canada
Yoshiharu Ishikawa	Nagoya University, Japan
Songnian Li	Ryerson University, Canada
Xiang Li	East China Normal University, China
Eleni Mangina	University College Dublin, Ireland
Wolfgang Reinhardt	Universität der Bundeswehr München, Germany
Markus Schneider	University of Florida, USA
Xiaohua Tong	Tongji University, China
Bin Wang	University of Calgary, Canada
Shuliang Wang	Wuhan University, China
Shuigeng Zhou	Fudan University, China

5th International Workshop on Data Quality in Data Integration System (DQDI 2012)

Workshop Co-organizers

Ke Deng   University of Queensland, Australia
Xiaochun Yang   Northeastern University, China
Shazia Sadiq   University of Queensland, Australia
Xiaofang Zhou   University of Queensland, Australia

Program Committee

Abrar Haider   University of South Australia, Australia
Adam Jatowt   Kyoto University, Japan
Hua Lu   Aalborg University, Denmark
Chaoyi Pang   CSIRO, Australia
Bela Stantic   Griffith University, Australia
John Wang   Griffith University, Australia
Kai Xu   Middlesex University, UK
Yu Zheng   Microsoft Research Asia, China

Table of Contents

Second International Workshop on Spatial Information Modeling, Management and Mining (SIM3 2012)

Spatial Indexing

Spatial Query Processing and Data Mining

Fifth International Workshop on Data Quality in Data Integration System (DQDI 2012)

Commercial SSD Products — Status Quo and Next

Bumsoo Kim

Indilinx Inc.
bumsoo@indilinx.com

Abstract. The speaker will give a short introduction of an SSD company "OCZ" and their products. By showing comprehensive products of one of the fast growing SSD companies, various technical topics and challenges will be touched.

Keywords: Flash Memory, SSD.

H. Yu et al. (Eds.): DASFAA Workshops 2012, LNCS 7240, p. 1, 2012.
© Springer-Verlag Berlin Heidelberg 2012

Improving Database Performance Using a Flash-Based Write Cache

Yi Ou and Theo Härder

University of Kaiserslautern
{ou,haerder}@cs.uni-kl.de

Abstract. The use of flash memory as a write cache for a database stored on magnetic disks has been so far largely ignored. In this paper, we explore how flash memory can be efficiently used for this purpose and how such a write cache can be implemented. We systematically study the design alternatives, algorithms, and techniques for the flash-based write cache and evaluate them using trace-driven simulations, covering the most typical database workloads.

1 Introduction

Flash memory[1] is popularly used in a variety of data storage devices, such as compact flash cards, secure digital cards, flash SSDs, flash PCIe cards, etc., primarily due to its non-volatility and high density. Flash-based storage devices (or flash devices for short) can be manufactured with some interesting properties, such as low-power consumption, small form factor, shock resistance, etc. which make them attractive to a large range of applications.

Due to the physical constraints of flash memory such as erase-before-program and wear-out of memory cells [1], flash devices typically implement an additional layer, the flash translation layer (FTL) [2], on top of the flash memory to support typical host-to-device interfaces.

The function of magnetic disks (HDDs), the currently dominating mass storage devices, relies on mechanical moving parts, which is one of the major threats to the device reliability and typically the bottleneck of the entire system performance. In contrast, flash devices do not contain mechanical moving parts. Therefore, they allow much faster random access (up to two orders of magnitude) and much higher reliability.

Flash memory and flash devices have received a lot of attention from the database research community due to the fact that existing database systems are designed and optimized for HDDs that have quite different performance characteristics than flash devices, e. g., flash SSDs [3].

To exploit the performance potential of flash memory and flash devices in a database environment, flash-specific query processing techniques [4,5], index structures [6,7,8], and buffer management algorithms [9,10,11] have been proposed. Efficient use of flash devices for update propagation [12], logging and recovery [13], and transaction processing [14] has also been studied.

---

[1] We focus on the NAND type flash memory due to its high density and low cost.

H. Yu et al. (Eds.): DASFAA Workshops 2012, LNCS 7240, pp. 2–13, 2012.
© Springer-Verlag Berlin Heidelberg 2012

Modern flash devices are much faster than HDDs, even for random write workloads. At the same time, they are also much more expensive than HDDs in terms of price per unit capacity. Some researchers realized that completely replacing HDDs with flash devices is not a cost-effective solution [15], instead, using flash as an intermediate tier between RAM-based memory and HDDs [16,17] can bridge the access-time gap between the two tiers without introducing much higher cost. However, it is doubtful whether such an abrupt architecture change will be widely accepted in practice, given the trend in industry to keep the whole working set in RAM for maximum performance [18,19].

1.1 Flash-Based Write Cache

In this paper, we study the use of flash as a write cache for databases based on HDDs. The flash-based write cache (or the flash-based second-tier cache) is a page cache layer between the RAM-based buffer pool (or first-tier cache) and the database stored in HDDs (the third tier). The accesses to all three tiers are via page-oriented interfaces, i. e., in units of database pages.

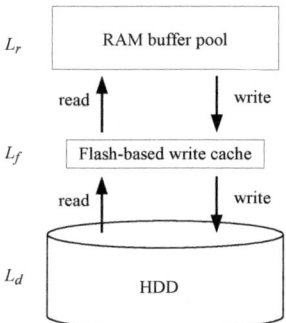

Fig. 1. Database storage system with flash-based write cache

The architecture of a three-tier database storage system are illustrated in Figure 1. It is very similar to the three-layer architecture (3LA) introduced in [17]. To be comparable, we adopt the notation of [17] to denote the RAM-based buffer pool as L_r, the flash-based write cache as L_f, and the third tier as L_d. Their capacities, in number of pages, are denoted as $|L_r|$, $|L_f|$, and $|L_d|$, respectively. As opposed to [17], which assumes that

$$|L_r| \leq |L_f| \leq |L_d| \tag{1}$$

we drop the constraint $|L_r| \leq |L_f|$ and assume that

$$|L_r| \ll |L_d| \text{ and } |L_r| \sim |L_f| \tag{2}$$

Note, according to Equation 2, $|L_r| > |L_f|$ is acceptable.

In practical systems, flash is used in different forms such as flash SSDs or flash PCIe cards. In our system, there is no constraint on which a specific form should be used. Instead, various forms of flash devices are abstracted into *page-oriented devices with flash performance characteristics*. We consider four different costs for device access in our system: the costs of flash read, flash sequential write, flash random write, and the cost of disk access, denoted as C_{fr}, C_{fw}, $C_{f\tilde{w}}$, and C_d, respectively.

In such a configuration, dirty[2] pages evicted from the RAM buffer pool are first written to the flash-based write cache and later propagated to HDDs. Depending on specific replacement policies, the number of disk writes can be substantially reduced in this way (see Section 3). Even as a write cache, L_f should ideally keep the "warm" pages, i.e., pages that are not so hot to be kept in L_r, but warmer than the remaining database pages. Hence, whenever a page request can be satisfied from L_f, a cost benefit of $C_d - C_{fr}$ can be obtained.

There are multiple benefits of using such a write cache: 1. Compared with a volatile write cache, it provides higher reliability and protection against data loss at power failure. 2. Due to the cost advantage of flash to RAM, the write cache can be much larger and potentially more efficient than volatile and expensive battery-backed RAM-based write cache (The same argument in favor of flash applies to energy saving [17]). 3. A dedicated write cache improves the write response time and offloads write workloads from HDDs and can asynchronously propagate them to HDDs, i.e., the read performance can also benefit from the write cache. 4. Potential page hits in the write cache reduce the number of expensive disk writes.

Using flash for a write cache, the wear-out problem of flash cells seems to be a concern. However, with proper wear-leveling techniques, the life time of flash memory becomes quite acceptable. For example, an SLC flash memory module typically has a write endurance of 100,000 program/erase cycles (Write endurance of one million cycles has already been reported [20]). With perfect wear-leveling, i.e., all flash pages are programmed and erased at equal frequency, 10 GB flash memory can have a life span of 27.4 years[3] under a daily write workload of 100 GB (factor 10 of its capacity).

1.2   Contribution

Our major contributions are:

- The use of flash devices as a write cache for a database stored on HDDs has been so far largely ignored. To the best of our knowledge, our work is the first one that considers this usage.
- We systematically study the algorithms and techniques for flash-based write caches and evaluate them using trace-driven simulations, covering the most typical database workloads.

---

[2] In contrast to its use in transactional contexts, we denote modified pages as "dirty", as long as they are not written to disk.

[3] $100000 \times 10/(100 \times 365) = 27.4$.

1.3   Organization

The remainder of this paper is organized as follows: Section 2 discusses related works. Section 3 presents and discusses the algorithms and techniques for the flash-based write cache. Section 4 reports our empirical study. The concluding remarks and future works are presented in Section 5.

2   Related Work

[16] is one of the pioneer works studying flash-aware multi-level caching. The authors identified three page-flow schemes in a three-level caching hierarchy with flash as the mid-tier and proposed flash-specific cost models for those schemes. In contrast, contribution [17] presented a detailed three-tier storage system design and performance analysis. In this study, the experiments have shown for certain range of applications, by reducing the amount of energy-hungry RAM-based memory and using a much larger amount of flash as the mid-tier, that system performance and energy efficiency can be both improved at the same time.

Canim et al. [21] proposed a temperature-aware replacement policy for managing the SSD-based mid-tier, based on the access statistics of disk regions (page groups). In [22], the authors studied three design alternatives of an SSD-based mid-tier, which differ mainly in the way how to deal with dirty pages evicted from the first-tier, e. g., write through or write back.

As a general assumption, all these approaches use a flash-based mid-tier being much larger than the first tier. As a consequence, both clean pages and dirty pages are cached in the mid-tier. In contrast, we focus on a configuration where the flash is used as a write cache, i. e., only dirty pages are cached in the mid-tier. More important, the above mentioned works only consider the cost asymmetry of reads and writes on flash, while the cost asymmetry between random and sequential flash writes are ignored, which can significantly impact the system performance, according to our experiments (see Section 4).

Using flash as a write cache has also been studied by Li et al. [23]. However, in their configuration, the database is completely stored in L_f and no L_d is considered, in contrast to the three-tier system (Figure 1) being studied by us. Their basic idea is to exploit the performance advantage of focused writes over random ones, by directing all the writes generated by L_r to a small logical flash area and reordering the writes so that they can be written back to their actual destinations on the same flash device, but in more efficient write patterns.

The authors of [24] and [25] have taken an approach that is somehow the "opposite" of ours. They consider the use of HDDs as the write cache for flash SSDs, based on the argument that the write performance of HDDs is better than that of some flash SSDs and HDDs don't have the wear-out problem. However, we believe that compared to HDDs, flash devices can be made much more reliable and they have a much higher performance potential. In fact, even the random write performance of many mid-range SSDs is now superior to that of the enterprise HDDs.

3   Algorithms and Techniques

Two design decisions are critical to the performance of flash-based caches:

- When should a page be admitted into the cache? This is specified by cache-admission strategies.
- How should admitted pages be written to the cache? This is specified by cache-writing strategies.

3.1   Cache-Admission Strategies

Following the architecture shown in Figure 1, there are only two cases where pages can be admitted into L_f: 1. *Admit-On-Read (AOR)*, i. e., when the read function of L_f is called; 2. *Admit-On-Write (AOW)*, i. e., when the write function of L_f is called.

The first case happens when a buffer fault occurs for page p in L_r. If p is not found in L_f either, it will be fetched from L_d and forwarded to L_r. After that, the newly fetched p can be admitted into L_f.

The second case happens when a dirty page p is evicted from L_r and the write function on L_f is called to write p back. Conceptually, pages admitted in this case are all dirty pages even if L_f is non-volatile, i. e., they need to be propagated to L_d at the latest, when they are evicted from L_f.

The GLB algorithm discussed in [17] actually uses a third cache-admission strategy, which we call *Admit-On-Eviction* (AOE), because it admits every page (either clean or dirty) evicted from L_r into L_f. However, AOE is not considered in this paper due to two problems: 1. It requires a write operation on L_f even on a cache hit in L_f (because the page currently hit has to be exchanged with a page from L_r); 2. It violates the transparency of L_f, because it requires for L_f an extension of the interface shown in Figure 1.

Cache-admission strategies are orthogonal to replacement polices, although classical second-tier cache algorithms such as LOC [17] and MQ [26] implicitly use AOR. In contrast, AOW is the cache-admission strategy used by our flash-based write cache.

3.2   Cache-Writing Strategies

According to the characteristics of flash memory, we consider two cache-writing strategies: *sequential cache write* (SCW) and *random cache write* (RCW). With SCW, pages are written to the flash media in a strictly sequential fashion (thus each write has a cost of C_{fw}, in contrast to $C_{f\tilde{w}}$ in case of RCW), as illustrated in Figure 2. If the flash-based cache has n pages and the most recently updated cache slot is $curr$, then the next cache slot to be used is given by $next = (curr+1)$ mod n. If an earlier version of a newly admitted page p is already in L_f, it has to be *invalidated*, i. e., if multiple versions of p exist in L_f, only the newest version is valid.

Because updates to the flash media are sequential and the slots are updated with equal frequency, SCW enjoys two advantages: *write performance and wear leveling without the need of an FTL*. However, due to the constraint of strict sequential writes, SCW does not allow much flexibility in the choice of replacement victims, which is always predetermined by the *next* pointer. Even if the page cached at the slot pointed to by *next* is a "warm" page, we have to write it back to the disk in order to make room for the page to be cached.

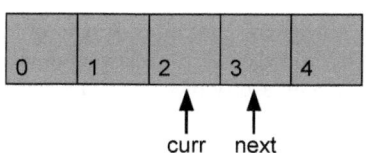

Fig. 2. Sequential cache write. Illustrated is a flash-based write cache with 5 pages. The slot number is marked at the bottom-left corner of the cache slots. Slot 2 is just written, the next slot to be used is slot 3.

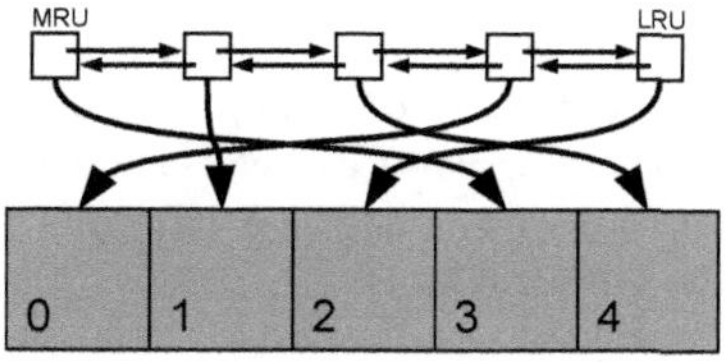

Fig. 3. Random cache write. In the example, the cache slots are ordered by their reference recency (as doubly linked list maintained in volatile memory), which implies the (random) order they are to be overwritten: 2, 0, 4, 1, 3.

Random cache write (RCW), as the name suggests, allows writing to the flash-based cache in a random fashion. If a earlier version of a newly admitted page p is already in L_f, it has to be *overwritten*. The cost of writing to the cache is higher than in the SCW case, and wear-leveling mechanisms such as FTL become necessary. However, in contrast to SCW, RCW does not impose any restriction on the replacement policy. For example, the cache slots can be ordered by their reference recency (as shown in Figure 3) or frequency, upon which the replacement decision can be made, as in the classical buffer management algorithms.

3.3   Track-Aware Algorithms

Seeking is the most expensive mechanical movement made by a magnetic disk – typically a few milliseconds for modern disks. If the information about which page belongs to which track, i.e., the page-to-track mapping function t that maps a logical page number a to its corresponding track number $t(a)$, is known to the flash-based write cache, it is possible to further improve the disk write performance by minimizing the number of seeks. Assume that the mapping function t is given by $t(a) = a/4$, then pages 0, 1, 2, and 3 have the track number 0, pages 4 to 7 have the track number 1, and so on. Cache algorithms making use of this track information are called *track-aware* algorithms.

Virtual track is an in-memory data structure used by the track-aware algorithms. It contains the pointers to the set of pages belonging to the same track. When a free cache slot is needed and all slots are currently occupied, a virtual

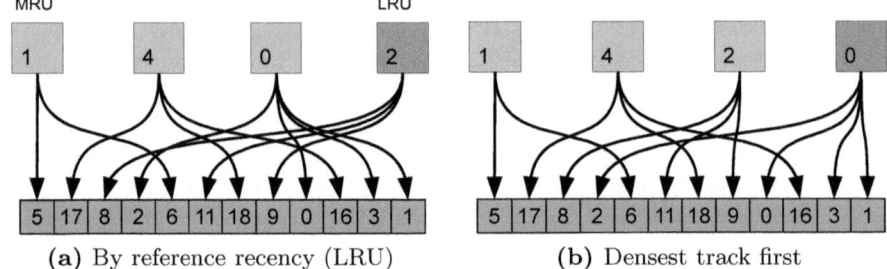

(a) By reference recency (LRU)   (b) Densest track first

Fig. 4. Virtual tracks (depicted as large rectangles with track number in the bottom-left corner) ordered by reference recency (a) and by density (b). Replacement victims are shown in green. Page numbers of the cached pages are depicted as the numbers inside the flash slots.

track can be chosen as the replacement victim, and the m pages pointed to by it are all flushed to disk at once, freeing m pages and requiring only one seek. This technique is called *coalesced flushing (CF)*.

The set of virtual tracks can be ordered by their reference recency, i. e., the time that a page belonging to the track is referenced (i. e., read or updated). When a replacement victim is needed, the least-recently-referenced virtual track is chosen, similar to the LRU replacement policy. Note if the page-to-track mapping function is $t(a) = a$ (one page per track), the described algorithm degenerates to the classical page-oriented LRU (see Figure 3). Therefore, we refer to this algorithm also as LRU whenever there is no ambiguity. An example runtime state of the algorithm is shown in Figure 4a.

Virtual tracks can also be ordered by their density, i. e., by the number of pointers they contain. The replacement victim is then the densest track, as illustrated in Figure 4b. This replacement policy is called *densest track first* (DTF).

Using the track information, we can further reduce the number of seeks using a *piggy-backing (PB)* technique. When the flash-based cache has to serve a read request for a page belonging to track t, a seek to t is very likely inevitable, but the cache can flush all pages pointed to by the corresponding virtual track without enforcing further seeks. Figure 5 illustrates a PB example.

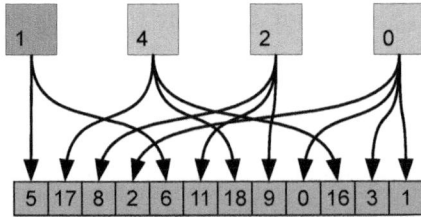

Fig. 5. Piggy-backing. In this sample, a read request for page 7 is to be served, thus a seek to track 1 is needed; in this case, the cache flushes pages 5 and 6 pointed to by virtual track 1.

3.4   Compatibility Matrix

The cache-write strategies SCW and RCW are orthogonal to the track-aware techniques. With SCW, it is also possible to perform coalesced flushing and piggy-backing. However, replacement policies, such as REC and DTF, are only compatible with RCW. Table 1 lists the compatibility relationships between the discussed techniques.

Table 1. Compatibility matrix of the techniques

	SCW	RCW
LRU, DTF, etc.		✓
CF	✓	✓
PB	✓	✓

Table 2. Device access costs

metric	high-end	low-end
C_{fr}	0.105 ms	0.165 ms
C_{fw}	0.106 ms	0.153 ms
$C_{f\tilde{w}}$	0.133 ms	7.972 ms
C_d	4.464 ms	

3.5   Logging and Recovery Implications

The data structures used by the algorithms discussed in this section, e. g., the table mapping logical page numbers to their physical locations, are stored in a small RAM area. To prevent from data loss in case of a system crash, the logical page number should be stored in the page (either in the payload or in the page header) to be able to restore the mapping table by a recovery procedure.

For SCW, it is possible that multiple versions of the same page exist in the cache at recovery. The solution is to store a version number in the page which increments whenever the page is updated. At recovery, only the version with the highest version number generates an entry in the mapping table. The log sequence number (LSN) [27] can serve as the version number. In this case, no extra space is required.

Because of the persistence of the mid-tier, all our 2-step update propagation approaches can be combined with the classical logging methods, the WAL principle (Write Ahead Log), and the recovery-oriented concepts (Atomic/NoAtomic, Steal/NoSteal, Force/NoForce) for mapping database changes from volatile to non-volatile storage [28]. As a consequence, the 2-step mechanism can not only be used to accelerate the propagation of data pages to disk, but can also be applied to log information. Such a practice automatically minimizes transaction latency caused by commit processing, i.e., the 2PC protocol, because all synchronous writes are first directed to the flash layer and not directly to HDDs.

4   Experiments

We implemented our algorithms and used trace-driven simulations to evaluate their performance and study their behavior under various workloads. Our test system consists of a disk layer supporting the block-device interface and collecting disk-access statistics, and a cache layer implementing the introduced algorithms and collecting flash-access statistics. Our traces contain the block-level accesses (physical page requests), collected with the help of the PostgreSQL database engine under TPC-C (100 warehouses) and TPC-H (scale factor: 10) benchmark workloads. The PostgreSQL engine was used to collect the buffer traces (logical page requests), which were then fed to and filtered by an LRU buffer pool of 10,000 pages (i. e., $|L_r| = 10,000$), and the resulting sequence of physical page requests make the block-level traces used in our experiments.

A test program parses the traces and generates block read and write requests, which are then served by the cache layer, either using the cached pages whenever possible or by accessing the disk layer if necessary. According to our three-tier storage architecture in Figure 1, only the bottom two tiers are used in our experiments. The observations made under the TPC-C workload are very similar to those made under the TPC-H workload. This means that our observations are not specific to the workload. For improved clarity, we choose to only report the experimental results collected using the TPC-C trace.

Based on our cost model introduced in Section 1.1 and with the variables n_{fr}, n_{fw}, $n_{f\tilde{w}}$, and n_d as the numbers for the related flash read, flash sequential write, flash random write, and disk accesses, we can define the performance metric *virtual execution time (v)* as:

$$v = n_{fr} \times C_{fr} + n_{fw} \times C_{fw} + n_{f\tilde{w}} \times C_{f\tilde{w}} + n_d \times C_d \qquad (3)$$

The actual values for device access costs, listed in Table 2, are obtained using device benchmarks on two SSDs (high-end and low-end) and a magnetic disk. According to our device benchmark, the high-end SSD has a very good random write performance: its average time of serving a random page write is only 25% slower than that of a sequential page write. In contrast, for the low-end SSD, the random writes are slower than the sequential writes by a factor of 50. The remarkable difference in the device performance characteristics can be explained by substantial differences in the proprietary FTL implementations.

4.1   AOR vs. AOW

We first compare AOR with AOW. Figure 6 shows their virtual execution times relative to the no-cache configuration. The replacement policy in both cases was LRU. The cache size $|L_f|$ was scaled by a factor of 4 from 1,000 to 16,000 pages. In this range, AOR suffers from the problem of duplicate caching. Hence, we can expect that AOW has a better performance. As indicated by the results, AOW clearly outperforms AOR, meaning that caching only dirty pages is quite efficient in our configuration, where the second-tier cache is of comparable size of the first-tier cache ($|L_r| = 10,000$, see also Equation 2). For this reason, this paper focus on AOW, and all algorithms evaluated in the remainder of the section use the AOW strategy.

4.2   SCW vs. RCW

To study the cache-write strategies SCW and RCW, we used the performance metrics both of the high-end and the low-end SSD. The difference in device

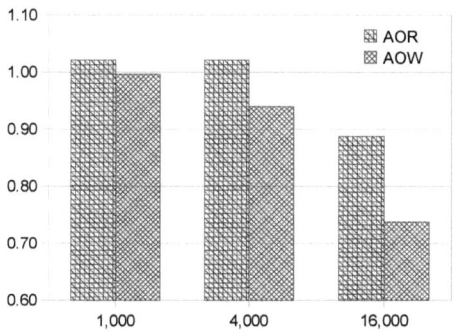

Fig. 6. Virtual execution times of AOR and AOW relative to the no-cache case

performance characteristics are reflected in our test results shown in Figure 7, where the virtual execution times of SCW and RCW with a cache of 4,000 pages relative to the no-cache configuration are shown. The replacement policy used in RCW was LRU. On the high-end SSD, RCW performance was superior because its higher hit ratio compensated the slightly higher cost of random flash writes. On the low-end SSD, however, RCW is much slower than SCW, because the latter only does sequential writes, which can be handled rather efficiently even by the low-end SSD.

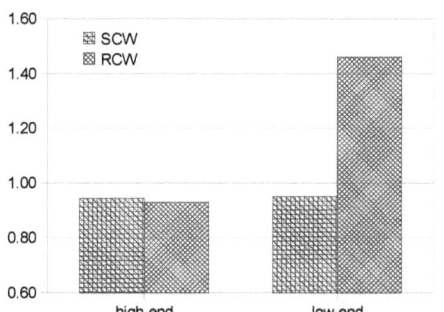

Fig. 7. Virtual execution times relative to the no-cache case

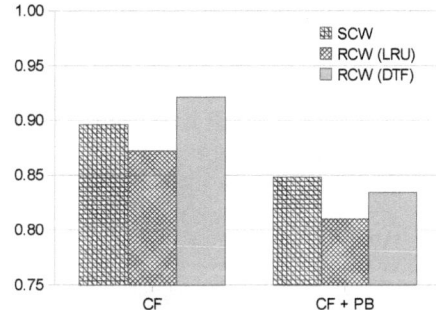

Fig. 8. Number of seeks relative to the non-track-aware LRU

4.3   CF and PB

We used the performance metrics of the high-end SSD to study the track-aware techniques discussed in Sect. 3.3. The goal of those techniques is to minimize the number of disk seeks. Figure 8 shows the numbers of seeks of various configurations relative to LRU (Admit-On-Write, without using track information) for a cache size of 4,000 pages. The page-to-track mapping function t used in the simulation was $t(a) = a/32$.

The results reveal that the track-aware technique CF clearly reduced the number of seeks (up to 12% for LRU), and the combination CF+PB achieved even

more significant improvements (up to 18% for LRU). Another observation is that
the simple replacement policy LRU, combined with the track-aware techniques,
achieved remarkably good performance.

5   Conclusion and Future Work

Based on our experimental results, we can conclude that:

- A small-sized (relative to the RAM buffer pool size) flash-based write cache
  can substantially improve storage system performance.
- Cache-writing strategies can significantly impact system performance, depending on the flash device implementation. For low-end flash devices with
  poor random write performance or raw flash memory, it is better to use SCW
  to handle the wear-leveling problem and random write problem natively in
  the database software.
- The page-to-track information, if available, can be used to further improve
  disk access performance.

The sector-to-track relation on modern disks can be much more sophisticate
than the mapping function used in our simulation. The fact that most enterprise
databases are hosted on RAID also adds complexity requiring further investigation. As future work, we plan to investigate the discussed algorithms and
techniques in a larger variety of configurations and on real devices. We will also
examine the correlation between the strategies and workload characteristics.

Acknowledgement. We are grateful to anonymous referees for valuable comments. This research is supported by the German Research Foundation and the
Carl Zeiss Foundation.

References

1. Gal, E., Toledo, S.: Algorithms and data structures for flash memories. ACM Computing Surveys 37(2), 138–163 (2005)
2. Chung, T.S., Park, D.J., Park, S., Lee, D.H., Lee, S.W., Song, H.J.: A survey of
   flash translation layer. Journal of Systems Architecture 55(5), 332–343 (2009)
3. Bouganim, L., Jónsson, B.T., Bonnet, P.: uFLIP: Understanding flash IO patterns.
   In: CIDR 2009 (2009)
4. Tsirogiannis, D., Harizopoulos, S., et al.: Query processing techniques for solid
   state drives. In: SIGMOD, pp. 59–72. ACM (2009)
5. Li, Y., On, S.T., Xu, J., Choi, B., Hu, H.: DigestJoin: Exploiting fast random reads
   for flash-based joins. In: MDM 2009, pp. 152–161. IEEE (2009)
6. Nath, S., Kansal, A.: FlashDB: dynamic self-tuning database for NAND flash. In:
   Int. Conf. on Information Processing in Sensor Networks, pp. 410–419 (2007)
7. Li, Y., He, B., Luo, Q., Yi, K.: Tree indexing on flash disks. In: ICDE 2009,
   pp. 1303–1306. IEEE (2009)
8. Yin, S., Pucheral, P., Meng, X.: A sequential indexing scheme for flash-based embedded systems. In: EDBT 2009, pp. 588–599. ACM (2009)

9. Park, S., Jung, D., et al.: CFLRU: a replacement algorithm for flash memory. In: CASES, pp. 234–241 (2006)
10. Ou, Y., Härder, T., et al.: CFDC: a flash-aware replacement policy for database buffer management. In: SIGMOD Workshop DaMoN, pp. 15–20 (2009)
11. Jin, P., Ou, Y., Härder, T., Li, Z.: AD-LRU: An efficient buffer replacement algorithm for flash-based databases. Data & Knowledge Eng. 72, 83–102 (2012)
12. Lee, S.W., Moon, B.: Design of flash-based DBMS: an in-page logging approach. In: SIGMOD 2007, pp. 55–66. ACM (2007)
13. Chen, S.: FlashLogging: exploiting flash devices for synchronous logging performance. In: SIGMOD 2009, pp. 73–86. ACM (2009)
14. On, S., Xu, J., Choi, B., Hu, H., He, B.: Flag Commit: Supporting efficient transaction recovery on flash-based DBMSs. IEEE Transactions on Knowledge and Data Engineering (99), 1–1 (2011)
15. Narayanan, D., Thereska, E., et al.: Migrating server storage to SSDs: analysis of tradeoffs. In: EuroSys, pp. 145–158. ACM (2009)
16. Koltsidas, I., Viglas, S.D.: The case for flash-aware multi-level caching. Technical Report (2009)
17. Ou, Y., Härder, T.: Trading Memory for Performance and Energy. In: Xu, J., Yu, G., Zhou, S., Unland, R. (eds.) DASFAA Workshops 2011. LNCS, vol. 6637, pp. 241–253. Springer, Heidelberg (2011)
18. Ousterhout, J., Agrawal, P., Erickson, D., et al.: The case for RAMClouds: scalable high-performance storage entirely in DRAM. ACM SIGOPS Operating Systems Review 43(4), 92–105 (2010)
19. Plattner, H.: A common database approach for OLTP and OLAP using an in-memory column database. In: SIGMOD 2009, pp. 1–2. ACM (2009)
20. Micron Technology, Inc. Micron collaborates with sun microsystems to extend lifespan of flash-based storage, achieves one million write cycles, http://investors.micron.com/releasedetail.cfm?ReleaseID=440650
21. Canim, M., Mihaila, G.A., et al.: SSD bufferpool extensions for database systems. In: VLDB, pp. 1435–1446 (2010)
22. Do, J., DeWitt, D.J., Zhang, D., Naughton, J.F., et al.: Turbocharging DBMS buffer pool using SSDs. In: SIGMOD 2011, pp. 1113–1124. ACM (2011)
23. Li, Y., Xu, J., Choi, B., Hu, H.: StableBuffer: optimizing write performance for dbms applications on flash devices. In: CIKM 2010, pp. 339–348. ACM, New York (2010)
24. Soundararajan, G., Prabhakaran, V., et al.: Extending SSD lifetimes with disk-based write caches. In: USENIX FAST 2010. USENIX Association (2010)
25. Yang, P., Jin, P., Yue, L.: Hybrid Storage with Disk Based Write Cache. In: Xu, J., Yu, G., Zhou, S., Unland, R. (eds.) DASFAA Workshops 2011. LNCS, vol. 6637, pp. 264–275. Springer, Heidelberg (2011)
26. Zhou, Y., Chen, Z., et al.: Second-level buffer cache management. IEEE Transactions on Parallel and Distributed Systems 15(6), 505–519 (2004)
27. Mohan, C., Haderle, D.J., et al.: ARIES: A transaction recovery method supporting fine-granularity locking and partial rollbacks using write-ahead logging. ACM Trans. Database Syst. 17(1), 94–162 (1992)
28. Härder, T., Reuter, A.: Principles of transaction-oriented database recovery. ACM Computing Surveys 15(4), 287–317 (1983)

h-Buffer: An Adaptive Buffer Management Scheme for Flash-Based Storage Devices

Rui Wang, Lihua Yue, Peiquan Jin, and Junjie Wang

School of Computer Science and Technology,
University of Science and Technology of China, 230027, Hefei, China
edenrui@mail.ustc.edu.cn

Abstract. Due to the limitations of flash memory, such as asymmetric I/O latencies and not-in-place update, there are two kinds of buffer replacement algorithms: page-clustered policy and group-clustered policy. That the former one organizes pages at page-level makes it easy to deal with hot pages, but shows a bad performance when the buffer size is large enough. The latter one organizes pages at group-level, which usually ignores the read request from the host as the RAM size inside SSDs (Solid State Disks) is limited. However, as the read/write latency for flash memory is about 1:10, and most of desk and server application programs are read-intensive, applying a small portion of buffer space for some hot clean pages will benefit most. In this paper, we propose such a buffer management scheme called h-Buffer with three lists. Applying less than 7.125% of the buffer size for clean pages, h-Buffer considers both the write and read requests by the adoption of a replacement policy, a write-back policy and a HL (hot list) compensating policy. Unlike certain existing algorithms, it does not only consider the recency and frequency of page references, but also interacts with the buffer capacities and FTL timely. Experiment results show that the erase count, write count, read count and run time of h-Buffer decrease 50% over traditional algorithms on average.

1 Introduction

Small size, shock resistance, low-power consumption and nonvolatile properties are some attractive features of flash memory [1][2], which lead to the wide use of it in our daily life, e.g. MP3 players, PDAs, cellular phones and SSDs. However, it has some limitations. Firstly, flash memory has asymmetric I/O speed and limited block erase count. From table 1 [3], a write operation is about ten times slower than a read operation, and an erase operation is about 8 times slower than a write operation. Secondly, an entire block erase and large data restoration will be performed when updating a byte in a page. Fortunately, because of the flash translation layer (FTL) [4][5] used in flash-based memory, the features of flash are transparent to file system. FTL translates logical page address in the file system to physical address used in flash memory.

The buffer algorithms should consider its particular characteristics while taking flash memory as storage devices. Recently, two kinds of buffer replacement algorithms were

H. Yu et al. (Eds.): DASFAA Workshops 2012, LNCS 7240, pp. 14–27, 2012.

designed based on the granularity of clustering, CFLRU [6], CCF-LRU [7], AD-LRU [8] and FOR [19] are the page-clustered algorithm (PCA), BPLRU [9], FAB [10], CLC [11] and l-Buffer [12] are the group-clustered the algorithm (GCA).

Table 1. The characteristics of Samsung K9XXG08UXA NAND flash memory [3]

Operation	Access time	Access granularity
Read	20 μs /page	Page (2KB)
Write	200 μs /page	Page (2KB)
Erase	1.5 ms /block	Block (128Kb = 64 pages)

In the PCA, it distinguishes hot pages and cold pages easily, and the replacement policy is simple as well. However, when the buffer size is proper, the PCA shows a worse performance compared to the GCA, because the former flushes dirty pages into flash memory page by page, but the latter is in the unit of a whole group, which will effectively decrease the count of write and erase operations.

However, the GCA owes a common problem, when the buffer size is too small, both the write and erase counts are extraordinary big. Because the buffer can only accommodate few groups and these groups are hard to grow bigger, then they will be flushed frequently and the write and erase counts will increase a lot in FTL. Moreover, as a group usually contains many pages, a hot page may pollute the whole group, which makes it hard to choose the best victim during the replacement operation. At last, most of the GCA used inside-SSDs just consider the writer request because of the limited buffer capacity, for instance BPLRU and l-Buffer. However, as most host and server applications are read-dominate, for example the real OLTP trace in Table 4, it is generated from a bank system, the read/write ratio is 77%/23%, it is better to allocate some portion of buffer space for hot clean pages.

In this paper, we propose a buffer management algorithm for flash-based storage, namely h-Buffer (Mixed Lists of Hybrid Granularities with Adaptive Policies). It chooses to manage hot pages at page-level, cold dirty pages at group-level, buffer some hot clean pages with little space and interact to the buffer capacities and host workload. The main contributions are summarized as follows:

(a) We present a novel structure to manage pages in the buffer with three lists of hybrid granularities, applying less than 7.125% of the buffer space for clean pages, h-Buffer considers both the write and read requests from the host (see Section 3.1).

(b) The replacement policy and write-back policy in h-Buffer are not fixed and can be dynamically adjusted according to the buffer capacities and FTL environment. Then h-Buffer can be applied to complex circumstance (see Section 3.2, 3.3).

(c) We conduct comparison experiments on synthetic traces and a real OLTP trace. The results show that h-Buffer outperforms all other competitors on both types of traces with respect to write count, erase count, run time and read count (see Section 4).

2   Related Work

In order to improve the performance of flash-based memory, the buffer replacement algorithms should not only consider the hit ratio, but also aim to reduce the write and erase counts to flash memory. Some famous PCA are CFLRU, CCF-LRU, AD-LRU and FOR. CFLRU partitions the page list into a working region and a priority region, and selects clean pages from the priority region as victim first. CCF-LRU uses a cold clean queue and a mixed queue to maintain buffer pages, it evicts out pages from the cold clean queue first. AD-LRU uses two LRU queues to capture both the recency and frequency of page references. FOR considers both the locality of read/write operation and the cost difference of read/write operations on the selection of a victim page. Because pages are organized at page-level, it is easy to distinguish hot pages and cold pages, then the selection of the victim page is simple. However, Kang et al. in [11] explained that managing pages at group-clustered level can decrease the number of extra write and erase operations efficiently with proper buffer size. As an evolution of CFLRU, CFDC [18] splits the priority region into a clean queue and a dirty queue. The dirty queue is composed of clusters and a cluster contains several pages. Then dirty pages can be flushed as a batch and the average time of single page flush can be reduced remarkably. However, the page order in the cluster does not correspond to page numbers, but to the time they entered the cluster. Unlike the group in the GCA, pages in the same cluster do not belong to the same flash block. In fact, because of the way of page clustering, as Yi Ou [18] said, CFDC flushes more pages than CFLRU. Then the number of write and erase operations to flash cannot be decreased.

In [9], Kim et al. proposed BPLRU only for write requests from the host. BPLRU uses three key techniques, block-level LRU, page padding, and LRU compensation, it always selects the least-recently updated group as the victim. However, if the victim group is small, many clean pages need to read from flash memory, which will cause many page writes during page padding. Jo et al. devised FAB [10] in which pages that belong to the same flash block are grouped together. In order to increase the chance of a *switch* merge [14] in FTL, FAB selects the group with the most pages as the victim. However, FAB is only suitable for a workload with sequential requests. Kang et al. proposed CLC in [11]. CLC evicts the largest one among the not recently updated groups. However, most of the victim groups contain un-sequential pages, which will cause a large number of full merge in FTL.

In [12], Li-Pin Chang proposed a write buffer management scheme called l-Buffer. Unlike former algorithms for flash-based storage devices, l-Buffer does not only consider the buffer capacities, but also monitors how the host workload stresses the FTL. To get the best I/O performance, the policies used in l-Buffer are dynamically adjusted according the state of FTL. When the merge cost in FTL is low, it replaces the victim for FTL logging, otherwise, it uses padding to produce large write bursts. However, as the use of a simulation helper program to adjust the two policies, l-Buffer will be interrupted regularly.

Above all, the PCA maintains pages at page-level and shows a better performance with a small buffer, but has a worse performance with a large buffer compared to the GCA. Most GCA ignores the read requests, such as BPLRU and l-Buffer. So they cannot benefit most desk and server application programs which are read-intensive. Therefore, a best flash replacement policy, the hot pages (including clean and dirty) should stay at page-level, cold dirty pages that belong to the same flash block should group together. Secondly, the buffer space for clean pages should not be too large. Thirdly, the policy should interact to the buffer capacities and FTL. To devise such a buffer replacement algorithm, we get the idea of using hybrid granularities of lists and classifying pages into three kinds, namely hot page, cold clean page, cold dirty page.

3   The h-Buffer Algorithm

In this section, we present a novel buffer replacement algorithm for NAND flash memory based storage system, which is called Mixed Lists of Hybrid Granularities with Adaptive Policies (h-Buffer). Applying a small portion of buffer space for clean pages, h-Buffer considers both the write and read requests from the host. h-Buffer employs three policies, namely a replacement policy, a write-back policy and a hot list compensating policy. h-Buffer selects victim page from CCL first, then victim group from CDGL. If the merge cost in FTL is high, h-Buffer selects the biggest group as the victim, otherwise the oldest group. If a victim group is choose, h-Buffer decides to log or to pad the group to flash memory according the circumstance of FTL used in flash memory.

3.1   Data Structure for h-Buffer

h-Buffer is composed of three lists in Fig.1, HL (hot list) and CCL (cold clean list) are page-clustered, CDGL (cold dirty group list) is group-clustered. Hot pages, cold clean pages and cold dirty pages stay in them separately. With the special structure, h-Buffer shows a good performance not only for erase count and write count but also for read count. The exact size of the three lists can be self-adjusted to different host workloads.

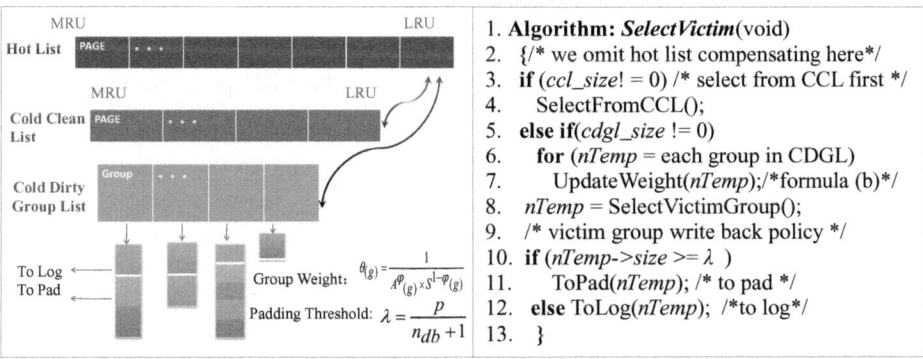

```
1. Algorithm: SelectVictim(void)
2.    {/* we omit hot list compensating here*/
3.    if (ccl_size! = 0) /* select from CCL first */
4.        SelectFromCCL();
5.    else if(cdgl_size != 0)
6.        for (nTemp = each group in CDGL)
7.            UpdateWeight(nTemp);/*formula (b)*/
8.    nTemp = SelectVictimGroup();
9.    /* victim group write back policy */
10.   if (nTemp->size >= λ )
11.       ToPad(nTemp); /* to pad */
12.   else ToLog(nTemp);  /*to log*/
13.   }
```

Group Weight: $\theta_{(g)} = \dfrac{1}{A^{\varphi}_{(g)} \times S^{1-\varphi}_{(g)}}$

Padding Threshold: $\lambda = \dfrac{p}{n_{db}+1}$

Fig. 1. The data structure of h-Buffer   **Fig. 2.** The victim selection algorithm in h-Buffer

If h-Buffer receives a read/write request, it will search the buffer first. On a hit, the page is moved to the MRU position of HL. Otherwise, h-Buffer allocates a new page cluster added to CCL/CDGL, and the page is written to the cluster from flash memory, then the page is read/overwritten. If free space in the buffer is not enough, h-Buffer will invoke the *SelectVictim* algorithm (Fig.2).

3.2 Replacement Policy

When free space in the buffer is not enough, the replacement algorithm is executed in Fig.2. Pages in the buffer can be classified into three types, namely cold clean page, cold dirty page and hot page. From Table 1, the cost of a write operation is about ten times higher than a read operation. The accurate ratio of read cost to write cost depends on the hidden FTL algorithm, which can only obtained by experiments. Thus it is easy to find out that the cost of evicting hot page (dirty or clean) is maximal and the cost of evicting cold clean page is minimal. Base on above analysis, we define expression as follow:

$$c_{cc} < c_{cd} < c_h \tag{a}$$

Here, the c_{cc}, c_{cd} and c_h are the cost of evicting a cold clean page, a cold dirty page and a hot page respectively.

According to formula *(a)*, the replacement policy in h-Buffer is described as follows:

(1) If CCL is not null, the page at the LRU position of CCL is selected as the victim page and is discarded.

(2) Otherwise, if CDGL is not null, h-Buffer updates the weight of each group $(\theta_{(g)})$ in CDGL according to formula *(b)*. The group with the smallest weight will be the victim and be flushed.

$$\theta_{(g)} = \frac{1}{A_{(g)}^{\varphi} \times S_{(g)}^{1-\varphi}} \tag{b}$$

(3) If CDGL is null, h-Buffer selects the page at the LRU position of HL as the victim.

In formula *(b)*, the $A_{(g)}$ and $S_{(g)}$ are the age and size of a group respectively. Now we consider how to calculate φ, we define n_{lb} and n_{db} be the average number of log blocks and data blocks during a full merge in FTL. The n_{db} reflects the association degree of log blocks, $n_{db} + 1$ block erases will be performed if a victim log block is selected. So the n_{db} can reflect the merge cost in FTL. We modify FTL to afford n_{db} to h-Buffer, whenever a victim group is selected, the value will be updated. When the merge cost in FTL is low, it is better to evict the oldest group in the buffer to allocate space for young and size-growing groups. Then we assign $\varphi = 1$, the oldest group will have the smallest weight and have more chance to evict out. Vice versa, it is better to evict the largest group with the purpose of increasing the chance of a *switch* operation in FTL. Storing small groups in the buffer can delay expensive full merge, and allows them to have more time to grow bigger to increase the chance of a padding operation. We assign $\varphi = 0$, the biggest group will have the smallest weight. Then the replacement policy can be adjusted dynamically according to the circumstance of FTL used in flash memory.

3.3 The Threshold of Logging or Padding On-Line Adaption in CDGL for FAST

If a victim group is selected, similar to the mechanism used in l-Buffer, h-Buffer decides whether to pad or log the victim. Let a flash block contains p pages, and r_{lb} be the space utilization of the log block during a full merge. Adopting FAST as the FTL used in flash memory for example, we know that FAST will choose the log block with the biggest space utilization as the victim block, so the space utilization is usually 100%. We define c_w and c_e be the time cost during a page write and a block erase operation. In a full merge, FTL rewrites $n_{db} \times p$ pages into flash memory and erases $n_{db} + n_{lb}$ blocks. So on average, to log $n_{db} \times p$ pages, the totally time cost is:

$$c_w \times (n_{db} \times p + n_{lb} \times r_{lb} \times p) + c_e \times (n_{db} + n_{lb}) \qquad (c)$$

Now we consider the padding operation, let λ be the average number of pages of the victim group, we will write $n_{lb} \times r_{lb} \times p$ pages. The totally time cost of padding is:

$$\frac{n_{lb} \times r_{lb} \times p}{\lambda} \times (c_w \times p + c_e) \qquad (d)$$

In the experiment, we use FAST as the FTL, then we get $r_{lb} = 100\%$ and $n_{lb} = 1$, let expression (c) and (d) be equal, we have the equation as follows:

$$\lambda = \frac{p}{n_{db} + 1} \qquad (e)$$

In h-Buffer, if the size of the victim group (vs) is bigger than λ, h-Buffer will execute a padding operation, read $p - vs$ pages from flash to the victim group, and then write the entire group to flash memory. Thus it will decrease the association degree of log blocks in FAST and increase the chance of a *partial* merge [15] or a *switch* merge. Otherwise, the $p - vs$ will be big, h-Buffer would just execute a log operation and flush the group to flash memory. Comprehensively, h-Buffer has better I/O performance.

Now we consider how to compute λ. h-Buffer uses an on-line computing method regularly session by session, it uses the write requests of the past session to compute the new λ, which is used for the next session. h-Buffer defines a constant *TIME_UPDATE_λ* as the session length, whenever the number of *TIME_UPDATE_λ* write requests passes, we update n_{db} for the next session.

3.4 Hot List Compensation

In h-Buffer, pages are firstly inserted into CCL or CDGL, and if they are referenced again, they will be moved to HL and become hot. However, if the host process transforms, the request sequence will change sharply and become irrelevant, so some pages in HL may have to wait a long time for a third reference and become cold. With time increasing, HL will become bigger and finally occupies the whole buffer. To avoid this phenomenon, we add a mechanism to shift these pages out from HL.

h-Buffer defines three parameters, namely *pagetime* , *cur_time* , *TIME_HL_OUT*. Each page in HL has the parameter *pagetime*, the *cur_time* is a global variable, whose initial value is zero, and the *TIME_HL_OUT* is a constant value. The *cur_time* increases by one each time a new page is added into the buffer. If the requested page is hit in the buffer, the page is moved to HL, and the *pagetime* of the page is changed to *cur_time*. As it is described in Fig.2, whenever the replacement operation happens, h-Buffer will examine the pages (*pTemp*) in HL from the LRU position. Once formula (*f*) is true, which means the page has not been referenced for a long time, then we move the page out from HL to accommodate space for new hot pages. The page is moved to CCL if it is clean, otherwise to CDGL. Then the size of HL will tend to be around the value of *TIME_HL_OUT*.

$$cur\_time - pTemp-> pagetime \geq TIME\_HL\_OUT \qquad (f)$$

4   Performance Evaluation

In this section, we compare the performance of h-Buffer with CFLRU, CCF-LRU, AD-LRU, FAB, and CLC. Because BPLRU and l-Buffer just consider the write request, we do not take them as comparison. We compare them with respect to erase count, merge count, write count, read count, and the run time. Then we test the influence of the two parameters $TIME\_HL\_OUT$ and $TIME\_UPDATE\_\lambda$. At last, we analyse the size of the three lists.

4.1   Experiment Setup

The experiments are conducted based on flash memory simulation framework, called Flash-DBSim [17], Flash-DBSim is a reusable and reconfigurable framework for the simulation-based evaluation of algorithms on flash disks. In our experiment, we simulate a 128MB Samsung K9K8G08U0 [3] NAND flash memory. The characteristics of the flash-based SSD are shown in Table 1 and the erasure limitation of blocks is 100000 cycles. We adopt FAST as the FTL algorithm with eight log blocks.

Table 2. The traces generated by DiskSim

Workload	Total references	Read/Write ratio	Locality
T7355	300,000	70%/30%	50%/50%
T7382	300,000	70%/30%	80%/20%
T1955	300,000	10%/90%	50%/50%
T9155	300,000	90%/10%	50%/50%
T1955-7382-9155	900,000	---	---

Table 3. The real OLTP trace

Attribute	Value
Total Buffer Requests	607391
Data Size	20 GB
Page Size	2048B
Duration	1 hour
Read/Write Ratio	77%/23%

In our experiment, we use two kinds of workloads. The first contains five synthesized traces generated by DiskSim [13]. In Table 2, the locality "80%/20%" means eighty percentages of requests focus on twenty percentages of total pages, and the read/write ratio "70%/30%" means 70% read requests contrast to 30% write requests. To simulate the complex environment of host process transforming, we join trace T1955, T7382 and T9155 together to generate trace T1955-7382-9155. The second workload is an OLTP trace in a real bank system supplied by Prof.Gerhard Weikum in Table 3. It contains 607391 page references to a CODASYL database with a total size of 20 Gigabytes. In CFLRU, we set the window size w to 0.1, which is the same as the value used in the paper of CCF-LRU. In CLC we set the partition parameter to 0.1.

4.2   Performance Evaluation on Synthesized Traces

Fig.3 shows the flash write count of each algorithm. The write count of h-Buffer is less than other five algorithms most of the time. When the buffer is smaller than 0.7MB, the write count of h-Buffer is a little higher than the PCA, but much smaller than the GCA. This is because when the buffer is too small, the GCA is hard to cluster pages, h-Buffer is composed of lists with hybrid granularities, so its performance would also be reduced. With the buffer size increasing, the write count of h-Buffer reduces sharply in the exponential form. While the write count of the PCA just decreases a little in the linear form, and the GCA decreases a lot but still more than h-Buffer. This is because the GCA can cluster pages that belong to the same block together, and write them to flash memory in the form of group. When the buffer size is bigger than 0.7MB, h-Buffer shows best, especially when the buffer size is 3MB, compared to the GCA and PCA, the write count of h-Buffer decreases 58% and 47.6% respectively. This is because in h-Buffer, the buffer is nearly full occupied by HL and CDGL (see Section 4.5). Secondly, h-Buffer evicts cold clean pages first, which need no flash write. Thirdly, h-Buffer selects the best victim group dynamically according the merge cost in FAST. Fourthly, it decides to log or pad the victim group according to the association of log blocks in FAST timely.

Fig.4 reflects the flash erase count of each algorithm, the line tendency is about the same as the write count. When the buffer size is smaller than 0.5MB, the erase count of h-Buffer is a little bigger than the PCA, but much lower than the GCA. That is because when the buffer size is too small, the CDGL list in the buffer is short and can only accommodate few groups. Then groups will be flushed frequently. However, h-Buffer performs best when the buffer size is bigger than 0.5MB. Especially at the point 3MB, compared to the GCA and PCA, the erase count of h-Buffer decreases 64.4% and 51.7% respectively. The reason is that the flash write count of h-Buffer which will cause an erase operation decreases a lot. Moreover, during the selection of a victim group, FAB and CLC may contain hot pages in the group. While in h-Buffer, no matter the victim is in CCL or CDGL, they are all cold.

Fig.5 shows the overall run time of various replacement algorithms, which is calculated as the sum of I/O time and memory time. The I/O time includes the time of all flash operations, such as read, write and erase. Each operation time is calculated by multiplying physical access time (Table 1) by the number of each operation. The memory time is the CPU time to run the algorithm. As flash I/O operations occupy most of the time, so the run time is mostly determined by the write count and erase count. In Fig.5, h-Buffer has the least run time except when the buffer size is smaller than 0.7MB. Especially when the buffer size is 3MB, compared to the GCA and PCA, the run time of h-Buffer decreases 62.5% and 49% respectively. The reason is the same as the case of write count and erase count.

Though most former algorithms do not consider read count to NAND flash memory, and some of them sacrifice read operation with the purpose of decreasing write count, such as BPLRU, CFLRU and CCF-LRU. In fact, as shown in Table (1), the time cost of a flash read is about ten times smaller than a flash write operation, so we cannot increase the read count too much. In Fig.6, h-Buffer has the least read count most of the time, except the situation that the buffer size is smaller than 0.7MB. When the buffer size is 3MB, the read count of h-Buffer decreases 49.5% than GCA and 40.2% than PCA.

To test the environment in the limit, we use trace 1955 and 9155, in which write request and read request are dominate respectively. Fig.7 shows the erase count of both the two environments. In the case of trace 1955, the PCA shows a terrible result than the GCA, and h-Buffer performs a little better than the GCA. The reason is that FAB, CLC and h-Buffer can flush dirty pages at the group-level, and as the read request from the host is too little, so the advantage of buffering hot clean pages cannot be sufficiently reflected in h-Buffer. However, when the read request is dominate, the PCA shows better performance than the GCA, and h-Buffer performs especially better than all of them. As most of the hot pages are clean, they can be organized at page-level in PCA and h-Buffer. Fig.8 describes the read count, in the case of 1955, h-Buffer performs about the same to the GCA, but much better than the PCA. That is because most of host requests are write, the advantage of the three lists cannot be reflected. However, when the read request is dominate, h-Buffer shows best.

Moreover, we add trace 1955-7382-9155 to simulate the environment of host processes transforming. We can see h-Buffer performs best most of the time with respect to the erase count and read count in Fig.9. As h-Buffer updates the parameter TIME_HL_OUT session by session, and the replacement policy and write-back policy is adaptive to the situation of FTL and buffer capacities.

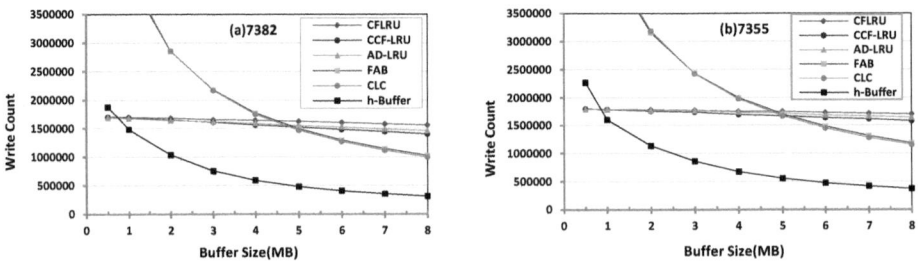

Fig. 3. Write count for trace 7382 and 7355 with different buffer size

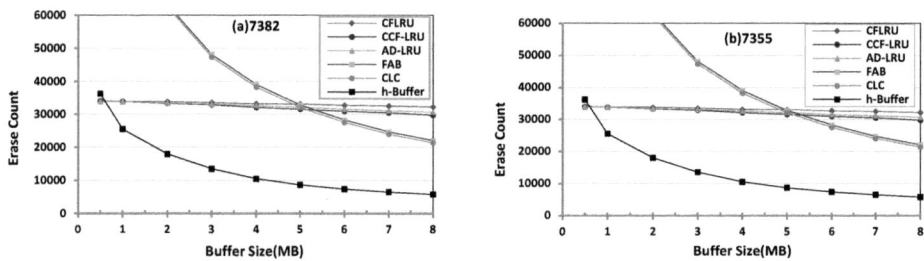

Fig. 4. Erase count for trace 7382 and 7355 with different buffer size

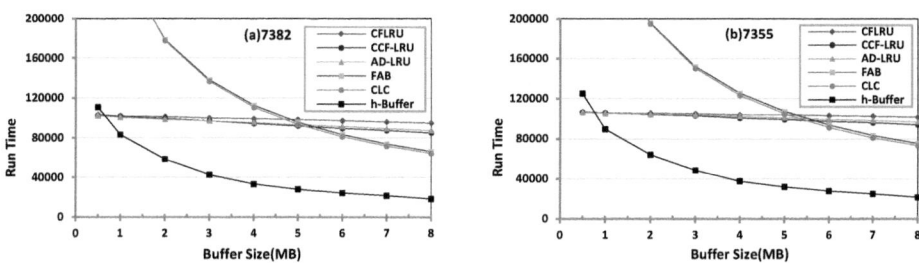

Fig. 5. Run time for trace 7382 and 7355 with different buffer size

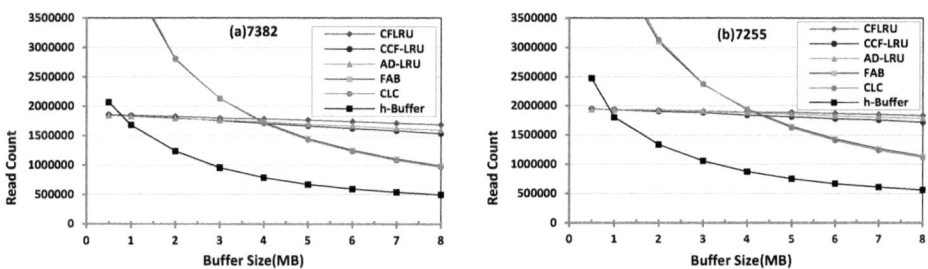

Fig. 6. Read count for trace 7382 and 7355 with different buffer size

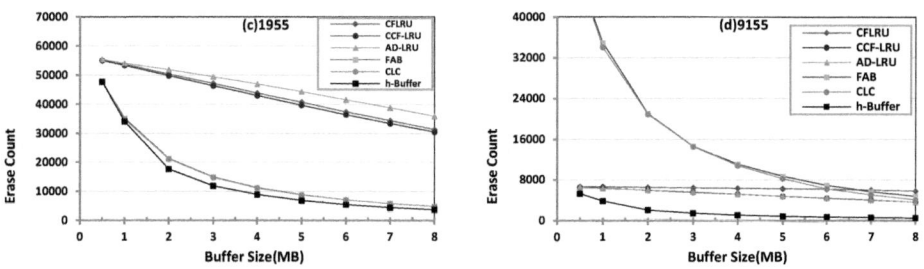

Fig. 7. Erase count for trace 1955 and 9155 with different buffer size

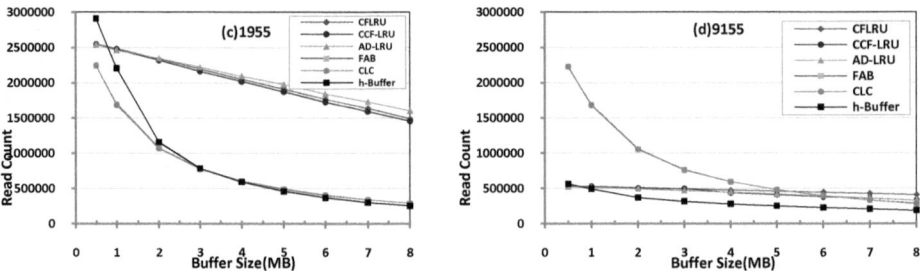

Fig. 8. Read count for trace 1955 and 9155 with different buffer size

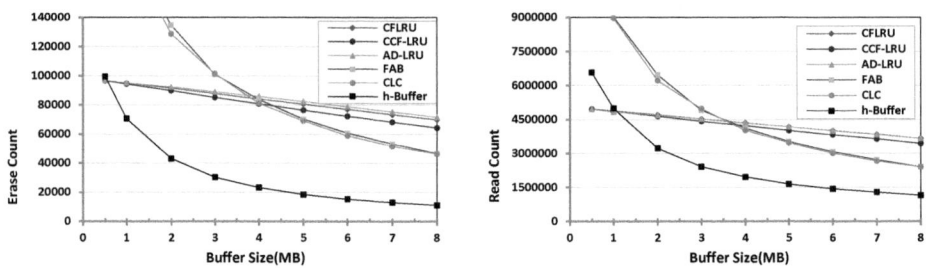

Fig. 9. Erase and read count for trace 1955-7382-9155 with different buffer size

4.3   Performance Evaluation on Real OLTP Trace

Running the OLTP trace, the erase count and write count of h-Buffer win the best most of the time according to Fig.10. When the buffer is smaller than 0.35MB, the erase count and write count of h-Buffer are both a little bigger than the PCA, but much smaller than the GCA. Compared to the first type of workload, h-Buffer performs better in the real OLTP trace. When the buffer size is 4MB, the erase count and write count of h-Buffer decrease 71.8% to the GCA, and 54.6% to the PCA on average. This is because the read and write locality are the same in the first type of workload, while the OLTP trace exhibits a read-intensive pattern in a real bank system and its read locality is a little higher compared with write locality, which can be concluded from Table 3.

Fig.11 shows the run time and read count of various policies, h-Buffer wins best most of the time except the situation when the buffer size is bigger than 0.35MB . However, there is a phenomenon worth attention when the buffer size reaches 28MB and grows bigger, the run time of CFLRU increases instead. The reason is that as the buffer size grows bigger, the write count and erase count decrease little, while it costs longer time to adjust the long LRU lists. Hence the searching overhead introduced by CFLRU has much impact on the run time.

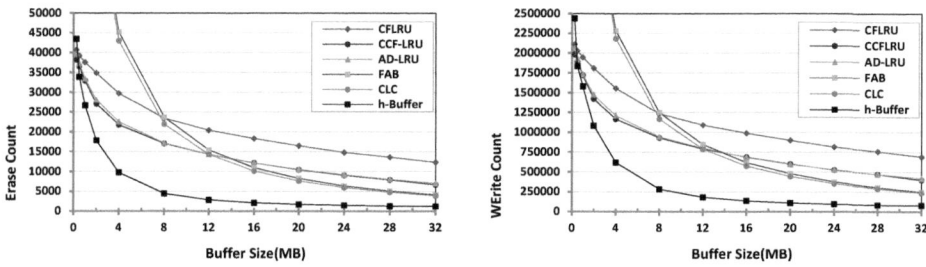

Fig. 10. Erase and write count for OLTP trace with different buffer size

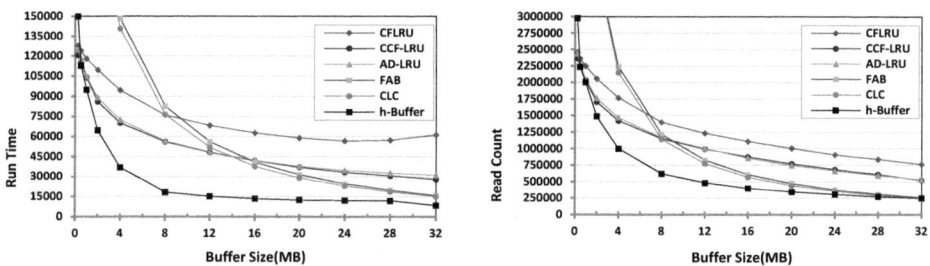

Fig. 11. Run time and read count for OLTP trace with different buffer size

4.4   Session Length

In section 3, we have discussed h-Buffer uses an on-line method to compute λ session by session by formula *(e)*. Whenever the number of $TIME\_UPDATE\_\lambda$ write requests passes, we update n_{db} for the next session. We use the OLTP trace with the buffer size 4MB. In Fig.12, when the value is 4000, h-Buffer gets the best performance. If the session length is small, h-Buffer will update n_{db} frequently, and the choice to log or to pad is more accurate, then the run time is smaller than the time when the session is big. However, we cannot update n_{db} so frequently, as the update will cost much CPU time. So the value 4000 performs best. If the host workload request invokes many merge operations, it is better to set $TIME\_UPDATE\_\lambda$ as a small value, otherwise a big one.

To the parameter $TIME\_HL\_OUT$, Fig.13 shows when it is equal to Buffsize*(1/64), h-Buffer has the least run time. We explain the phenomenon as this, according to formula *(f)*, the size of HL tend to be close to $TIME\_HL\_OUT$. When the value is big, then few pages is needed to move out from HL, the size of HL will be big and the size of CDGL will be small corresponding, so h-Buffer will have the tendency to the LRU list and do not shows a good performance. However, we cannot maintain $TIME\_HL\_OUT$ too small, as in that case, the size of HL will be too small, some hot pages will be transferred to CCL and CDGL and have bigger chance to swap out. If the host workload has few processes or the request sequences do not change sharply, $TIME\_HL\_OUT$ can be a big value, then the size of HL will maintain big enough to increase the page hit ratio. Otherwise, $TIME\_HL\_OUT$ should be small. With the parameter $TIME\_HL\_OUT$, h-Buffer can work well with difference kinds of application environment.

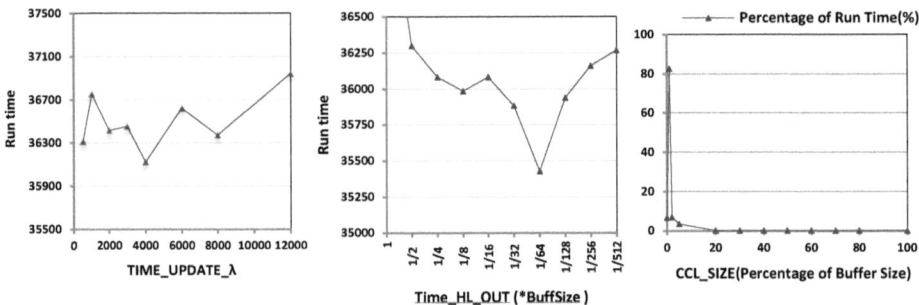

Fig. 12. Run time of OLTP for different session with buffer size 4MB

Fig. 13. Run time of OLTP for different TIME_HL_OUT with buffer size 4MB

Fig. 14. The percentage of run time for different size of CCL

4.5   The Size Analysis of HL,CCL and CDGL

As the RAM size in SSDs is limited, we need to decrease the number of cold clean pages to accommodate enough space for hot pages and cold dirty groups as much as possible. Taking the OLTP trace for example, we measure the size of CCL during the runtime of h-Buffer. Fig.14 shows that the time occupy 89.26% of the whole runtime when the size of CCL is smaller than 1% of the buffer size, and the value comes up to 99.45% when the size of CCL is smaller than 4% of the buffer size. So the CCL only occupies smaller than 4% of the buffer size about all of the time. h-Buffer contains a hot list compensating policy to move pages that may become cold again out. In the past section, we have measured that when the size of HL is 1/32 to 1/128 of the buffer size, h-Buffer performs best, so the hot list occupies 3.125% to 0.78% of the buffer size. Above all, the cold clean pages and hot clean pages just occupy smaller than 7.125% of the whole size.

5   Conclusions and Future Works

In this paper, we proposed an efficient buffer algorithm for flash-based NAND memory, namely h-Buffer. h-Buffer maintains three lists with hybrid granularities in the buffer. h-Buffer contains a replacement policy, a flash write-back policy and a hot list compensating policy. h-Buffer not only considers the recency and frequency of page references, but also interacts to the buffer capacities and the FTL used in flash memory. The policies are adaptive, if the merge cost is high in FTL, it selects the biggest group and let small ones staying longer to grow bigger to become candidates for padding, otherwise, the oldest group is selected to accommodate space for new pages. Whenever a victim group is choose, h-Buffer decides to log or to pad the group intelligently according to λ. h-Buffer adds a mechanism to move pages in HL that may become cold sometime later, then h-Buffer can be applied to complex circumstance. Experiment results show that h-Buffer performs best among its competitors most of the time.

Acknowledgments. We are grateful to Prof. Gerhard Weikum for the provision of the real OLTP trace. This work is supported by National Natural Science Foundation of China under the granted no. 60833005 and 61073039.

Reference

1. Wu, C.H., Kuo, T.W.: An Adaptive Two-Level Management for the Flash Translation Layer in Embedded Systems. In: Proc. of IEEE/ACM ICCAD, pp. 601–606 (2006)
2. Xiang, X., Yue, L., et al.: A Reliable B-Tree Implementation over Flash Memory. In: Proc. of ACM SAC, pp. 1487–1491 (2008)
3. Samsung Electronics, K9XXG08UXA.1G X8 Bit/ 2G X8 Bit/ 4G X8 Bit NAND Flash Memory (2006)
4. Intel Corporation, Understanding the Flash Translation Layer (FTL) Specification, Technical Report AP-684 (1998)
5. Kim, J., Kim, J.M., Noh, S.H., et al.: A Space-Efficient Flash Translation Layer for Compact-Flash Systems. IEEE Trans. on Consumer Electronics 48(2), 366–375 (2002)
6. Park, S.-Y., Jung, D., Kang, J.-U., Kim, J.-S., Lee, J.: CFLRU: A Replacement Algorithm for Flash Memory. In: CASES 2006, pp. 234–241. ACM (2006)
7. Li, Z., Jin, P., et al.: CCF-LRU: A New Buffer Replacement Algorithm for Flash Memory. IEEE Transactions on Consumer Electronics, 1351–1359 (2009)
8. Jin, P., Ou, Y., Harder, T., Li, Z.: AD-LRU: An Efficient Buffer Replacement Algorithm for Flash-Based Databases. Data & Knowledge Engineering (2011)
9. Kim, H., Ahn, S.: BPLRU: A Buffer Management Scheme for Improving Random Writes in Flash Storage. In: Proc. Sixth USENIX Conf. File and Storage Technologies, pp. 239–252. FAST (2008)
10. Jo, H., Kang, J.-U., Park, S.-Y., et al.: FAB: Flash-Aware Buffer Management Policy for Portable Media Players. IEEE Trans. Consumer Electronics, 485–493 (2006)
11. Kang, S., Park, S., Jung, H., et al.: Performance Trade-Offs in Using NVRAM Write Buffer for Flash Memory-Based Storage Devices. IEEE Trans. Computers, 744–758 (2009)
12. Chang, L.-P., Su, Y.-C.: Plugging Versus Logging: A New Approach to Write Buffer Management for Solid-State Disks. In: DAC, pp. 23–28. ACM (2011)
13. Bucy, J.S., Schindler, J., et al.: The DiskSim Simulation Environment Version 4.0 Reference Manual, Carnegie Mellon University Technical Report (2008)
14. Kim, J., Kim, J.M., Noh, S.H., et al.: A Space- Efficient Flash Translation Layer for Compact Flash Systems. IEEE Trans. Consumer Electronics, 366–375 (2002)
15. Lee, S.-W., Park, D.-J., Chung, T.-S., et al.: A Log Buffer Based Flash Translation Layer Using Fully Associative Sector Translation. ACM Trans. Embedded Computing Systems, 436–453 (2007)
16. Park, C., Cheon, W., Kang, J., Roh, K., Cho, W., Kim, J.-S.: A reconfgurable ftl architecture for nand fash-based applications. ACM Trans. Embed. Comput. Syst. 7(4), 1–23 (2008)
17. Jin, P., Su, X., Li, Z., Yue, L.: A Flexible Simulation Environment for Flash-aware Algorithms. In: Proc. of CIKM 2009, demo. ACM Press (2009)
18. Ou, Y., Harder, T., Jin, P.: CFDC: a flash-aware replacement policy for database buffer management. In: Science And Technology (DaMoN), pp. 15–20. ACM (2009)
19. Lv, Y., Cui, B., He, B., et al.: Operation-Aware Buffer Management in Flash-based Systems. In: SIGMOD, pp. 12–16. ACM (2011)

A Study of Space Reclamation on Flash-Based Append-only Storage Management

Yulei Fan, Wei Cao, and Xiaofeng Meng

School of Information, Renmin University of China, Beijing, China
{fyl815,caowei.cn,xfmeng}@ruc.edu.cn

Abstract. Flash disks exhibit many different IO characteristics from magnetic disks. Flash-based databases should consider such traits of flash disks as faster access performance but discrepant between reads and writes, and between sequential and random accesses etc. To overcome the poor performance of random writes on flash memory, log-based and append-only storage management had been proposed by replacing random write with sequential write; this paper studies append-only storage management(ASM) with effective space reclamation methods, which work for both traditional relational database systems and key-value database systems built on flash devices. In this paper, we detail two kinds of record layouts used for space reclamation and then propose three space reclamation algorithms based on them. Our experiments show that the append-only storage management with bitmap index-based space reclamation algorithm(BI-ASM) has a factor of 32 of spatial improvement and factors from 23 to 44 of temporal improvement.

Keywords: Flash-based DBMS, Storage Management, Append-only, Space Reclamation.

1 Introduction

With its capacity increasing and price dropping fast, flash memory is being rapidly deployed as secondary data storage for more kinds of devices. For example, sensors and mobile devices(PDAs, MP3 players, mobile phones, digital cameras, etc), portable computers(individual PCs, etc) and enterprise servers(such as search engine servers, etc), and so on.

Because of its electronic limitations, such as erase-before-write constraints, limited life span, etc, flash-based storage management systems have to be carefully designed to fully exploit flash memory's beneficial merits, e.g. low access latency, low power consumption, higher shock resistance etc. Although there is a flash translation layer(FTL) as the most important component inside flash Solid State Disks(SSD) to realize address mapping, wear-leveling, space reclamation and so on, SSD-based storage management systems still encounter the problem of inferior performance, especially for random writes. So the trend of using flash memory, or more specifically, SSDs, as popular storage media, makes us rethink storage management of database systems, regardless of traditional

H. Yu et al. (Eds.): DASFAA Workshops 2012, LNCS 7240, pp. 28–39, 2012.

relational database systems or key-value database systems. Much research work suggests improving performance by reducing random write operations, i.e. adopting log-based storage management, append-only storage management(ASM) or the alike. Log-based storage management and ASM replace random writes by sequential writes.

Sequential writes will consume free space of flash devices quickly. Due to relatively limited capacity of flash memory, another important problem, space reclamation, becomes more prominent in the circumstances of flash memory which prevalently adopts out-of-place updates and favors append-only storage management for higher performance in flash-based database systems. Much previous research has been done on FTL, buffer management, index design, and so on, but, to the best of our knowledge, no research has ever elaborated on space reclamation. So in this paper, our main focus is to define and detail the space reclamation problem for flash-based database management with append-only storage management and then explore three alternative space reclamation algorithms to discover the optimized one both spatially and temporally.

In this paper, we identify the problem of space reclamation in the ASM background for flash-based storage management in section 2. Then in section 3, we describe the architecture of ASM with space reclamation and simultaneously define "space reclamation" and two kinds of record layout explicitly, namely, version-link-list-based record layout and delete-record-based record layout. According to record layout and data structure used for space reclamation, we proposed three kinds of space reclamation algorithms, i.e. version-link-list-based space reclamation, sorted-array-based space reclamation and bitmap-index-based space reclamation in section 4. In section 5, we compare and analyze three space reclamation algorithms both theoretically and experimentally on spacial and temporal efficiency. In the last section 6, we conclude the whole paper. The contribution of our work is to study the strong and weak points of three space reclamation algorithms in ASM and to find an optimized one in the general sense.

2   Flash Memory and Flash-Based Storage Management

2.1   Flash Memory and SSDs

Flash memory comes in two flavors: NOR flash and NAND flash. NOR flash is mainly used for storing running codes, but NAND flash is considered suitable for data storage. Henceforth, if not pointed out explicitly, we mean NAND flash by saying flash.

Flash memory mainly has four important characteristics: 1)Flash memory has **no mechanical latency** in data accessing; the latency is only linearly proportional to the amount of data transferred; 2)Flash memory has the characteristic named erase-before-write; to avoid frequent expensive erasures, **out-place updates** are preferred to in-place ones; 3)Another important characteristic of flash memory is **asymmetric read, write and erasure speeds**; 4)There is an important property, limited life span, because it wears out and becomes unreliable after **limited number of writes** on flash memory.

SSDs are normally built on an array of flash memory packages for providing higher bandwidth than flash memory. As an important component in SSDs, flash translation layer(FTL) is used to emulate a hard disk and expose an array of logical blocks to the upper-level components. Functions of FTL mainly include address mapping, garbage collection and wear leveling. According to mapping address granularity, FTL algorithms can be divided into three types[1]: page-level mapping[2, 3], block-level mapping[4–6] and hybrid mapping with page and block granularity[7]. Recently, more novel FTL algorithms have appeared to solve the problems when using flash memory or SSDs for newer and bigger applications.

2.2   Flash-Based Storage Management

Due to the unique physical characteristics of flash memory, data management techniques, especially those in the lower levels of systems, which are implementation-dependent(e.g. storage techniques), deserve thorough redesigning. To address these limitations, many flash-specific techniques concerning storage management have been developed.

In flash-based storage management, there are three kinds of strategies, which are hybrid storage management, log-based storage management and append-only storage management(ASM).

Partition Attributes Across(PAX[8]), typical hybrid storage management of N-ary Storage Model(NSM[9]) and Decomposition Storage Model(DSM[10]), was proposed to solve querying processing of flash-based databases, especially scan and join operations, but did not consider update operations. Because of no description of update operation in [8], there is no information about space reclamation.

Log-based storage management(e.g. In-page logging(IPL)[11] and Differential IPL[12]), reduces random writes by log-based sequential writes. As shown in [11, 12], changes made to a data page in data area are buffered in main memory as logs, and then the change logs are written sector by sector to the log area in flash memory sequentially. The difference between [11] and [12] is the ratio of the size of data area to the size of the log area in a block. Because of limited space in an erase unit, when data areas or log areas have no free space for new data or log sectors, "merging data pages and their log sectors to reclaim data space is triggered by storage manager." Index storage can also adopt IPL storage model, such as B+-trees[13, 14]. But for IPL, when querying, it's necessary to apply all necessary logs on the data before getting the current values for the results. This has a negative impact on the latency of query processing.

But for key-value data, to meet application requirements of high throughput and low response latency, back-ends of big websites mainly focus on the third kind of storage management, ASM, such as FAWN[15] and FlashStore[16]. FAWN uses SSDs as secondary storage devices replacing magnetic disks and part of memory for bridging the gap between CPU and IO devices, etc. FlashStore, SkimpyStash[17] and ChunkStash[18] uses SSDs as a write cache for magnetic disks. In [15–18], data pages are only appended on flash memory and a hash table is used to index the valid data records on flash SSDs. When there is no free space on flash SSDs, space reclamation triggered by the storage manager reads all data pages and then check the validity of every record by looking it up in the hash table. Because a hash entry can index multiple records with the same hash key value in the form of a linked list, [17] solved the problem of oversize of the hash table in [16], but space reclamation in [17] pays multiple flash accesses to check the validity of each record, so the overhead of space reclamation is very high.

Due to relatively limited capacity of flash memory, space reclamation has an important impact on space utilization. Even though space reclamation, as a background processing procedure, is an internal facility that shouldnt be the concern of users, it pervasively influences the user applications, because it need main memory and the time for finishing up space reclamation. So it is important to discover the optimized space reclamation algorithm both spatially and temporally.

3   Structure of ASM with Space Reclamation

In this section, we describe and detail the two components of our topic: append-only storage management(ASM) and space reclamation management(Section 3.1). And then we propose two kinds of record layout used for space reclamation(Section 3.2). Figure 1 gives an overview of ASM with space reclamation.

3.1   ASM with Space Reclamation

ASM with space reclamation is composed of a read buffer, a vacuum buffer and a write buffer in main memory and data storage on flash devices, as shown in figure 1.

Read Buffer is a fixed-size main memory buffer caching recently read items. Because read buffer caches data which will not be updated in place, when page replacement is desirable upon a read operation, the least recently used page will be chosen as the evicted page, without having to be flushed to flash SSDs.

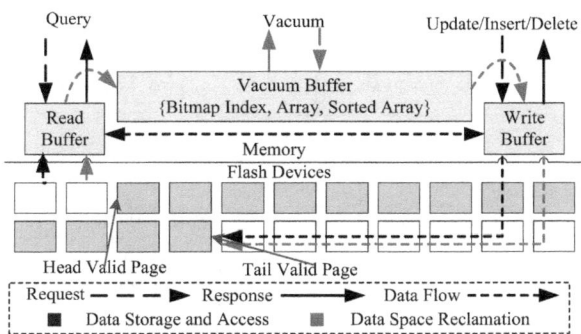

Fig. 1. Overview of ASM with space reclamation management

Write Buffer is a fixed-size main memory buffer caching recently written items produced by insert, update and delete operations. When one write buffer page is filled with data, a write to a flash SSD will happen. To guarantee durability of data, we can alternatively set a configurable timeout interval to force the writes to flash SSDs. The write buffer is sized to 2-3 times the flash page size so that inserts, updates and deletes can still go on when part of the buffer is being written to a flash SSD.

Flash Devices provides persistent storage for user's data which have been appended on flash SSDs with the write to flash being in the units of page. The pages on flash are used in a bidirectional circular linked list order. When the free space on flash SSDs is less than a configurable threshold α, space reclamation is triggered by space reclamation management.

Space reclamation is the procedure of freeing invalid data records' flash storage space and defragmenting valid data records' flash storage space. Space reclamation management works on the basis of vacuum buffer in main memory.

Vacuum Buffer is an unfixed-size data structure that buffers vacuum information used for freeing flash storage space. The data structure can alternatively be a bitmap index, a sorted array, a regular array, a linked list, or any other data structures which can be used for space reclamation.

Every space reclamation algorithm can be divided three steps: 1) Reversely scanning the data on *flash devices* and simultaneously reading every page into the *read buffer*; 2) Validating every data record by searching the data structure in the *vacuum buffer*; 3) Writing valid data records to the *write buffer* to be later flushed out to *flash devices* when a write buffer page is full.

3.2   Record Layout

For ASM with space reclamation, we can use different space reclamations for different record layouts. In this section, we briefly describe the two kinds of record layout: version-link-list-based record layout and delete-record-based record layout. Figure 2 gives an overview and examples of two kinds of record layout.

Figure 2(a) shows the version-link-list-based record layout(short for VLL-RL). For VLL-RL, we need to change the data record schema by adding an auxiliary attribute *PVRA*, which holds the address of previous version of the record. For the

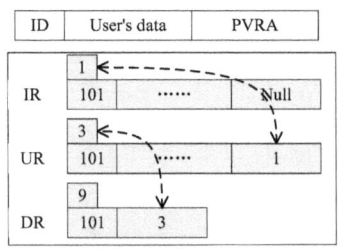

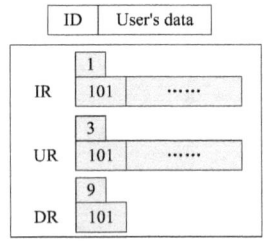

(a). Version Linked List-based Record Layout   (b). Delete Record-based Record Layout

Fig. 2. Overview of two kinds of record layout

record produced by an insert operation(short for IR), the *PVRA* is set as *NULL*.
For the record produced by an update operation(short for UR), the *PVRA* is set
as the original address of the record being updated. For the record produced by
a delete operation(short for DR), the *PVRA* is set as the original address of the
record being deleted. To differ DRs from URs, a DR has no user data. For example,
firstly, we insert a new record(IR) whose address is '1' and *PVRA* is *NULL* as in
figure 2(a). And later we want to update this IR by producing a new record(UR)
whose address is '3' and *PVRA* of the UR is the same as the address of the IR,
1. At last, we want to delete the record by producing a new record(DR) whose
address is '9' and *PVRA* of the DR is the same as the address of the UR, which is
3. And there is no user's data in the DR.

Figure 2(b) shows the delete-record-based record layout(short for DR-RL).
The difference between VLL-RL and DR-RL is whether there is an auxiliary
attribute *PVRA* in a record(IR, UR, or DR). There is no *PVRA* in any record
of DR-RL. For example, in figure 2(b), there is not *PVRA* in the IR, UR or
DR. But other parts of the IR, UR and DR are the same as the corresponding
records in VLL-RL.

4 Algorithms of Space Reclamation

According to record layout and data structure in vacuum buffer, we mainly
proposed three space reclamation algorithms: version-link-list-based space recla-
mation(Section 4.1), sorted-array-based space reclamation(Section 4.2) and
bitmap-index-based space reclamation(Section 4.3).

4.1 Version-Link-List-Based Space Reclamation

For ASM with version-link-list-based space reclamation(VLL-ASM), the records
including IR, UR and DR are produced in the form of VLL-RL. In this subsection, we propose version-link-list-based space reclamation algorithm based on
VLL-RL shown in algorithm 1. The size of the sorted array can be increased
dynamically and the elements in it must be sorted in ascending or descending
order of the key, denoted as ID. The sorted array can be placed in the vacuum
buffer. Note CR is the record being processed currently.

At first, we use *Intialization_Sorted_Array*() function to initialize a sorted
array to store the *PVRA* values and save *tail_id* into *temp*. And then we use the
first *for* loop and *Read_Page_Flash*() function to reversely read page by page
into the read buffer from flash devices. In the filtering phase, for every record of
one read buffer page, we get the *CRA* and *PVRA*. if the *PVRA* is not *NULL*,
we must insert the *PVRA* into sorted array for processing the older version of
CR in the future and update the *PVRA* of CR as *NULL*. And then we use
Is_Exist_Sorted_Array() function to decide whether CR should be write back
to flash SSDs or be deleted completely. If *CRA* exists in the sorted array, CR is an
old version of the record which can be discarded and the *CRA* can be deleted from
the sorted array; otherwise, we must check whether CR is DR by record layout.

Algorithm 1. Vacuum ($head\_id, tail\_id$)

Input   : $head\_id$ is ID of head valid page, $tail\_id$ is ID of tail valid page
Output: $head\_id, tail\_id$
begin
> $sorted\_array = Intialization\_Sorted\_Array(0); \ temp = tail\_id;$
> **for** $page\_id$ *from* $temp$ *to* $head\_id$ **do**
>> $page = Read\_Page\_Flash(page\_id);$
>> **for** *each record in page* **do**
>>> $CRA = address \ of \ CR; \ PVRA = address \ of \ previous \ version \ of \ CR;$
>>> $flag\_cra = Is\_Exist\_Sorted\_Array(CRA);$
>>> **if** $PVRA! = NULL$ **then**
>>>> $Insert\_Sorted\_Array(PVRA); \ Update\_Record(PVRA, NULL);$
>>>
>>> **if** $flag\_cra == false$ **then**
>>>> **if** CR *is not* DR **then**
>>>>> $Write\_Record\_Buffer\_Flash(tail\_id, CR);$
>>>
>>> **else**
>>>> $Delete\_Sorted\_Array(CRA);$
>
> $Delete\_Flash\_Page(head\_id, temp); \ head\_id = temp + 1;$
end

If CR is not a DR, CR is the newest version of the record that need be later appended to flash devices through write buffer by $Write\_Record\_Buffer\_Flash()$ function. At last, after processing pages from $head\_id$ to $tail\_id$, we must free these pages on flash devices and then set $head\_id$ again.

4.2   Sorted-Array-Based Space Reclamation

For ASM with sorted-array-based space reclamation(SA-ASM), the records including IR, UR and DR are produced in the form of DR-RL. In this subsection, we propose sorted-array-based space reclamation algorithm 2 based on DR-RL. The difference from VLL-ASM is that the sorted array keeps the *ID* encountered when running space reclamation but not the *PVRA*.

As shown in Algorithm 2, the procedures of reading data pages from flash devices into read buffer, appending write buffer page to flash devices and deleting all pages processed are same as these in VLL-ASM. The differences from VLL-ASM is the procedure of filtering data records. In SA-ASM, for every record in read buffer, firstly, we check whether the CR is a DR. If the CR is a DR, the CR can be ignored and simultaneously the ID of CR must be inserted in sorted array by $Insert\_Sorted\_Array()$ function. If the CR is not a DR, we must check whether the ID of CR had been encountered through $Is\_Exist\_Sorted\_Array()$ function. If the ID of CR is not in sorted array, the CR is encountered for the first time, so the ID of CR should be insert in sorted array and should be flushed out to flash devices through the write buffer; otherwise, we had seen the newer version record before processing CR, so the CR can be ignored.

Algorithm 2. Vacuum ($head\_id, tail\_id$)

Input : $head\_id$ is ID of head valid page, $tail\_id$ is ID of tail valid page
Output: $head\_id, tail\_id$
begin

> $sorted\_array = Intialization\_Sorted\_Array(0); \ temp = tail\_id;$
> **for** $page\_id$ from $temp$ to $head\_id$ **do**
>> $page = Read\_Page\_Flash(page\_id);$
>> **for** each record in page **do**
>>> **if** CR is not DR **then**
>>>> $flag\_id = Is\_Exist\_Sorted\_Array(ID);$
>>>> **if** $flag\_id == false$ **then**
>>>>> $Insert\_Sorted\_Array(ID);$
>>>>> $Write\_Record\_Buffer\_Flash(tail\_id, CR);$
>>>
>>> **else**
>>>> $Insert\_Sorted\_Array(ID);$
>
> $Delete\_Flash\_Page(head\_id, temp); \ head\_id = temp + 1;$

end

4.3 Bitmap-Index-Based Space Reclamation

Considering the speedup of checking and the save of the main memory space, we propose bitmap-index-based space reclamation algorithm 3 based on DR-RL. Every possible value of the *ID* has a unique bit in bitmap for indicating its validity, which is to ensure there is no collision.

Algorithm 3. Vacuum ($head\_id, tail\_id, moid$)

---

input : $head\_id$ is ID of head valid page, $tail\_id$ is ID of tail valid page, $moid$
    is the maximum of all OIDs
output: $head\_id, tail\_id$
begin

> $bitmap = Intialization\_Bitmap(moid, 0); \ temp = tail\_id;$
> **for** $page\_id$ from $temp$ to $head\_id$ **do**
>> $page = Read\_Page\_Flash(page\_id);$
>> **for** each record in page **do**
>>> **if** CR is not DR **then**
>>>> $id\_bit = Find\_Bitmap(ID);$
>>>> **if** $id\_bit == 0$ **then**
>>>>> $Set\_Bitmap(ID, 1); \ Write\_Record\_Buffer\_Flash(tail\_id, CR);$
>>>
>>> **else**
>>>> $Set\_Bitmap(ID, 1);$
>
> $Delete\_Flash\_Page(head\_id, temp); \ head\_id = temp + 1;$

end

In algorithm 3, we initialize the bitmap as 00...00. For every record, if the bit corresponding to the ID of the CR is 1, the CR is the newest version data record or a DR; otherwise, the CR is an invalid data record. In the filtering phase, for every record, firstly, we check whether the CR is a DR through the record layout. If the CR is a DR, the CR can be ignored and the bit in bitmap must be set to 1 by *Set_Bitmap()* function. If the CR is not a DR, we must check whether the bit corresponding to the ID of the CR is 1 or 0 by *Find_Bitmap()* function. If the bit is 0, the CR is encountered for the first time, so the bit corresponding to the ID of the CR must be set to 1 and should be flushed out to flash devices through the write buffer; otherwise, we had seen the newer version record before processing the CR, so the CR can be ignored.

5   Experiments

5.1   Simulation Setup and Performance Metrics

We conducted a simulation experimental study on a PC running Windows7 with an Intel Quad 2.4GHz CPU and 2GB memory. There are two kinds of storage media which are Intel 2.5" 3Gb/s SATA SSD 80G 5V(Short for SSD) and Hitach 3.5" SATA 3Gb/s 250G 7200rpm(Short for HDD). We implemented three append-only storage methods with version-link-list-based(VLL-ASM), sorted-array-based(SA-ASM) and bitmap-index-based(BI-ASM) space reclamation algorithms using Microsoft Visual Studio 2008. For the dataset, we utilize table *item* and *stock* in TPC-C benchmark, and then extract part from TPC-C dataset as IRs, and then produce URs and DRs with the ratio of insert records, update records and delete records as 100 : 100 : 5 by running update operations and delete operations randomly.

We consider the following performance metrics to compare the performance of the different methods:(1) the size of data space on flash SSDs, which shows the spatial efficiency of append-only storage management;(2) the running time of space reclamation, which shows the temporal efficiency of space reclamation algorithms;(3) the size of vacuum information in memory, which shows the memory utilization of space reclamation algorithms.

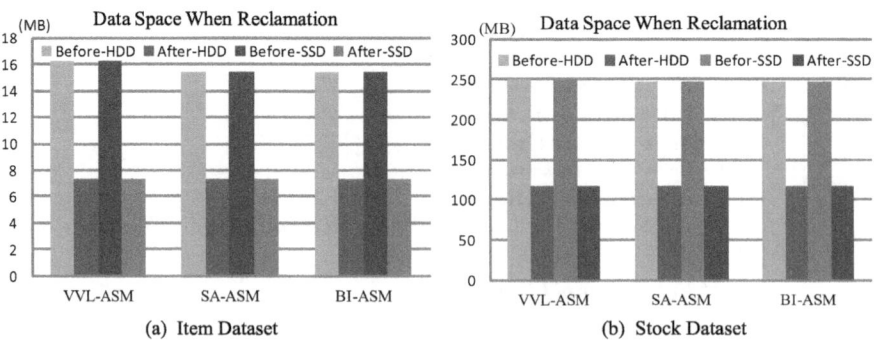

Fig. 3. Data space on flash SSDs of every ASM

5.2   Performance Comparison and Analysis

Evaluation results about the size of data space storing table *item* and *stock* on SSD and HDD are shown in Figure 3. There is sharp difference between the before- and after- space reclamation cases because there are many invalid versions of some records after update or delete operations, but after space reclamation, there is only one version for every record. There is no difference between HDD and SSD used to store the dataset *item* and *stock* in either before- or after- space reclamation cases, because there are identical records on SSD and HDD. But VLL-ASM has a bigger data space size than the other two because for every record in VLL-ASM, an additional attribute value storing *PVRA* for checking the older version data takes up more space. For SA-ASM and BI-ASM, the data space size is identical for every circumstances because they adopt the same record layout: DR-RL.

The running time of space reclamation is shown in Figure 4. For VLL-ASM and SA-ASM, the running time of space reclamation is longer than that of BI-ASM because we can directly locate a bit in bitmap through hash function. For table *item*, the running time of space reclamation in VLL-ASM is longer than that in SA-ASM showed in figure 4(a). But for table *stock*, it is reversely

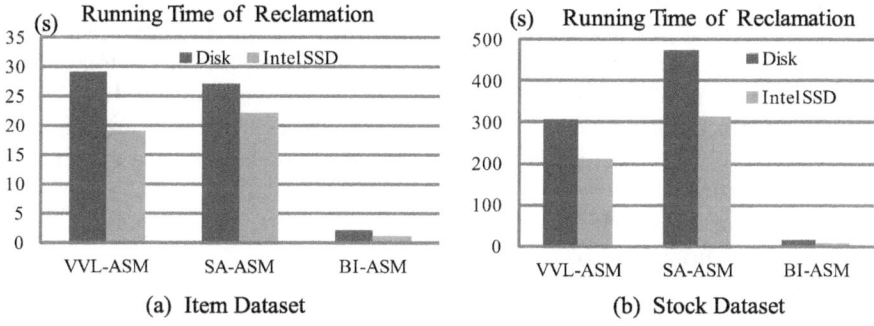

Fig. 4. Running time of every space reclamation

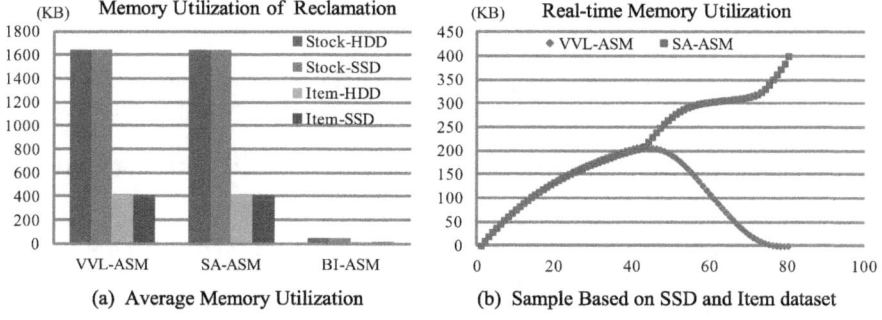

Fig. 5. Memory utilization of every space reclamation

shown in figure 4(b). Because there is sharp difference between the two sorted arrays used for VLL-ASM and SA-ASM shown in figure 5(b). The sizes of the sorted arrays affect the runnning time of $Is\_Exist\_Sorted\_Array()$ function in algorithm 1 and 2. So factors of 23 to 44 temporal improvement by BI-ASM was achieved by running on table *item* and *stock*.

The memory utilization of the vacuum information is shown in Figure 5(a). Average size of the memory utilized in VLL-ASM and SA-ASM is bigger than that in BI-ASM because BI-ASM maintains a bitmap in main memory but VLL-ASM and SA-ASM utilize sorted array in main memory. So a factor of 32 spatial improvement by BI-ASM was achieved by running on table *item* and *stock*. In Figure 5(b), the real-time size of the sorted arrays in VLL-ASM and SA-ASM varied with the running time, so we extracted 80 data sample points and drew the curve. For VLL-ASM, sorted array keeps the addresses of the unprocessed invalid records recently, so the curve starts from 0 and then reaches a highest point, and then returns to 0 at the end of space reclamation. For SA-ASM, sorted array keeps all the encountered *OID*s since the algorithm began, so the curve starts from 0 to the highest point in a monotonically increasing way.

6   Conclusions and Future Work

More and more flash-based database systems adopt append-only storage management, but an important module in append-only storage management is space reclamation management. Some data structures, such as bitmap indexes and arrays, etc, can be used as vacuum information for space reclamation. In this paper, we propose two kinds of record layout and three space reclamation algorithms based on them. BI-ASM can speed up the space reclamation and save main memory. In future work, we will study other data structure and the compressed bitmap that can reduce the size of vacuum information in main memory but not increase the running time of space reclamation.

Acknowledgement. This research was partially supported by the grants from the Natural Science Foundation of China (No.60833005, 61070055).

References

1. Tae-Sun, C., Dong-Joo, P., Sangwon, P., Dong-Ho, L., Sang-Won, L., Ha-Joo, S.: A survey of Flash Transalation Layer. Journal of Systems Architecture - Embedded Systems Design (JSA) 55, 332–343 (2009)
2. Intel: Understanding the flash translation layer specification. In: Application Note AP-684 (December 1998)
3. Sungjin, L., Dongkun, S., Young-Jin, K., Jihong, K.: Last: locality-aware sector translation for nand flash memory-based storage systems. ACM SIGOPS Operating Systems Review 42, 36–42 (2008)
4. Jesung, K., Jong-Min, K., Sam, H.N., Sang Lyul, M., Yookun, C.: A Space-Efficient Flash Translation Layer for Compact-Flash Systems. IEEE Transactions on Consumer Electronics 48, 366–375 (2002)

5. Jeong-Uk, K., Heeseung, J., Jinsoo, K., Joonwon, L.: A superblock-based flash translation layer for NAND flash memory. In: Proceedings of the 6th ACM & IEEE International Conference on Embedded Software (EMSOFT 2006), pp. 117–161. ACM & IEEE Press, Seoul (2006)
6. Lee, S.-W., Choi, W.-K., Park, D.-J.: FAST: An Efficient Flash Translation Layer for Flash Memory. In: Zhou, X., Sokolsky, O., Yan, L., Jung, E.-S., Shao, Z., Mu, Y., Lee, D.C., Kim, D.Y., Jeong, Y.-S., Xu, C.-Z. (eds.) EUC Workshops 2006. LNCS, vol. 4097, pp. 879–887. Springer, Heidelberg (2006)
7. Chin-Hsien, W., Tei-Wei, K.: An Adaptive Two-Level Management for the Flash Translation Layer in embedded Systems. In: The International Conference on Computer-Aided Design (ICCAD 2006), pp. 601–606. IEEE Press, San Jose (2006)
8. Mehul, A.S., Stavros, H., Janet, L.: Fast scans and joins using flash drives. In: SIGMOD Workshop DaMoN, pp. 17–24. ACM Press, Vancouver (2008)
9. Raghu, R., Johannes, G.: Database Management Systems. In: WCB, 2nd edn., McGraw-Hill (2000)
10. George, P.C., Setrag, K.: A Decomposition Storage Model. In: Proceedings of the ACM SIGMOD International Conference on Management of Data (SIGMOD 2085), pp. 268–279. ACM Press, Austin (1985)
11. Sang-Won, L., Bongki, M.: Design of flash-based DBMS: an in-page logging approach. In: Proceedings of the ACM SIGMOD International Conference on Management of Data (SIGMOD 2007), pp. 55–66. ACM Press, Beijing (2007)
12. Yi-Reun, K., Kyu-Young, W., Il-Yeol, S.: Page-differential logging: an efficient and DBMS-independent approach for storing data into flash memory. In: Proceedings of the ACM SIGMOD International Conference on Management of Data (SIGMOD 2010), pp. 363–374. ACM Press, Indianapolis (2010)
13. Na, G.-J., Moon, B., Lee, S.-W.: In-Page Logging B-Tree for Flash Memory. In: Zhou, X., Yokota, H., Deng, K., Liu, Q. (eds.) DASFAA 2009. LNCS, vol. 5463, pp. 755–758. Springer, Heidelberg (2009)
14. Gap-Joo, N., Sang-Won, L., Bongki, M.: Dynamic in-page logging for flash-aware B-tre index. In: Proceedings of the 18th ACM Conference on Information and Knowledge Management(CIKM 2009), pp. 1485–1488. ACM Press, Hong Kong (2009)
15. David, G.A., Jason, F., Michael, K., Amar, P., Lawrence, T., Vijay, V.: FAWN: a fast array of wimpy nodes. In: Proceedings of the 23rd ACM Symposium on Operating Systems Principles (SOSP 2011), vol. 54(7), pp. 101–109. ACM Press, Cascais (2011)
16. Biplob, D., Sudipta, S., Jin, L.: FlashStore: High throughput persistent key-value store. In: Proceedings of the VLDB Endowment (VLDB 2010), vol. 3, pp. 1414–1425. ACM Press, Singapore (2010)
17. Biplob, D., Sudipta, S., Jin, L.: SkimpyStash: RAM space skimpy key-value store on flash-based storage. In: Proceedings of the ACM SIGMOD International Conference on Management of Data (SIGMOD 2011), pp. 25–36. ACM Press, Athens (2011)
18. Biplob, D., Sudipta, S., Jin, L.: ChunkStash: Speeding up Inline Storage Deduplication using Flash Memory. In: 2010 USENIX Conference on USENIX Annual Technical Conference (ATC), p. 16. ACM Press, Berkeley (2010)
19. Duck-Ho, B., Ji-Woong, C., Sang-Wook, K.: Clustering and Non-clustering Effects in Flash Memory Databases. In: Database and Expert Systems Applications, DEXA, International Workshops, pp. 4–8. IEEE Press, Linz (2009)

A Dual-Grained FTL for Flash Memory

Junjie Wang, Lihua Yue, Peiquan Jin, and Rui Wang

School of Computer Science and Technology,
University of Science and Technology of China, 230027, Hefei, China
wjj107@mail.ustc.edu.cn

Abstract. Flash memory has been widely used in both embedded devices and
enterprise storage devices, due to its specific characteristics such as small size,
light weight, high speed, shock resistance, and less energy consumption. However, in order to deal with the special limitation of flash memory, i.e., erase-
before-write, an intermediate software layer called *flash translation layer* (FTL)
was employed in modern flash-based disks to map logical page addresses from
the file system to physical page addresses used in flash memory. However,
most existing FTL schemes suffer from the overhead of small random writes
and merge operations, especially full merges. In this paper, we proposed a novel
FTL named DGFTL (Dual-Grained FTL), which divides flash memory into two
regions, namely a page region and a block region. DGFTL uses new algorithms
to manage the dual-grained flash memory and can effectively transfer small
random writes into sequential ones. This leads to more efficient switch merges
and less costly full merges and partial merges. Our experimental results show
that DGFTL reduces the count of erase operation by more than 50% over some
existing flash memory management techniques.

1   Introduction

NAND flash memory has been widely used as storage medium for embedded systems
and portable devices in recent years, due to its specific characteristics, such as small
size, lightweight, non-volatile, high speed, shock resistance, and less energy consumption [1]. However, NAND flash exhibits some different characteristics compared
with magnetic disks. The most important one is that, NAND flash memory is a kind
of *write-once and bulk erase* medium, which is organized in terms of blocks and each
block consists of a fixed number of pages. There are three basic operations for NAND
flash memory, namely read, write and erase operation, where the read and write operations are performed in a page unit but the erase operation is in a block granularity.
Moreover, a block should be erased first if one of its pages needs rewriting. This characteristic is called *erase-before-write*.

In order to cope with the erase-before-write nature of flash memory in the file system, an intermediate software layer called *flash translation layer* (FTL) was employed
[2]. FTL maps logical page addresses from the file system to physical page addresses
used in flash memory. At present, log-buffer-based FTL, also called hybrid-level FTL
[4], is typically used in commercial flash-based disks (also called solid state drives
(SSD)), in which a log buffer is used to save the logs recording the updates occurring

H. Yu et al. (Eds.): DASFAA Workshops 2012, LNCS 7240, pp. 40–52, 2012.

on flash memory. In other words, according to the hybrid-level FTL, flash memory is divided into log blocks for saving logs and data blocks for saving data. However, the hybrid-level FTL introduces new overhead called merge operations [3], which refers that we have to apply the logs in log blocks to the data blocks to produce valid data whenever the log blocks are exhausted. In a merge operation, firstly a victim log block should be selected, and then the logs in the log block are applied to the corresponding data block to generate new data pages, which are then written to a free data block. Finally, the old log block and data block are erased and labeled free.

The performance of log-buffer-based FTL is sensitive to the association of log block, which means the number of data blocks associated with a log block. There are two major techniques to deal with the association, namely BAST and FAST. BAST proposed to associate one data block with one log block, while FAST assigned N data blocks to one log block. The BAST scheme will lead to low space efficiency of flash memory in case that there are a small number of writes occurring on the data block associated with the log block. The FAST scheme is more suitable to this occasion. However, it will result in a lot of merge operations when a great number of writes occur on the associated N data blocks.

In this paper, we propose a novel FTL called DGFTL (Dual-Grained FTL). DGFTL is also a log-buffer-based FTL. However, unlike BAST and FAST, the data blocks in DGFTL are organized with two grains, i.e., a page region and a block region. The main contributions of DGFTL are summarized as follows:

(a) We propose DGFTL, a new FTL scheme with dual grains, in which the data blocks of flash memory are divided into a page region and a block region. The page region is used to cache random page writes, which will finally be evicted and moved to the block region but with a sequential pattern. In DGFTL, we also develop a bitmap-table-based way to enable uniform addresses mapping for both page region and block region (see Section 3).

(b) We conduct comparison experiments on synthetic traces and real TPC-C traces to measure the performance of DGFTL and other competitors including BAST and FAST, in terms of different metrics such as erase count and elapsed time. The results show that our scheme outperforms all the other competitors (see Section 4).

2   Background and Related Work

2.1   Flash Translation Layer(FTL)

FTL is used to encapsulate flash memory into a block device like traditional magnetic disk. FTL is helpful to make flash memory practical in enterprise applications, as file systems can deal with flash-based disks (also known as SSDs) in the same way as they treat magnetic disks.

According to the granularity of address mapping, FTL can be classified into three groups, i.e., page-level FTL, block-level FTL, and hybrid-level FTL.

Page-level FTL maps the logical page number (LPN) into a physical page number (PPN) in flash memory. Therefore, a logical page can be written by the *out-of-place* scheme, which means a logical page can be written to any physical page in flash memory.

In the block-level FTL, the address mapping is implemented on the basis of block granularity. Firstly, a logical page number (LPN) is divided into a logical block number (LBN) and a page offset. Then the LBN is translated to a physical block number (PBN) and finally the page is written at a fixed location in the block which is determined by the page offset. Compared with the page-level FTL, the address mapping table in block-level FTL is much smaller and can be stored in the buffer. However, when a page in a block needs to be updated (rewritten), the other pages in the same block as well as the latest version of the page have to be migrated into a free block. The high cost of page migration will result in poor write performance.

The hybrid-level FTL employs a combination of page-level mapping and block-level mapping. All the physical blocks were divided into data blocks and log blocks. The log blocks are maintained to serve the update information of the page. When all the log blocks are exhausted the merge operation is needed. There are three types of merge operations, namely *full*, *partial* and *switch merge* [3]. The partial merge and switch merge can be done only when the pages in the data block are sequentially updated. While full merge requires many page copies and block erases, partial and switch merges have low overhead for garbage collection.

The design criterion of hybrid-mapping FTL is to avoid full merges and do partial or switch merges whenever possible. However, full merges are difficult to avoid as I/O requests are usually with different patterns. As a consequence, how to reduce or even avoid full merges has been a critical issue for all hybrid-mapping FTL schemes.

2.2   BAST and FAST

BAST and FAST are two representatives of hybrid-level FTLs. BAST and FAST are both hybrid-level FTLs except that they use different block association policies, i.e., 1:1 log block mapping in BAST [3] and 1: N log block mapping in FAST [4].

The block association policy means how many data blocks a log block can be used for. In the 1:1 scheme, a log block is allocated for only one data block. The 1:1 log block mapping of BAST can invoke frequent log block merges. Furthermore, the log blocks in BAST would show very low space utilization when they are replaced from the log buffer. If the write request pattern is random, the 1:1 mapping scheme shows poor performance since frequent log block merges are inevitable. Such a phenomenon where most write requests invoke a block merge is called log block thrashing.

To prevent the log block thrashing problem, the 1: N mapping scheme of FAST was proposed. In 1: N scheme, a log block can be used for multiple data blocks at a time. Using the 1: N mapping, we can prevent the log block thrashing problem. However, the problem of 1: N mapping is its high block associability, where the block associability means how many data blocks are associated with a log block. This means that FAST scheme requires a large cost per block merge though it invokes a small number of block merge. The maximum block associability is same to the number of pages in a block.

However, FAST has a higher cost that BAST when reclaiming a single log block. That is because every log block in FAST is associated with N data blocks, so when a log block is reclaimed, $N+1$ erase operations will be triggered. FAST also maintains a *sequential log block* to store the sequential writes so as to reduce full merge operations, but it only uses a simple hash function on logical page number to determine whether a page write is sequential   or not. This will lead to a lot of partial merges.

Compared with FAST, DGFTL scheme uses a page region to collect all the random writes, and conducts an in-buffer clustering step to transfer random writes into sequential ones. This technique can avoid frequent partial merges in FAST. For example, suppose a block contains four pages numbered 0, 1, 2 and 3, and there are five updates sequentially focusing on page 0, 1, 4, 2, 3, FAST will put those logs into the sequential log block as well as the random log block and finally triggers a lot of partial merges on both sequential and random log blocks. However, in DGFTL, the page region will buffer all the five updates and re-organize 0, 1, 2 and 3 into a cluster, which is then written to the block region. This ensures that we can perform a switch merge to apply those updates, as page 0, 1, 2 and 3 are located in the same block.

2.3   Other FTL Schemes

There are also some other FTL schemes proposed in recent years, such as Superblock FTL [5], SAST (Set-Associative Sector Translation) [6], LAST (Locality-Aware Sector Translation) [7], and LazyFTL [14].

Both Superblock FTL and SAST are depending on the block association policy, i.e., N: K log block mapping, which means N data blocks share (at most) K log blocks. The difference is that Superblock FTL keeps a page-level mapping in the spare areas of the super-block while SAST restricts the number of log blocks and maintains the page-level mapping table in the buffer. However, both Superblock FTL and SAST use some pre-tuned parameters and those parameters are not suitable for different access patterns.

Unlike Superblock FTL and SAST, LAST divides LBAs into several segments to fully utilize the log blocks and keep the reclamation overhead as low as possible. By exploiting the locality of storage access patterns, LAST places write requests into different segments. However, the reclamation overhead of a random log block is very high because of the high association of the log block. LazyFTL divides the entire flash memory into four segments, one of which is used to store mapping information.

3   Dual-Grained FTL (DGFTL)

In this section, we will discuss the basic idea and details of DGFTL. The motivation of DGFTL is to reduce full merges and partial merges by introducing a page region to cluster random page writes into sequential ones. This will result in more switch merges and reduce full and partial merges. In the following we will introduce the basic idea of our scheme, and then the detailed algorithm for region partitioning and mapping.

3.1   Basic Idea

The typical workload in real applications is a mixture of random writes and sequential writes. Random writes often result in a large number of full merge operations. Furthermore, a series of sequential writes will be interrupted by a random write and in turn a lot of partial merges will be introduced. Therefore, the motivation of DGFTL is to detect the random writes that appear in a series of sequential writes so that more switch merges are expected to be performed. For this purpose, DGFTL divides the data blocks in flash memory into two regions, namely a page region (PR) and a block region (BR), as illustrated in Fig.1. Random writes are first placed in the page region. They are then clustered into sequential writes and written to the block region. The page region is accessed in a page granularity and the page-level address mapping is applied. Besides, the page region is not associated with log blocks. This is because the page-level mapping scheme is efficient to reclaim a block, no matter what kind of write pattern occurs. However, the block region is associated with a log region (as shown in Fig.1). Each log block in the log region can be associated with multiple data blocks in the block region, which is similar with FAST [4]. Since the block region only absorbs sequential writes from the page region, the number of data blocks associated with a log block is very small, which will decrease the overhead of garbage collection in the block region. Hence, through the dual-grained regions, DGFTL can transfer random writes into sequential ones and avoid the merge operation caused by random write.

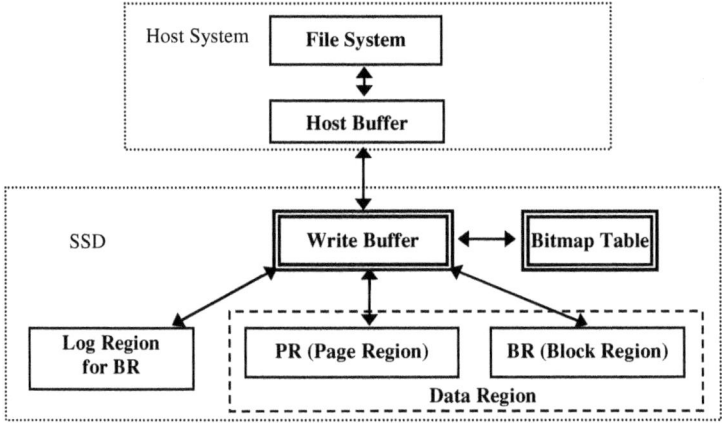

Fig. 1. Overall architecture of DGFTL

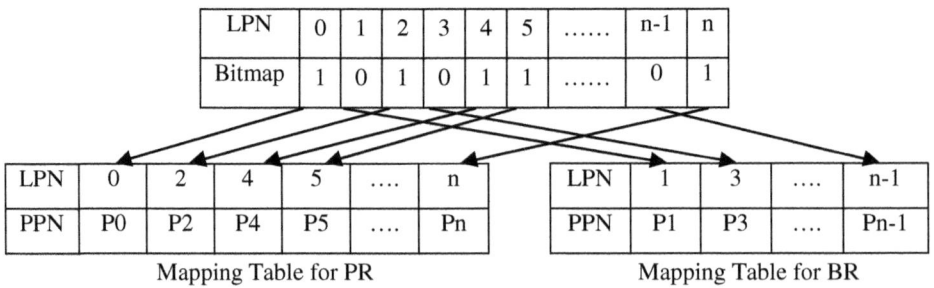

Fig. 2. Bitmap table for address mapping in DGFTL

Moreover, the page region shares the same address space with the block region. To accomplish the uniform addressing goal, we introduce a bitmap table in DGFTL to maintain the address mapping information, as shown in Fig.2. Each page is assigned a bitmap indicating its current status, where a 1-bit means that the corresponding page has been written into the page region and a 0-bit indicates that the corresponding page has been written into the block region. Based on the bitmap table, we can quickly determine the target region that the latest version of a page is located in.

3.2   Page Write in DGFTL

In this section, we will explain how DGFTL deal with page write requests. A page write is first sent to the page region, where all the page writes are clustered into sequential ones and written into the block region.

When a page write request arrives, the first step is to determine which region the page should be written in. In DGFTL, sequential writes are directly written to the block region, while random writes should be absorbed by the page region. Therefore, the type of write request should be recognized at first. In the previous work a locality detecting policy was proposed [7]. This policy determines the locality type by comparing the size of each request with a threshold value. However, this policy does not consider the impact of write buffer. There are also some previous works proposed to utilize the write buffer to detect the pattern of write requests [8, 12, and 13]. They all employ a cluster-based approach, in which the pages in the buffer are grouped into several clusters and the pages in the same cluster will be written into flash memory sequentially. This is much like a sequential write.

We also adopt the cluster-based approach in DGFTL. Our approach is based on a previous algorithm called CLC [8]. CLC is a representative cluster-based buffer management policy. It takes into account both temporal locality and cluster size when selecting a victim. In DGFTL, we use an improved CLC buffer management policy, which is named CLC-t. In the CLC-t scheme, we add a variable threshold to control which region the victim cluster should be written to. A threshold parameter NT is introduced in CLC-t. When a victim cluster is to be written into flash memory, we first compare the size of cluster with NT. If the size is bigger than NT then the cluster will be written into the block region, otherwise the victim cluster will be placed in the page region.

The CLC-t algorithm determines the right region (page region or block region) for a page write request. After that, the page will be written into the page region or the block region. Fig.3 shows the detailed algorithm of page write. Note the inputted parameter *flag* is determined by the CLC-t algorithm. In particular, it is determined on the basis of NT. Basically, *flag* is one means that the corresponding page should be written into the page region, and otherwise the page should be written into the block region. When *flag* is one, first we need to judge whether there is a free page in the page region. If there is no free page, a migration operation will be triggered to reclaim a block in the page region. The migration operation aims to move the pages in the page region to the block region. The principle of migration operation includes the following two rules:

(a) The migrated pages in the page region should be within the same logical block.

(b) The logical block that the migrated pages are located in contains the maximum count of pages in the page region.

This process will be easily implemented with the help of the bitmap table. Based on the migration operation, DGFTL aggregates the pages in the page region into clusters. Besides, if there is already an old version of the page but it is not located in the page region, we have to perform some additional operations to alter the bitmap table and modify the address mapping information.

On the other hand, if *flag* is zero, the page should be written into the block region, this process is similar with that in FAST [4], except that in DGFTL we have to perform some additional operations to ensure the data consistency (as shown in Line 22 to 25 in the algorithm).

Algorithm 1: write(LPN, data, flag)   /*data are logically written to the page LPN, and flag would be decided by our buffer algorithm*/

```
1:  LBN:=LPN div PagesPerBlock;
2:  offset:=LBN mod PagesPerBlock;
3:  if flag is nonzero    /*data should be written to page region(PR)*/
4:        if there are no empty pages in PR region
5:              first perform a migration operation, and then select the victim block which has
              the maximum dirty pages;
6:              move the valid pages to the reserved block and modify the map information;
7:              erase the victim block;
8:              put the victim block into the free block pool and get the least erase block as a
              new reserved block;
9:        end if
10:       get an empty page and write the data, update PMT ;
11:       if Bitmap[LPN]==0
12:             delete corresponding map information of page LPN in the BMT or SMT;
              /*PMT means the page-level mapping table for PR region, BMT means the
              block-level mapping table for BR, and SMT is the mapping table for the log
              block of log region*/
13:             set Bitmap[LPN] to one;
14:       end if
15:  else
16:       PBN:=getPbnFromBMT(LBN); /*get PBN from BMT */
17:       if a collision occurs at offset of the data block PBN
18:             place the data to the appropriate log block in the region of BR;
19:       else
20:             write data at offset in the data block of PBN;
21:       end if
22:       if Bitmap[LPN]==1
23:             delete corresponding map information of page LPN in the PMT;
24:             set Bitmap[LPN] to zero;
25:       end if
26:  end if
```

Fig. 3. The algorithm of write operation in DGFTL

4   Performace Evaluation

In this section, we compare the performance of DGFTL with other two FTLs, i.e., BAST and FAST, with respect to number of erase operations and total elapsed time.

4.1   Evaluation Methodology

In order to make a comparison with three policies that introduced in the front, we performed the evaluation on Flash-DBSim [9]. Flash-DBSim is a reusable and reconfigurable framework for the simulation-based evaluation of algorithms on flash disks. It can be regarded as a reconfigurable SSD. In our experiment, we simulate a 1GB NAND flash device with 64 pages per block and 2KB per page. In the experiment, we refer to the Samsung K9XXG08UXA flash chip in the Flash-DBSim. The detailed parameters of the selected flash chips are listed in Table 1.

Table 1. The characteristics of NAND flash memory [10]

Operation	Access Time	Access Granularity
read	80 μs/pages	Page(2KB)
write	200 μs/pages	Page(2KB)
erase	1.5 ms/block	Block(128KB=64pages)

In the experiments we use three types of traces. The first one is a one-hour OLTP trace in a real bank system, as shown in Table 2. The trace contains 607391 page references focusing on a CODASYL database with a total size of 20 Gigabyte. The second type of trace is generated by DiskSim [11]. Table 3 shows the details about this type of traces. The locality "50%/50%" means fifty percentages of requests focusing on fifty percentages of total pages. The last type of trace is a TPC-C trace which was collected by executing undo operations on an Oracle database engine. Note that all of those two types of traces introduced before are write only requests. Moreover, in order to get a simple description, the log block in BAST, the log block in FAST, and the block in the page region and log block in the block region of DGFTL are all called *page-level mapping blocks* (PB).

Table 2. The real OLTP traces

Attribute	Value
Total Buffer Requests	607391
Data Size	20 GB
Page Size	2048B
Duration	One hour
Total Different Pages Accessed	51870
Read / Write Ratio	77% / 23%

Table 3. Two types of synthetic traces (write only)

Type	Total References	Different Pages Accessed	Locality
T1	300,000	42441	50%/50%
T2	300,000	49637	80%/20%

For each given trace, the simulator counts the number of read, write, and erase operations FTL generates, and calculates the total elapsed time using the following formula, and that the time spend by three operations are depicted in the Table 1.

$$Total\_elapse\_time = read\_count * read\_time + write\_count * write\_time + erase\_count * erase\_time$$

For the buffer management policy, we use CLC for BAST and FAST and CLC-t for DGFTL.

4.2   Influence of NT

Fig.4 shows the effect of the threshold variable NT in CLC-t. Here we use the real OLTP trace, and set the page-level mapping blocks to 32. The X axis means the value of NT and Y axis indicates the total elapsed time. Fig.4 shows that the elapsed time first decreases but then increases when NT exceeds 24. That is because a lot of small-sized clusters are written into the page region when NT is less than 24. But as the value of NT grows, many pages with sequential access property will be wrongly placed in the page region. In such case we have to move data from the page region to the block region which produces a lot of extra reads and writes. However, as the size of buffer increases, the effect of clustering will be much better in DGFTL. In the remainder of our experiment the value of NT is set to 24 and the buffer size is set to 1 MB.

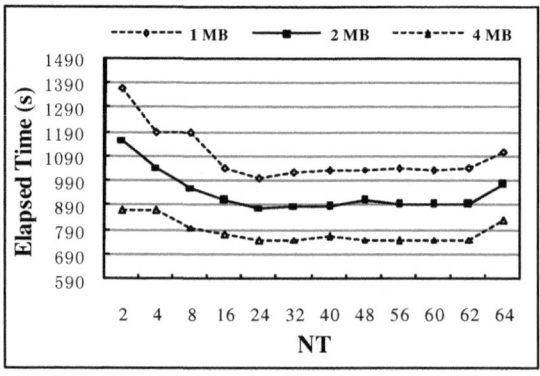

Fig. 4. Performance comparison by different NT

4.3   Performance Comparison

The results of our simulation are shown in Table 4 and 5. The performance metrics we used are the number of total erase count and elapsed time. Compared to read/write operations, the erase operation is more time consuming and, therefore, the efficiency of an FTL scheme mainly depends on how many erase operations it can avoid. In the experiment, we test the impact of the number of page-level mapping blocks by increasing the number of page-level mapping blocks from 32 to 256 for each configuration. In addition, the number of log blocks in the block region of DGFTL is set to 8 for every configuration.

Table 4. The erase count with different count of PB

Trace type	The real OLTP trace				T2			
Size of PB	256	128	64	32	256	128	64	32
BAST	**162743**	219338	250172	263442	**97193**	139607	165659	178409
FAST	**30590**	36936	47870	66018	**22659**	25562	31769	45695
DGFTL	**15957**	21170	32869	57542	**10448**	15389	24576	41885
Trace type	T1				TPC-C			
Size of PB	256	128	64	32	256	128	64	32
BAST	**110642**	152643	180552	198010	**50142**	60171	103375	166193
FAST	**23882**	26565	33755	49092	**47542**	51226	55834	61615
DGFTL	**11282**	16949	27509	45958	**16557**	18846	19689	22889

Table 5. The elapsed time with different count of PB (second)

Trace type	The real OLTP trace				T2			
Size of PB	256	128	64	32	256	128	64	32
BAST	**1319**	1917	2404	2669	**883**	1323	1652	1843
FAST	**426**	527	707	997	**312**	357	467	696
DGFTL	**289**	374	573	1001	**192**	272	426	735
Trace type	T1				TPC-C			
Size of PB	256	128	64	32	256	128	64	32
BAST	**1012**	1469	1818	2085	**902**	1082	1104	1743
FAST	**330**	376	502	753	**691**	738	802	889
DGFTL	**207**	299	479	808	**249**	289	305	336

Overall, DGFTL exhibits shorter garbage collection overhead than other schemes. Especially, as the count of page-level mapping blocks increased the superiority of DGFTL became more obvious. The DGFTL scheme outperforms FAST by reducing the number of erase operations by more than 50% over the whole trace when the page-level mapping blocks increase to 256.

BAST shows the largest count of erase operations compared to other FTLs. As described in section 2.2, for the block thrashing problem BAST cannot efficiently handle random write patterns. Therefore, there is a large number of full merge operation caused by random writes. However, our proposed scheme would place the small random write request to the page region which could dramatically decrease the amount of full merge operations.

Table 6. The association of Log Block

Trace type	The real OLTP trace				T2			
Size of PB	256	128	64	32	256	128	64	32
FAST	2.43	2.9	3.73	5.12	3.09	3.41	4.19	5.98
DGFTL	1.98	2.38	3.09	4.26	1.60	1.99	2.76	4.20
Trace type	T1				TPC-C			
Size of PB	256	128	64	32	256	128	64	32
FAST	3.28	3.57	4.49	6.48	1.87	2	2.17	2.38
DGFTL	1.59	2	2.85	4.38	1.15	1.23	1.27	1.34

FAST exhibits a better garbage collection performance than BAST by efficiently remove the block thrashing problem. However, it cannot outperform DGFTL because of its log block's high association which could cause a higher cost for garbage collection especially for random writes. For DGFTL, random writes were clustered into sequential writes and then migrated into block region which effectively lower the association of log block. In the Table 6, the association of the log block is less than 2 when the size of page-level mapping blocks increase to 256. It's very simple to find that the association of log block is much less than FAST at every configuration. Therefore, much less time was spending for DGFTL to reclaim a log block.

However, there is a little bit more total run time for DGFTL when the page-level mapping block is set to 32 in the beginning. This phenomenon was lead by the migration operation, this operation is a kind of extra operation which is not exist in the BAST and FAST, especially when the size of the page region is too small this operation will be frequently called and bring about more overhead.

Extraordinary, for the real OLTP trace with benchmark TPC-C DGFTL shows a much more improvement, as shown in Table 5 and 6. The access patterns of TPC-C are categorized into two types: random writes with high temporal locality and sequential writes. Because these two types of write requests are simultaneously issued from the file system, they arrive at the FTL layer in mixed patterns of random and sequential accesses. DGFTL could efficiently separates out random and sequential accesses and places the different kind of access to different region. However, the sequential writes will be interrupting by the random writes which will decrease the number of partial or switch merge operation in the scheme of FAST, but increase full merge operation. Therefore, FAST shows a poor performance.

One interesting result is that the performance of DGFTL is better than BAST and FAST as the number of page-level mapping block increased. In Fig.5, this result was got by real OLTP trace, others have a similar result, the X axis means every time the number of the page-level mapping block increased, and Y axis means the proportion of reduction in the number of erase operations compared with the configuration that the page-level mapping block is set to 32. In Fig.5 we can find that when the size of page-level mapping block is increased from 32 to 256 there is about 72.27% reduction of erase operations, but at the same situation only 53.66% and 38.22% for FAST and BAST. Therefore, DGFTL is much more sensitive to the increase of page-level mapping blocks than BAST and FAST. So in order to get a better performance we can assign more blocks to the page region.

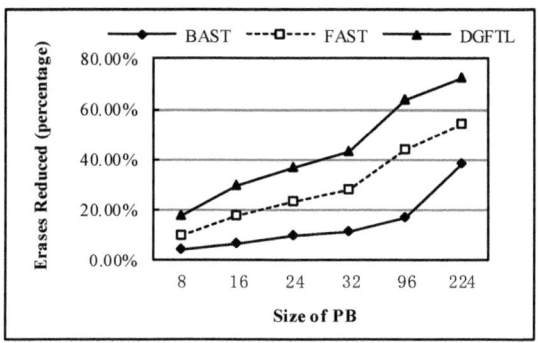

Fig. 5. The performance improved with PB increased

5   Conclusions

In this paper, we proposed a novel FTL scheme, called DGFTL, which outperforms the well-known BAST and FAST. DGFTL can effectively decrease the association of log block through the introduction of threshold of NT. Meanwhile, by the assistance of the CLC-t algorithm DGFTL can efficiently extract the sequential writes from the random writes and isolate them into different regions. The dual regions together with the migration operation and the designed write operation can reduce many unnecessary merge operations and improve the performance of garbage collection. In the future, we will design a new data structure for the page region, which will place the page from the same logical block into the same physical block as much as possible. This design could further decrease the number of read, write and erase operations when we move pages from the page region to the block region.

Acknowledgement. This paper is supported by the National Science Foundation of China (No. 60833005 and No. 61073039).

References

[1] Wikipedia: Flash memory (2010),
    http://en.wikipedia.org/wiki/Flashmemory
[2] Intel Corporation: Understanding the Flash Translation Layer (FTL) Specification. Technical report (1992)

[3] Kim, J., Kim, J.M., Noh, S.H., et al.: A Space-efficient Flash Translation Layer for Compact Flash Systems. IEEE Transactions on Consumer Electronics 48(2), 366–375 (2002)

[4] Lee, S.-W., Park, D.-J., Chung, T.-S., et al.: A Log Buffer Based Flash Translation Layer using Fully Associative Sector Translation. ACM Transactions on Embedded Computing Systems 6(3), article 18 (2007)

[5] Kang, J.-U., Jo, H., Kim, J.-S., et al.: A Superblock-based Flash Translation Layer for NAND Flash Memory. In: EMSOFT 2006 (2006)

[6] Park, C., Cheon, W., Kang, J., et al.: A Reconfigurable FTL (Flash Translation Layer) Architecture for NAND Flash-Based Applications. ACM Transactions on Embedded Computing Systems 7(4), article 38 (2008)

[7] Lee, S., Shin, D., Kim, Y.-J., et al.: LAST: Locality-Aware Sector Translation for NAND Flash Memory-Based Storage Systems. ACM SIGOPS Operating Systems Review 42(6), 36–42 (2008)

[8] Kang, S., Park, S., Jung, H., et al.: Performance Trade-Offs in Using NVRAM Write Buffer for Flash Memory-Based Storage Devices. IEEE Trans. Computers 58(6), 744–758 (2009)

[9] Jin, P.Q., Su, X., Li, Z., et al.: A Flexible Simulation Environment for Flash-aware Algorithms. In: Proc. of CIKM 2009, demo. ACM Press (2009)

[10] Lee, S.-W., Moon, B.: Design of Flash-Based DBMS: An In-Page Logging Approach. In: SIGMOD 2007, Beijing, China (June 2007)

[11] Bucy, J.S., Schindler, J., Schlosser, S.W., et al.: The DiskSim Simulation Environment Version 4.0 Reference Manual. Carnegie Mellon University Technical Report (2008)

[12] Kim, H., Ahn, S.: BPLRU: A Buffer Management Scheme for Improving Random Writes in Flash Storage. In: Proc. Sixth USENIX Conf. File and Storage Technologies (2008)

[13] Zhao, H., Jin, P.Q., Yang, P.Y., et al.: BPCLC: An Efficient Write Buffer Management Scheme for Flash-Based Solid State Disks. International Journal of Digital Content Technology and its Applications 4(6), 123–133 (2010)

[14] Ma, D.Z., Feng, J.H., Li, G.L.: LazyFTL: A Page-level Flash Translation Layer Optimized for NAND Flash Memory. In: SIGMOD 2011 (2011)

Impact of Storage Technology on the Efficiency of Cluster-Based High-Dimensional Index Creation

Gylfi Þór Gudmundsson[1], Laurent Amsaleg[1,2], and Björn Þór Jónsson[3]

[1] INRIA, Rennes, France
{gylfi.gudmundsson,laurent.amsaleg}@inria.fr
[2] CNRS, Rennes, France
[3] School of Computer Science, Reykjavík University, Iceland
bjorn@ru.is

Abstract. The scale of multimedia data collections is expanding at a very fast rate. In order to cope with this growth, the high-dimensional indexing methods used for content-based multimedia retrieval must adapt gracefully to secondary storage. Recent progress in storage technology, however, means that algorithm designers must now cope with a spectrum of secondary storage solutions, ranging from traditional magnetic hard drives to state-of-the-art solid state disks. We study the impact of storage technology on a simple, prototypical high-dimensional indexing method for large scale query processing. We show that while the algorithm implementation deeply impacts the performance of the indexing method, the choice of underlying storage technology is equally important.

1 Introduction

Multimedia data collections are now extremely large and algorithms performing content-based retrieval must deal with secondary storage. Magnetic disks have been around for decades, but their performance, aside from capacity, has not improved significantly. Better storage performance, and improved reliability, has been achieved, however, by grouping many disks together and striping data as in the Redundant Array of Independent Disks approach. Recently, Solid-State Disks (SSDs) have emerged as a disruptive storage technique based on memory cells on chips. Their storage capacity grows quickly and they outperform magnetic approaches. It is therefore of high interest to study what impact secondary storage technologies can have on the design and performance of high-dimensional indexing algorithms that are a core component of content-based retrieval approaches. This paper is an initial investigation in that direction.

1.1 Content-Based Retrieval and High-Dimensional Indexing

Retrieving multimedia documents based on their *content* means that the search analyzes the actual content of the documents rather than metadata associated with the documents, such as keywords, tags, and free text descriptions [11]. For images, the term content might, for example, refer to colors, shapes, texture, or

H. Yu et al. (Eds.): DASFAA Workshops 2012, LNCS 7240, pp. 53–64, 2012.
© Springer-Verlag Berlin Heidelberg 2012

any other information, possibly very fine-grained, that can be derived from the image itself [7]. Users can then submit a photo to the system to search for multimedia material that is visually similar. For music, information such as pitch, energy and timbral features can be used for such content-based retrieval. Searching by content is needed for such applications as image copyright enforcement, face recognition, query by humming, or automatic image classification [5].

Content-based retrieval systems are frequently built from two primary software components: the first component automatically extracts some low level features from the multimedia material; while the second component builds an index over this database of features allowing for efficient search and retrieval. For images, the computer vision literature includes numerous feature extraction techniques. The visual properties of each image are encoded as a set of numerical values, which define high-dimensional vectors. There is one such vector per image when the description is global, for example when a color histogram encodes the color diversity of the entire image. Some applications require a much more fine-grained description; in this case multiple features are extracted from small regions in each image [12]. Typically, the texture and/or the shape that are observed in each region of interest are somehow encoded in a vector. Overall, the similarity between images is determined by computing the distance between the vector(s) extracted from the query image and the ones kept in the database. As vectors are high dimensional, repeatedly executing this distance function (typically Euclidean) to run a *k-nearest neighbor* search is CPU intensive.

Efficient retrieval is facilitated by high-dimensional indexing techniques that prune the search space (e.g, see [15]). Most techniques from the abundant literature partition the database of features extracted from the collection of images into cells and maintain a tree of cell representatives. The search then takes a query vector, traverses the tree according to the closest representative at each level, fetches bottom leafs containing database features, computes distances and returns the k nearest-neighbors found. At a large scale, when indexing a few hundred millions vectors or more, approximate search schemes returning near-neighbors (possibly not the nearest) must be used for efficiency, potentially trading result quality for response time [1,6,10].

1.2   Contribution

We have created two very different implementations of a rather simple, yet very effective, high-dimensional indexing strategy relying on clustering to partition the database of features [8]. These implementations differ in the way they access secondary storage, emphasizing small vs. large I/Os and random vs. sequential I/Os. We have then run these implementations on a machine connected to various magnetic storage devices, as well as various SSD devices, and accurately logged their respective performance.

This study focuses on the efficiency of the index creation, which is the most disk-intensive phase, far more intensive than the search phase. Not only must the entire data collection be read from secondary storage during index creation, and then eventually written back to secondary storage again, but a gigantic

number of CPU intensive distance calculations between vectors are also required
to cluster them. High-dimensional indexes are typically created in a bulk manner:
all vectors to index are known before the process starts and the index tree
structure, as well as the bottom leafs of the tree, are all created in one go. From
a traditional DBMS perspective, this process can be seen as being analogous to
a sort-merge process with very CPU-intensive comparison function calls.

With our detailed analysis, we show that a good understanding of the underlying hardware is required at design time to get optimal performance. In
particular, as the devil is in the details, we show that the theoretical behavior of
storage devices may differ much from what is observed in the battlefield and that
even members of a presumably homogeneous family of storage solutions may be-
have quite differently. We also show that balancing efforts on software versus
hardware is not always obvious: it is key to clearly evaluate the cost of paying a
very skilled programmer to nicely tune and debug a complex piece of code versus
producing a naive implementation and throwing at it efficient hardware.

1.3 The Case for Reality

This investigation is done using real algorithms working on real data with standard production hardware. Working with various technical specifications does
not accurately reflect the complex layers of interactions between application,
operating system and I/O devices (disks, network) while processing real data.

We implemented a cluster-based high-dimensional indexing algorithm that is
prototypical of many approaches published in the literature. The algorithm is
approximate in order to cope with truly large high-dimensional collections. It has
an initial off-line phase that builds an index. This is a very demanding process:
the entire raw data collection is read from disk and vectors that are close in the
feature space are clustered together and then written back to disk. This process
is essentially I/O bound.

The search phase is quite different, but it is also prototypical of what has
been published. From a query submitted by a user, a few candidate clusters are
identified using the index, fetched from disks, and then the CPU is heavily used
for scanning the clusters in search of similar vectors.

We also use a real data collection made of more than 110 million local SIFT
descriptors [12] computed on real images randomly downloaded from Flickr.
This descriptor collection is clearly not made of independent and identically
distributed random variables as is the case for most synthetic benchmarks; it
has some very dense areas, while some others are very sparse, which together
strongly challenge the indexing and retrieval algorithms. Observing the behavior
of indexing and searching real data is known to give more valuable insights in
general [16], and this certainly extends to the impact of various storage solutions.

1.4 Overview of the Paper

This paper is structured as follows. Section 2 briefly reviews the state of art in
secondary storage techniques. Section 3 gives an overview of the high dimen-

sional indexing algorithm we use in this paper, while Section 4 details the two
implementations which stress differently the underlying hardware. Section 5 then
describes the experiments we performed and Section 6 concludes the paper.

2   Secondary Storage Review

We now briefly review the state-of-the-art in storage technology, focusing on the
aspects that are most relevant for our work.

2.1   Magnetic Disks

Magnetic disks are the standard for cheap secondary storage. During the last
decade, only the capacity of the disks has dramatically increased; the latest is
the terabyte platter announced by Seagate in May 2011. Their read/write performance has improved little as their mechanical parts are inherently slow. This
impacts the performance of small and random operations which are significantly
slower than large sequential ones. Accessing data that is contiguous on disk is
therefore key to disk performance. Note that sophisticated embedded software
in the disk controller tries to minimize the costs of reading and writing (e.g.,
reordering accesses, enforcing contiguity, writing asynchronously, caching) but
programmers have little or no control over these decisions.

2.2   Solid-State Disks

SSD technology is based on flash memory chips. Two types of memory cells are
used: single-level chips (SLC), which are fast and durable, and multi-level chips
(MLC) that take more space and are not as durable, but are cheaper. Recently
Intel, with its 710 Series SSD, introduced new MLC technology that is nearly on
par with SLC in endurance. This new SSD is targeting the needs of the enterprise
with both endurance and capacity.

The memory modules are typically arranged in 128KB blocks. Since there are
no slow mechanical parts, reading from an SSD is extremely fast and sequential or
random reads are equally fast. In contrast, writing is more costly as it sometimes
requires a special erase operation done at the level of an entire 128KB block.
The cost of writes is therefore not uniform and write performance is typically
unpredictable from a programmer's point of view. In addition to minimizing
write costs, internal controlling algorithms do wear-leveling to extend the life
span of the chips. SSD performance has been extensively studied (e.g., see [4]).

With the release of the SATAIII standard, the potential transfer rate doubled,
from 300MB/sec to 600MB/sec. In turn, the SSD vendors released 500+MB/sec.
capable disks for the public market. The enterprise market, however, has shown
more restraint in this area and focused on durability and capacity. For example,
the Intel 710 Series disks have only 270MB/s read capability, and 170MB/s write
capability, far below the capacity of SATAIII.

2.3   Hybrid Disk: Magnetic Disk and SSD in One Device

Seamlessly combining SSD and magnetic disk technology into a single device has long been expected. The first such disk, Momentus XT, was introduced by Seagate in 2010. It is basically a 250-500GB 7200rpm magnetic disk that has a single 4GB SLC cell embedded. However, as there is only one SSD cell, the read/write capacity is limited. Furthermore, the SSD cell is only used for read caching. With such limitations, the Momentus XT is primarily targeting the laptop market where it provides power savings and reduces boot- and loading time. One can fully expect, however, to see rapid advances in this area.

2.4   Network Attached Storage (NAS)

A NAS is typically a quite large secondary storage solution made available over a network. It usually contain an array of disks, operating in parallel thanks to advanced RAID controllers, and made available through a network connected file-servers or dedicated hardware directly connected to the network. The performance of the NAS can be very hard to evaluate as there are many layers of hardware, caching and communication, each with its own bottlenecks. Often the network links between the server and the clients limit the throughput as the links are typically shared by many clients.

2.5   Interactions with the OS

Many sophisticated routines trying to reduce the costs of accessing secondary storage exist in all operating systems. Prefetching is a common technique: instead of reading few bytes at a time from the disk, more data than asked for is brought into RAM in the hope that data needed later will therefore already be in memory, thus avoiding subsequent disk accesses. Reads are blocking operations and the requesting process can only be resumed once the data is in memory, but writes might be handled asynchronously as there is effectively no need to wait for the data to reach the disk. Overall, the operating system uses buffers for I/Os and fills or flushes them when it so desires, trying to overlap the I/O and CPU load as much as possible. If reads or writes are issued too rapidly, there is little overlapping and performance degrades.

3   Extended Cluster Pruning

To study the impact of secondary storage technologies on high-dimensional indexing, we implemented a prototypical index creation scheme built on the Cluster Pruning algorithm [6], as extended by [8]. Cluster Pruning is quite representative of the core principles underpinning many of the quantification-based high-dimensional indexing algorithms that perform very well [17,14,9].

Overall, the extended Cluster Pruning algorithm (eCP) is very much related to the well-known k-means approach. Like k-means, eCP is an unstructured quantifier, thus coping quite well with the true distribution of data in the high-dimensional space [13]. The extension in eCP is to make the algorithm more I/O friendly as the database is assumed to be too large to fit in memory and must therefore reside on secondary storage.

3.1   Index Construction

eCP starts by randomly selecting C points from the data collection, which are used as representatives of the C clusters that eCP will eventually build. The cluster count C is typically chosen such that the average cluster size will fit within one disk I/O, which is key to secondary storage performance. Then the remaining points from the data collections are read, one after the other, and assigned to the closest cluster representative (resulting in Voronoi cells).

When the data collection is large, the representatives are organized in a multi-level hierarchy. This accelerates the assignment step as finding the representative that is closest to a point then has logarithmic complexity (instead of linear complexity). Eventually, once all the raw collection has been processed, then eCP has created C clusters stored sequentially on disks, as well as a tree of representatives which is also kept on disks.

The tree structure is extremely small compared to the clusters as only a few nodes are required to form that tree. The tree is built according to a hierarchical clustering principle where the points used at each level of the tree are representatives of the points stored at the level below. This does not, however, provide total ordering of the clusters and thus this is not a B^+-tree-like index.

3.2   Searching the Index

When searching, the query point is compared to the nodes in the tree structure to find the closest cluster representative. Then the corresponding cluster is accessed from the disk, fetched into memory, and the distances between the query point and all the points in that cluster are computed to get the k nearest neighbors.

The search is approximate as some of the true nearest neighbors may be outside the Voronoi cell under scrutiny. Experience from different application domains has shown that the quality results of eCP can be improved by searching in more than one cell, because this returns better neighbors. In this case, however, more cells must be fetched from secondary storage and more distance computations performed.

3.3   Result Quality of eCP

Experiments have shown shows that eCP returns good quality results despite its approximate behavior [6,8]. Two main reasons can explain this. First, the high-dimensional indexing process produces Voronoi cells that nicely preserve the notion of neighborhood in the feature space. The search process is therefore likely to find the actual nearest neighbors in the clusters identified at query time. Second, most modern image recognition technique use local description of images where a single image is typically described using several hundred high-dimensional vectors. At search time, the many vectors extracted from the query image are all used one after the other to probe the index and get back k neighbors. What is returned for every single query vector is eventually aggregated (typically by voting) to identify the most similar images. Because there is so much redundancy in the description, missing the correct neighbors for some of the query vectors has indeed little impact of the final result.

The cluster count, and the corresponding average cluster size, obviously influences heavily the CPU cost for the search phase as the number of distance calculations is linked to the cardinality of clusters. It also impacts the performance of the index creation as it determines the total number of clusters that must be created to hold the entire collection, and thus influences the number of nodes in the index tree as well as its height and width. This, in turn, impacts the number of distance calculations done to find the cluster into which each point has to be assigned. Experiments with various cluster sizes indicate that using 64-128KB as the average cluster size gives the best performance overall [8].

4   Index Creation Policies

We have designed two implementations of the eCP index creation algorithm that have quite different access patterns to secondary storage. They differ during their *assignment phase*, when assigning vectors to their clusters, and also during their *merging phase*, when forming the final file that is used during the subsequent searches. We have not changed the search process of eCP at all.

Both index creation policies start by building their in-memory index tree by picking leaders from the raw collection. Then they allocate a buffer, called *in-buff*, for reading the raw data collection in large pieces. They then iterate through the raw collection via this buffer, filling it with many not-yet-indexed vectors. The index is used to quickly identify the leader that is the closest to each vector in *in-buff*, representing the cluster that the vector must be assigned to. Once all vectors in *in-buff* have been processed, one of the two policies described below is used to transfer the contents of the buffer to secondary storage.

4.1   Policy 1: TempFiles (TF)

This first policy uses temporary files, one for each cluster. Each temporary file contains all the vectors assigned so far to that cluster. When called, the TF policy loops through the representatives, appending to each temporary file all vectors in *in-buff* assigned to that cluster. When appending to a cluster, its associated temporary file is opened, appended to and closed, as they are too numerous to remain open. When all vectors from *in-buff* have been written to disk, a new large piece from the raw collection is read into *in-buff*, and eCP continues. After having assigned all vectors from the raw collection, all these temporary cluster files are then concatenated into a single final file by reading them sequentially from disk and writing to the final file.

Table 1. Key storage device performance indicators

Disk	Type	Specified Ave. Seek Time	Specified Ave. Rot. Latency	Specified Cache Size	Measured Seq. R/W Thr.put.
Seagate	Magnetic	11.0 ms	4.16 ms	8 MB	46/40 MB/s
Fujitsu	Magnetic	11.5 ms	4.17 ms	16 MB	68/53 MB/s
SuperTalent	SSD	<1 ms	-	Unknown	124/34 MB/s
Intel	SSD	<1 ms	-	16 MB	220/66 MB/s

In terms of access patterns, TF performs, at cluster assignment time, large sequential reads to fill *in-buff* with new vectors as well as many small random writes, one per cluster, every time all the vectors in *in-buff* have been processed. When creating the final file, it also performs cluster-sized sequential reads (one per cluster, typically 128KB) as well as large sequential writes for the final file.

4.2   Policy 2: ChunkFiles (CF)

This second policy generally follows a sort-merge principle. When called, CF sorts *in-buff* on increasing values of the leader identifiers. It then creates a new chunk file on disk and flushes *in-buff* into that chunk file before closing it. It then reads another large piece from the raw collection into *in-buff* and continues. After having processed all vectors from the raw collection, CF merges all the sorted chunk files using a typical secondary storage merging process.

In terms of access patterns, CF performs, at cluster assignment time, large sequential reads (typically 128MB) to fill *in-buff* and large sequential writes when creating each chunk file. When creating the final file, it performs many small random reads to get data from all the chunk files as needed and large sequential writes for the final file.

5   Experiments

5.1   Experimental Setup

In our experiments we used a collection of more than 110 million SIFT descriptors [12] of 128 dimensions extracted from 100,000 images randomly downloaded from Flickr. This collection is about 14.5GB. We reused the parameters from [8] that were found to work best, i.e., the depth of the index was 3 and the average cluster size was 128KB, resulting in 111,424 clusters on secondary storage. Note that clusters are not equally filled as the true distribution of vectors in space is not balanced (30% of the clusters are smaller than 64KB, while 21% are larger than 192KB). In all experiments the size of *in-buff*, and thus each chunk file, is 128MB. Note that this is much larger than the cluster size.

Experiments were run on a Dell Precision T3400, 3GHz Intel E6400 dual core CPU with 6MB cache and 4GB RAM (only one core was used). For all disks we use the ext3 file system and Debian OS. We tested two magnetic

Table 2. Performance of eCP index creation policies, single drive setups

Disk	Total Time I/O+CPU (s)		Assignment I/O (s)		Merging I/O (s)	
	TF	CF	TF	CF	TF	CF
Seagate	43,144	12,949	12,556	548	18,829	1,299
Fujitsu	32,895	12,689	9,975	388	11,145	1,207
SuperTalent	32,540	11,528	17,120	149	3,529	236
Intel	14,164	11,398	2,028	46	402	244
NAS	22,335	14,564	5,314	610	2,349	202

disks: 3.5" Seagate Barracuda 7200.10 and 2.5" Fujitsu MHZ2160BJ. Both are 7200 rpm disks with similar seek time and rotational latency. We also used two SSDs: SuperTalent FTM28GL25H and Intel X-25M, type SSDSA2MH080G1GC. Finally, we used a NAS 3070 from NetApp. Table 1 provides more details on the single drives. The three first columns are filled using vendor figures, while the last column shows sustained observed sequential read and write performance. Accurately measuring the performance of the NAS is much more complicated.

We then ran two different experiments. In the first experiment we used a single drive: the file containing the raw collection, the temporary files/chunk files, and the final cluster file were all stored on a single disk. In this case some reads and writes overlap in time and compete for the disk. This causes slower performance as enforcing truly sequential accesses is much more difficult.

In the second experiment we used two drives. In this case, the raw collection was kept on one drive and the temporary files/chunk files were stored on another drive, eliminating any competition between reads and writes at assignment time. Similarly, the final file and the temporary files/chunk files were stored on different drives; it is sufficient to put the final file on the first drive where the raw collection is to eliminate any competition at merging time. We now detail the performance measurements for these two experiments.

5.2   Single Drive Experiment

Table 2 shows the performance measurements when using the single drive setup. The total (wall clock) time includes the time for I/Os as well as for executing the many distance calculations on the CPU. The CPU usage is almost identical for both TF and CF and equal to 11,000 seconds on average, divided into 10,930 seconds for assignment and 70 seconds for merging. The second and third columns show the overall time it takes to perform the assignment of vectors to leaders and the final merging. These times include the time spent on I/Os but exclude the almost constant CPU costs.

Overall, focusing on the total time, regardless of the device, the first key observation is that CF always outperforms TF. The TF policy repeatedly opens, writes to, and then closes clusters, forcing the OS to flush data on disks using synchronized blocking writes. TF appends data to many relatively small files

Table 3. Performance of eCP index creation policies, two drive setups

	Total Time I/O+CPU (s)		Assignment I/O (s)		Merging I/O (s)	
Two Drive Setup	TF	CF	TF	CF	TF	CF
Fujitsu-Intel-Fujitsu	13,467	11,640	1,977	370	208	220
Intel-Intel'-Intel	13,484	11,301	1,666	67	180	188

(111,424 files of 128KB), in contrast to CF which writes only once to each of fewer but much larger files (109 files of 128MB). The performance of TF differs much from CF with magnetic storage devices as many arm movements are done. Interestingly, for TF, the SuperTalent SSD performs poorly—unfortunately, not all SSDs are equal, as reported in [4]. In contrast, the Intel SSD completely outperforms all the other setups, showing that it handles random reads and writes very well.

Turning to the assignment phase, Table 2 shows that CF spends very little time waiting for I/O. With CF we observed much overlapping between CPU computations and disk requests thanks to OS and device optimizations which keep the processor (usefully) busy while waiting for I/O completion. This explains the very small times for CF, in particular with the Intel SSD which proves to handle competing reads and writes very well. With TF, the assignment phase is CPU bound, suggesting a look at parallelism.

The merging phase for TF is costly due to the multitude of (relatively) small file accesses compared to CF. Merging for CF also greatly benefits from the prefetching done by the OS: the few large files are brought into memory before the data gets processed, reducing I/O cost. Prefetching is less profitable for merging with TF as many small files are involved.

5.3   Two Drive Experiment

By using separate physical drives for the reading and writing, competition for the disk is potentially eliminated. We observed that the larger costs occur when writing the assigned vectors to disk and then reading them back as in both cases many random accesses are performed; using an SSD is therefore ideal to speed up indexing. We defined two setups: First, we kept the input and final output on the Fujitsu (the magnetic disk with the best observed performance) but used an Intel SSD for the intermediate files. The second setup used two identical Intel SSDs. The performance measurements for these setups are reported in Table 3.

The table shows that using the SSD for costly random operations provides dramatic total time improvements, regardless of the type of the other device. With the Fujitsu-Intel-Fujitsu setup, the Intel 66MB/s write speed matches closely the Fujitsu 68MB/s read capacity. Replacing this magnetic device with an SSD does not help much as their total times are very similar. One key lesson is that it is not necessary to put SSDs everywhere, which could be terribly expensive, but to use them solely where random accesses are massively needed. This greatly reduces costs, both in terms of performance and money. Note that it is the CPU cost that dominates the time for CF, with I/Os being relatively cheap. The TF policy, however, suffers again from the multitude of small files.

Turning to assignment, CF again outperforms TF since it is dealing with a multitude of small files with blocking write accesses. CF with Fujitsu-Intel-Fujitsu is limited by the time it takes to read the data from the magnetic source: it has a lot of CPU to do once the *in-buff* buffer gets filled and the disk is not accessed again for some time, long enough to have the disk entering a power-saving mode, typically reducing its rotational speed. This, in turn, increases the cost of the next data request. The Intel-Intel'-Intel setup has no such problems and its performance is extremely good. It turns out to be slightly above what was observed in Table 2; the reasons are unclear, but some fluctuations have been observed. Note, however, that 67 seconds are insignificant with respect to the total time of more than 13 thousand seconds.

The quite small values during merging, for both policies and both setups, show the improvements from the lack of competition between reads and writes as they are directed to different drives. Even SSDs suffer from I/O competition.

5.4   Other Results

We also checked the impact of a larger *in-buff* on the performance, both for the single drive and two drive setups. As expected, enlarging *in-buff* speeds up TF as large *in-buff* reduces the number of random writes that are necessary. In contrast, larger writes slow down CF because the OS is better at overlapping CPU and I/Os when performing writes in bursts as large writes overwhelm buffers.

SSD performance degradation over time has been reported in other studies, especially for certain IO patterns (e.g., see [2,3]). We therefore monitored the performance of our SSDs over time but did not find any significant performance change. Our pattern of always ending with a large sequential write may work to the advantage of the SSDs by preventing such degredation.

6   Conclusion

We have created two very different implementations of a rather simple, yet very effective, cluster-based high-dimensional indexing strategy. These implementations differ in the way they access secondary storage, emphasizing small vs. large I/Os or random vs. sequential I/Os. We have then run these implementations on a machine connected to various magnetic storage devices, as well as various SSDs devices, and measured the performance.

Our results show that the secondary storage devices used for large scale high-dimensional indexing are key to performance. On the one hand, a carefully crafted implementation can get good performance when using traditional magnetic devices. On the other hand, simpler implementations, potentially saving RAM, can perform very well when high performance SSDs devices are used, as they cope very well with random accesses. SSDs, however, are not the magical answer to all performance problems: their capacity is still limited; their price is so far very high, although this will probably quickly change; and their observed performance varies tremendously from one model to the other.

But generalizing this result, one should carefully evaluate the cost of an extremely sophisticated implementation versus buying efficient storage devices and placing them along the performance-critical paths.

Acknowledgement. This work was partly achieved as part of the Quaero Project, funded by OSEO, French State agency for innovation.

References

1. Andoni, A., Indyk, P.: Near-optimal hashing algorithms for approximate nearest neighbor in high dimensions. Commun. ACM 51, 117–122 (2008)
2. Athanassoulis, M., Ailamaki, A., Chen, S., Gibbons, P.B., Stoica, R.: Flash in a dbms: Where and how? IEEE Data Eng. Bull. 33(4), 28–34 (2010)
3. Bonnet, P., Bouganim, L.: Flash device support for database management. In: CIDR, pp. 1–8 (2011), `www.crdrdb.org`
4. Bouganim, L., Jónsson, B.T., Bonnet, P.: uFLIP: Understanding flash IO patterns. In: Proc. CIDR (2009)
5. Casey, M., Veltkamp, R., Goto, M., Leman, M., Rhodes, C., Slaney, M.: Content-based music information retrieval: Current directions and future challenges. Proceedings of the IEEE 96(4), 668–696 (2008)
6. Chierichetti, F., Panconesi, A., Raghavan, P., Sozio, M., Tiberi, A., Upfal, E.: Finding near neighbors through cluster pruning. In: Proc. PODS (2007)
7. Datta, R., Joshi, D., Li, J., Wang, J.Z.: Image retrieval: Ideas, influences, and trends of the new age. ACM Comput. Surv. 40, 5:1–5:60 (2008)
8. Gudmundsson, G., Jónsson, B.T., Amsaleg, L.: A large-scale performance study of cluster-based high-dimensional indexing. In: Proc. ACMMM–Workshop on Very-Large-Scale Multimedia Corpus, Mining and Retrieval (2010)
9. Jégou, H., Douze, M., Schmid, C.: Product quantization for nearest neighbor search. IEEE TPAMI 33(1), 117–128 (2011)
10. Lejsek, H., Ásmundsson, F.H., Jónsson, B.T., Amsaleg, L.: NV-Tree: An efficient disk-based index for approximate search in very large high-dimensional collections. IEEE Trans. Pattern Anal. Mach. Intell. 31, 869–883 (2009)
11. Lew, M.S., Sebe, N., Djeraba, C., Jain, R.: Content-based multimedia information retrieval: State of the art and challenges. ACM Trans. Multimedia Comput. Commun. Appl. 2, 1–19 (2006)
12. Lowe, D.G.: Distinctive image features from scale-invariant keypoints. International Journal of Computer Vision 60(2) (2004)
13. Paulevé, L., Jégou, H., Amsaleg, L.: Locality sensitive hashing: A comparison of hash function types and querying mechanisms. Pattern Recognition Letters 31(11), 1348–1358 (2010)
14. Philbin, J., Chum, O., Isard, M., Sivic, J., Zisserman, A.: Lost in quantization: Improving particular object retrieval in large scale image databases. In: Proc. CVPR (2008)
15. Samet, H.: Foundations of Multidimensional and Metric Data Structures. Morgan Kaufmann Publishers Inc., San Francisco (2005)
16. Shaft, U., Ramakrishnan, R.: Theory of nearest neighbors indexability. ACM TODS 31(3), 814–838 (2006)
17. Sivic, J., Zisserman, A.: Video Google: A text retrieval approach to object matching in videos. In: Proc. ICCV (2003)

Implementation of the Aggregated R-Tree
over Flash Memory

Maciej Pawlik and Wojciech Macyna

Institute of Mathematics and Computer Science
Wrocław University of Technology
Poland
`146350@student.pwr.wroc.pl,`
`wojciech.macyna@pwr.wroc.pl`

Abstract. Flash memory becomes the very popular storage device. Almost every kind of hand-held devices use flash memory because of its shock - resistance, power economy and non-volatile nature. Recently more attention has been paid to the data storage in flash memory. Due to the different architecture, the implementation of the B-tree and R-tree indexes on the solid discs cannot be applied to flash memory directly. In this paper we propose the efficient implementation of the aggregated R-tree index. In our approach we separate the R-tree meta data and the aggregated data into different sectors of flash memory. We also calculate the number of read and write operations and compare it with the standard R-tree implementation. Our proposition is particularly effective, since the R-tree structure is quite stable and the aggregated values change frequently. The experiments confirm the effectiveness of our implementations.

Keywords: flash memory, aggregated values, aR-tree.

1 Introduction

Flash discs or Solid State Drives (SSDs) are considered as an alternative to the magnetic discs. Almost every kind of hand-held devices use flash memory because of its shock - resistance, power consumption and non-volatile nature. In recent years, the capacity of flash memory has grown rapidly, so that the devices are able to store the huge amount of data. Flash memory tends to replace the magnetic discs in many areas. It is hard to imagine sensor networks, embedded systems or hand-held devices without flash memory.

Unlike the magnetic discs, the flash discs or SSDs have no mechanic movement overhead. As a result, the seek and rotational delays are no dominant I/O costs. Due to this fact, the random reads in flash memory are faster than in a magnetic disc. On the other hand, the random writes to flash memory are much slower than reads. Moreover, the write operation is much more energy consuming than the read operation. This asymmetry has an impact on data storage in flash memory.

In this article, we consider the effective storage of the spatial data over flash memory based on an R-tree index. The main aim of using that index is to speed up the access to the spatial objects in a database. The idea of the R-tree is similar to the B-tree. Every

H. Yu et al. (Eds.): DASFAA Workshops 2012, LNCS 7240, pp. 65–72, 2012.

entry in the R-tree corresponds to the rectangle of the considered area and has pointers to the smaller rectangles inside it. To find the objects residing in the given area, the traverse from the root to the appropriate rectangle in the R-tree is needed.

The concept of the R-tree index was proposed in [1]. Since then many variants of the R-trees have been developed. Recently the aggregated R-tree (aR-tree) index has attracted attention. In that index, the aggregated values, connected with the rectangles, are not calculated every time but they are stored in the aR-tree explicitly.

In the disc storage system, the operations, such as: insert, delete and rebalancing on the R-tree, are implemented efficiently. Sectors are read and written to the same location. In case of flash memory - it is not possible. The blocks of flash memory could not be overwritten, unless they are erased first. The block consists of the fixed number of pages, where data may be written. The frequent erasing of the same blocks in flash memory can quickly deteriorate them. This is caused by the fact that the number of erase operations is limited and depended on the type of flash memory. To reach the even sector usage, the flash memory vendors use the Flash Translation Layer (FTL). The different kinds of FTL are described in more detail in: [2], [3], [4]. Taking into consideration all these restrictions, we claim that the implementation of the aR-tree index over the magnetic disc could not be directly applied to flash memory.

In [5] authors propose the efficient implementation of the R-tree structure. However, they don't consider storing aggregated values connected with the R-tree nodes.

In this paper we propose the extension of the R-tree implementation which deals with the aggregated values. Our work may be treated as the first step towards effective storage of the aggregated values associated with the tree index in the flash memory context. The paper is organized as follows. In the next section we formulate the problem. In the third section we propose our aR - tree implementation. In the forth section we make some calculations connected with our approach and compare it with the standard R-tree implementation. After that, we describe the experiments, which have been done to confirm our methods empirically. In the last section we summarize the paper and describe conclusions and possible extensions of our work.

2   Problem Formulation

2.1   R-Tree Description

The R-tree is the structure similar to the B-tree. There are several implementations of the B-tree over flash memory [6] [7]. The main purpose of the R-tree index is to manage the spatial data very efficiently. A common operation, which speeds up the index is, for example, the searching of all spatial objects in the particular area. To find all the objects residing in the rectangle A (see figure 1(a)), the R-tree node containing this rectangle must be found (see figure 1(b)).

There are two kinds of nodes in the R-tree: the leaf nodes and the internal nodes. Besides, there is one root node. Every internal node contains pointers to the child nodes. The leaf node points to the objects in the database. Figure 1(b) shows the example of the R-tree index. There are four nodes in the R-tree: the root node R and three child nodes: A, B, C, which contain the spatial objects residing inside them.

The aR-tree is the extension of the R-tree structure. It enables storing the aggregated values for the entries inside the tree nodes. Sometimes the calculation of the aggregated values each time may be too complex in the real applications. In such situations, it may be better to store these values inside the R-tree explicitly. Figure 1(c) shows the example of aR-tree, where the value connected with the entry is the sum of the values from the entries of the child node. The more detailed description of the aR-tree can be found in [8].

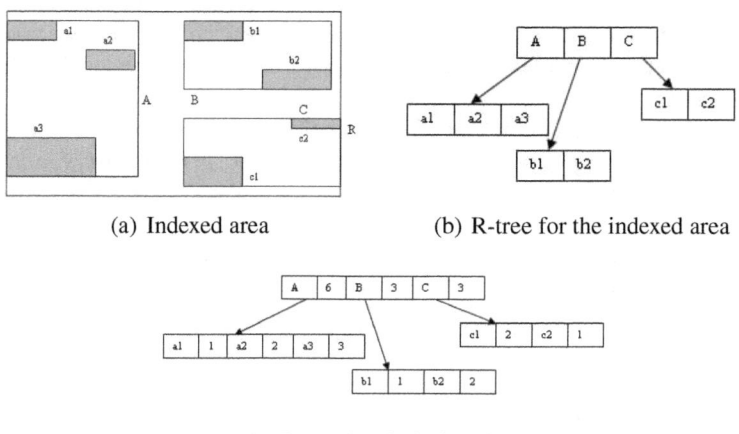

(a) Indexed area   (b) R-tree for the indexed area

(c) aR-tree for the indexed area

Fig. 1. Spatial indexes

3 Proposed Implementations

3.1 Overview

The aR-tree over flash memory should be implemented independently of the FTL. The purpose of the implementation is to maximize efficiency and to minimize the number of writes and energy consumption. In this section we propose our aR-tree implementation idea and compare it with the Original AR Method described in [5]. Both methods are write optimal.

3.2 Original AR Method

In this section we present our first implementation, which is based directly on the proposition described in [5].

When the new spatial object is created, it is written to the reservation buffer firstly. Each object has two parts: meta data and data. The meta data of the object contain data that are necessary for building the aR-tree index. The meta data are stored in the structure called an index unit. The index unit consists of the following components: $dataptr, parentnode, nextnode, id, minimalboundingbox, opflag$ and $agg-value$. $dataptr, parentnode, nextnode, id, minimalboundingbox$ are: a pointer to data, a pointer to the parent node, a pointer to the child node, a pointer to the aR-tree node containing the object and the minimal bounding box, respectively. $opflag$ represents the corresponding operation, i.e. an insertion, a deletion or an update. $agg - value$ denotes the list of the aggregated values.

If the index item changes, it is registered in the reservation buffer firstly. If it is full, its content is flushed to the flash memory. The architecture of flash memory influences on the fact that index units may be scattered over various sectors of flash memory.

To facilitate an access to the aR-tree, the additional mapping is needed. A node translation table is adopted to maintain the index unit and the corresponding sectors of the memory. The node translation table is an array, which for each aR-tree node contains the number of the sectors, where the index units belong to that node reside. The following example illustrates this idea.

Let's assume that the reservation buffer can hold up to four objects and the objects, $a4, b3, b4, c3$, are inserted into the spacial database (fig. 2(a)). If the reservation buffer is full, the inserted objects are transformed to the index units: $I1, I2, I3, I4$, respectively. Then, the new created index units are written to the sectors of the flash memory. Please note that the object $a4$ belongs to the node A, the objects $b3, b4$ to the node B and the object $c3$ to the node C.

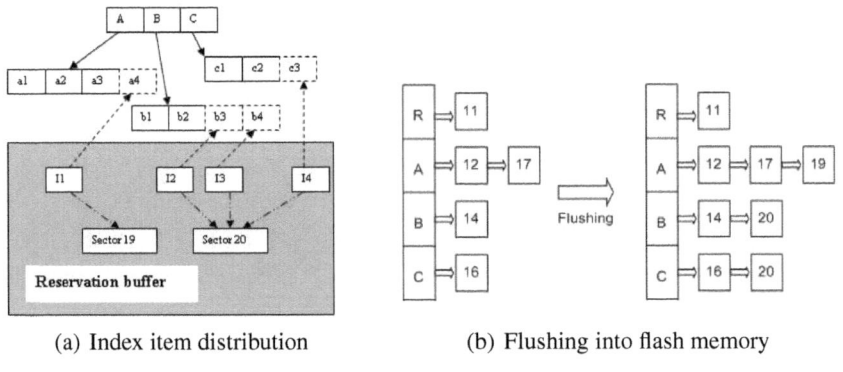

(a) Index item distribution   (b) Flushing into flash memory

Fig. 2. R-tree writing process

Now let's assume that each memory sector is able to maintain up to three index units. Then, the index units $I2, I3, I4$ may be stored in the first sector (for instance sector 20) and the index unit $I1$ in the other sector (for example sector 19), as it is shown on figure 2(a). After that, the new sectors are added to the node transition table. The figure 2(b) shows the node transition table before (left side) and after the inserting of the new objects (right side). It is easy to see that after flushing to flash memory, the sectors 19 and 20 are added to the node transition table. This structure is very helpful for reconstructing of the aR-tree. For example, if we want to reconstruct the node A, we need to visit only the sectors with numbers: 12, 17 and 19.

The implementation packs the index units into the sectors of flash memory. The goal is to reach the minimal number of the written sectors. The problem is NP-hard and may be reduced to the Bin-Packing problem [5]. In this method, the FIRST-FIT algorithm is used.

3.3 Proposed AR Method

Original AR Method is a simple extension of the R-tree implementation proposed in [5]. To the index item we added the $agg-value$, which contains the list of aggregated values connected with this index item.

However, this approach would not be efficient, since the aR-tree structure is stable and the aggregated values change frequently. Every update of the aggregated value would affect the modification of the whole index item, what could be impractical.

In Proposed AR Method the data connected with the index item and the aggregated values are separated to the different sectors. Thus, if the aggregated value is changed, there is no need to rewrite all other values of the index item.

If the aggregated values are modified, the changes are written to the reservation buffer. If it is full, its content is flushed to the aggregated table. It is a simple structure, which consists of: the identifier of the entry of the aR-tree and the aggregated value (see fig. 3). Because the number of the entries may be high, the aggregated values may be scattered over many sectors of the flash memory. To facilitate the access to the values connected with the node, the index table is used. It binds every node with the list of the sectors, which contain the aggregated values (the fragments of the aggregated table) for that node.

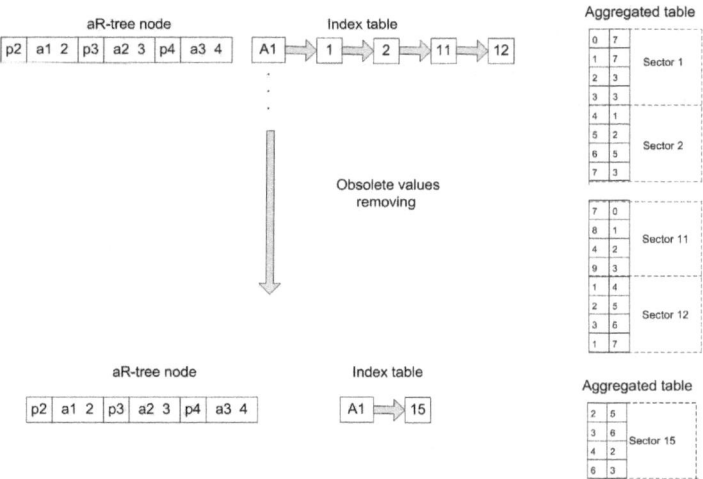

Fig. 3. Compacting of the aggregate index

The index table contains for each node the list of sectors, in which the aggregated values for this node are stored. If the aggregated value is modified, the new value is inserted into the aggregated table. It may add the new entry into the index table, since the value is written into the new sector. The old value is not immediately deleted. As a consequence, the aggregated table and the index table may grow. Thus, it is necessary to create the mechanism, which removes the obsolete values from the aggregated table and compacts the index table. To do that, we define the constant C, which denotes the maximal number of sectors where the aggregated values of one aR-tree nodes are written. Figure 3 shows the way of erasing of the obsolete values for $C = 4$. The node $A1$ consists of three spatial objects: $a1$, $a2$ and $a3$. The aggregated values connected with these objects are written to the sectors: 1, 2, 11 and 12. The very important issue is related to the order of the sectors in the index table. Note that sectors 1 and 12 contain the aggregated value for the entry 3. The actual value is 6 because the sector 12 is after the sector 1 in the index table. After the compacting process, all the actual aggregated values are written in the sector 15.

4   System Analysis

In this section we calculate the number of reads and writes for two implemented methods: Original AR Method (OARM) and Proposed AR Method (PARM). Our calculations are similar to [5] and are presented in more detail in [9] . In this section, we assume that the aR-tree structure is stable and only the aggregated values change.

In case of OARM, the number of reads and writes, needed for the updating of k aggregated values, may be bounded by $R_{1-Agg} = O(k * H * C)$ and $W_{1-Agg} = O(2 * (\frac{k*H}{f})))$, respectively. H denotes the height of the aR-tree, C is the maximal number of sectors connected with one node and f denotes the number of index units, which one memory sector can contain.

On the other hand, PARM needs $R_{2-Agg} = O(2 * k * H * (C + 1))$ reads and $W_{2-Agg} = O(2 * (\frac{k*H}{r}))$ writes, where r denotes the number of records, which can contain one memory sector.

According to our calculations, we can derive the following conclusions:

$$R_{1-Agg} \leq R_{2-Agg} \tag{1}$$

This is because PARM additionally reads from the aggregated table, which is placed in the different sectors than the index items.

$$W_{2-Agg} \leq W_{1-Agg} \tag{2}$$

$$O(2 * (\frac{k * H}{r})) < O(2 * (\frac{k * H}{f}))) \tag{3}$$

The condition (3) holds because: $f \ll r$. One memory sector may contain much more aggregated values than index items.

5   Experiments

In this section, we shall discuss some simulations, which confirm the effectiveness of the proposed methods. We compared two implementations: Original aR-Tree Method (OARM) and Proposed aR-Tree Method (PARM). Our experiments were conducted on the 16 GB flash memory where the read and write access time is: 54 μs and 400 μs, respectively.

In this experiment we created two aR-trees using each of the above described strategy. Every aR-tree consisted of 200 minimal bounding rectangles (MBR). Then we modified 200000 times the aggregated values connected with the randomly chosen MBRs without changing the aR-tree structure.

Figure 4 shows the number of reads and writes for both methods. We see that PARM needs more read operations to perform the modification of the aggregated values, because it requires to scan the sectors containing the index item data and the aggregate table. On the contrary, OARM only needs to read the index item data. However, PARM is more effective than OARM, as far as the number of writes is considered. Such a situation is caused by the fact that in OARM the aggregated values are inside the index item, therefore, the changing of those values implies the rewriting of the whole index item. In case of PARM only these blocks are changed, where the aggregated values reside.

Besides, we measured the evaluation time for both methods. According to the experimental results, PARM outperforms OARM. It is strongly connected with the number of reads and writes. The write operation needs more time to be performed. Besides, the erase operation is often triggered in OARM, what slows down this method.

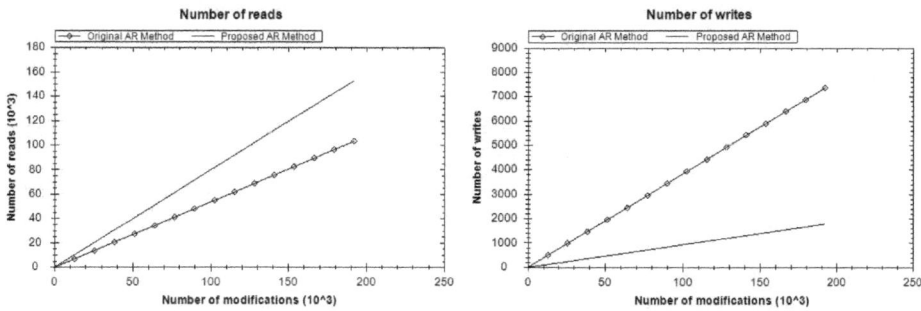

Fig. 4. The charts show: the number of reads, the number of writes and the evaluation time for each implementation

6 Conclusions and Future Work

In this paper, we proposed some approaches to the implementation of the aR-tree index over flash memory. The aR-tree index may be considered as the extension of the R-tree index. Apart from the typical components using for the storing of the spatial objects, it contains the precalculated aggregated values inside. The implementation of the aR-tree over flash memory must consider the limitations of that memory and cannot be directly adopted from the magnetic discs environment.

We compared two implementations. In Original AR Method, we use index items with the aggregated values. If the entry of the aR-tree node changes, it is written to the reservation buffer and if the reservation buffer is full, all entries are flushed into the flash memory. In Proposed AR Method, we split the aR-tree structure and aggregated values into the different sectors of the flash memory. This method is very effective, since the aR-tree structure changes rarely and the aggregated values frequently.

We calculated and compared the number of writes and reads for all methods. To confirm our results we carried out several experiments.

In the future work we are going to deal with the view materialization problem. It is because not all the aggregated values should be stored in flash memory. Therefore, the question is, which values should be materialized to make the implementation energy efficient.

References

1. Guttman, A.: R-trees: A Dynamic Index Structure for Spatial Searching. In: International Conference on Management of Data, pp. 47–57. ACM (1984) 66
2. Park, C., Cheon, W., Kang, J., Roh, K., Cho, W., Kim, J.S.: A Reconfigurable FTL (flash translation layer) Architecture for NAND Flash-based Applications. ACM Trans. Embed. Comput. Syst. 7(4), 1–23 (2008) 66
3. Cho, H., Shin, D., Eom, Y.I.: Kast: K-associative Sector Translation for NAND Flash Memory in Real-time Systems. In: DATE 2009, pp. 507–512 (2009) 66
4. Gal, E., Toledo, S.: Algorithms and Data Structures for Flash Memories. ACM Computing Surveys 37 (2005) 66
5. Wu, C.H., Chang, L.P., Kuo, T.W.: An Efficient R-tree Implementation over Flash-memory Storage Systems. In: GIS, pp. 17–24. ACM (2003) 66, 67, 68, 69, 70
6. Nath, S., Kansal, A.: Flashdb: Dynamic Self-tuning Database for NAND Flash. In: IPSN, pp. 410–419. ACM (2007) 66
7. Wu, C.H., Kuo, T.W., Chang, L.P.: An Efficient B-tree Layer Implementation for Flash-memory Storage systems. ACM Trans. Embedded Comput. Syst. 6(3) (2007) 66
8. Papadias, D., Kalnis, P., Zhang, J., Tao, Y.: Efficient OLAP Operations in Spatial Data Warehouses. In: Jensen, C.S., Schneider, M., Seeger, B., Tsotras, V.J. (eds.) SSTD 2001. LNCS, vol. 2121, pp. 443–459. Springer, Heidelberg (2001) 67
9. Pawlik, M., Macyna, W.: Implementation of the Aggregated R-tree over Flash Memory. Technical Report, Wroclaw University of Technology (2011) 70

A Flash-Based Decomposition Storage Model

Qingling Cao, Zhichao Liang, Yulei Fan, and Xiaofeng Meng

School of Information, Renmin University of China, Beijing, China
{qinglingcao,zhichaoliang,fy1815,xfmeng}@ruc.edu.cn

Abstract. The traditional HDD-based columnar storage is an important technology to improve the performance of query-intensive database. However, some features of HDD weaken the advantages of columnar storage. In this paper, we study the advantages of SSD over HDD on columnar storage and propose a new columnar storage model based on SSD, named Flash Based Decomposition Storage Model(FBDSM). FBDSM stores each attribute of a table in a column and each column has a log table to store its recently updated data, which will be merged with original data periodically. In this way, FBDSM not only provides efficient update, but also almost has no negative impact on query performance.

Keywords: DSM,columnar storage,flash memory.

1 Introduction

In 1985, columnar storage was proposed in [1] and Sybase launched Sybase IQ in 2004, which is the first commercial column-store mainly for on-line analysis and data mining[3]. Subsequently, C-store[2], an open source write-optimized column-store was released and SQL Server implemented columnar storage technology in its version 11.0[4]. Nowadays, many workloads, like data warehouse, require read-optimized systems, so the column-store storage systems have been widely used. Comparing with its row-store counterpart for query performance, column-store enjoys some distinguish features such as better compression utilization and low I/O cost[5][6]. However, as a storage technology optimized for query-intensive, random read operations are inevitable. Consequently, the poor random read performance of HDD may become the bottleneck in column-store storage systems.

1.1 SSD Properties

SSD is an electronic equipment with small size, light weight, good shock resistance and particularly excellent access performance. Physically, SSD consists of some flash chips, and each flash chip is composed of certain number of blocks. Block is made up of a certain number of pages, usually 32, 64 or 128 pages. Page is the unit to read and write, and block is the unit to erase. A block can't be overwrite until erased. Erase is a time-consuming operation and a block will become unreliable after a certain times of erase. These features cause that small size write

H. Yu et al. (Eds.): DASFAA Workshops 2012, LNCS 7240, pp. 73–80, 2012.

on SSD is not so convenient, and in-place update is not supported.Without mechanical latency, SSD exhibits much better random read throughput than HDD.
Much work have been done to make better use of SSD based on its special features [7][8][9]. Accordingly, many papers indicate that SSD is much better than
HDD in columnar storage.

1.2   Our Approach: FBDSM

Considering the properties of SSD, we proposed a new column storage model,
Flash-based Decomposition Storage Model(FBDSM), which not only suits the
characteristics of SSD, but also provides efficient update with little negative
affect on read performance. Just like the column-store, all attributes in a table
is stored decomposed in FBDSM as column. Every column has two storage
structures: one is primary table(PT), to store data that is loaded or inserted and
another is a vice table just like log, named as log table(LT), which is designed
to store update(UPDATE and DELETE) data. In this way, update could be
postponed in the LT until the merge of PT and LT.

The rest of this paper is organized as follows. In Section 2 we introduce some
existing columnar storage models and in section 3, we present our storage model
FBDSM. We illustrate query processing differences between FBDSM and traditional columnar storage model in section 4 and give our experiment and performance analysis in section 5. Section 6 presents our conclusion finally.

attr0	attr1	attr2
V00	V01	V02
V10	V11	V12
V20	V21	V22

TID	attr0
0	V00
1	V10
2	V20

TID	attr1
0	V01
1	V11
2	V21

TID	attr2
0	V02
1	V12
2	V22

Fig. 1. The traditional table and its corresponding DSM storage model

2   Related Work

The first column storage model was described in [1] as DSM(Decomposition Storage Model). The DSM need to support only the simple binary relation between
two objects. The first object is a fixed length attribute called surrogate, which is
similar to TID(tuple identity) to identify a tuple. The second object could be a
fixed or variable value attribute. As figure 1 shows, the traditional table on the
left is stored as three binary table on the right in DSM. For this DSM storage
model, it is very easy to recognize a tuple by surrogate, but attaching a surrogate
column for every attribute wastes much storage space. Some column-stores, like
KDB, MonetDB/X100 [10] and c-store [2], keep data in entry sequence. Each attribute of a tuple have the same relative location in columns, which means that
they have the same calculated TID. In this way, there is no need to store TID.

Insertion could just be put at the end of the columns. However, it is expensive to update a few attributes in a tuple at the same time. Generally, it will be substituted by an insertion after a deletion.

3   FBDSM

As we mentioned ahead, there is only comprised columnar storage on SSD recently, which hinders the columnar storage to exert its advantages on SSD. So we put forward a novel columnar storage model FBDSM that fits for SSD. We introduce the storage structure of FBDSM in section 3.1, and present how to deal with INSERT, DELETE and UPDATE in section 3.2 and 3.3. Finally, in section 3.4 we show how to integrate data of attribute column and LT.

3.1   Storage Structure of FBDSM

Flash-based Decomposition Storage Model (FBDSM) is a columnar storage model designed to make full use of SSD. SSD, for its distinctive features as mentioned above, has some special read and write mechanisms in practical applications. Since the write unit on SSD is page, it is very inconvenient to do small-size update. Moreover, it is also difficult to keep entry sequence on flash since it supports no in-place update. To solve this dilemma, we put forward a new storage model, FBDSM. Just like column-store, all attributes in a table is stored decomposed as column in FBDSM. Every column has two storage structures. One is primary table(PT), to store loaded or inserted data while another is a vice table serves just like log, which is named as log table(LT) and used to store updated(UPDATE and DELETE) data. LT consists two attributes, TID and the same attribute as column, in which TID identifies the data in the column. TID identifies the updated data in the column, and the other attribute is the new data. In figure 1, the table on the right stored as FBDSM is presented in figure 2.

attr0	TID	attr0	attr1	TID	attr1	Attr2	TID	attr2
V00			V01			V02		
V10			V11			V12		
V20			V21			V22		

Fig. 2. Table in figure 1 stored as FBDSM

3.2   INSERT

In FBDSM, any loaded or inserted data is put at the end of PT. So these new added tuples are kept entry sequence. Figure 3 shows table T with two columns 9 tuples loaded, and 2 tuples inserted at the end of the PT. Here we assume that every block consists four attribute values in columns, and consists 3 tuples in LT. Each block consists two pages. Then attributes A and B are both stored in 3 blocks.

A		TID	A		B		TID	B
#0	A0			#0	B0			
#1	A1			#1	B1			
#2	A2			#2	B2			
#3	A3			#3	B3			
#4	A4			#4	B4			
#5	A5			#5	B5			
#6	A6			#6	B6			
#7	A7			#7	B7			
#8	A8			#8	B8			
#9	A9			#9	B9			
#10	A10			#10	B10			
#11	A11		Insert	#11	B11			

Fig. 3. Table T after load data and insert two tuples

3.3   DELETE and UPDATE

To delete a tuple, it should compute its TID at first, and then insert every LT of the table a record <TID, NULL>. To update a tuple, find the TID of the attribute to be updated at first, then add the TID and the new value to the corresponding LT, leaving the old value in the column as it is. Figure 4 shows table T after some DELETE and UPDATE operations. The deleted or old value in a column will not be deleted until merged. Since there is no flag in column to mark the invalid, and the data in LT may be updated again as <2,A2'> in LT of A, thus it should search the LT from tail to head in query processing and the first value encountered of the same TID is the latest.

A		TID	A	B		TID	B
#0	A0	6	A6'	#0	B0	8	B8'
#1	A1	2	A2'	#1	B1	10	NULL
#2	A2	4	A4'	#2	B2	5	A5'
#3	A3	10	NULL	#3	B3	9	B9'
#4	A4	2	A2''	#4	B4		
#5	A5	11	A11'	#5	B5		
#6	A6			#6	B6		
#7	A7			#7	B7		
#8	A8			#8	B8		
#9	A9			#9	B9		
#10	A10			#10	B10		
#11	A11			#11	B11		

Fig. 4. Table T after some UPDATE and DELETE operations

3.4 Merge of Data

For update-intensive workloads, many "holes" that store invalid data will appear in blocks of PTs after many UPDATE and DELETE operations, thus it is very necessary to merge the new data in LTs into PTs in time. Here we set 1/2 as the threshold to trigger the merge operation, namely if the invalid data occupied half of a block, the block can be merged. Once the merge is triggered, a background procedure will be woke up to merge the target blocks in batch once. The blocks in the PT are merged with the new data in its LT and written into new blocks, hence the data in old blocks of columns certainly will not be used in the future. Then these blocks can be recycled. The new blocks may have free space, since the merged blocks may have deleted data. Certainly, the cost to move data followed ahead is unaffordable. We leave the free space as it is, until the the blocks next to it is merged, and write the data into its free space.

Then, how to deal with these blocks in LTs? After merge, some new data have been integrated into columns. If we delete all of them from the LTs thoroughly, it may lead to blocks rewrite frequently. To avoid this, there is a threshold for blocks of LTs. For example, if more than 2/3 of the data in the block has been merged into its column, than this block can be rewritten leaving only valid data. For free space, it takes the same mechanism as in column. Every LT is also assigned a data structure, we name it as Bit_Flag_Map(BFM). Every bit in BFM shows if the corresponding value in its LT is valid data. The BFM is stored in SSD and is changed after merge. So those new data of LT that are inserted after merge is not shown in BFM. Their validation should be judged by scanning from tail. Figure 5 shows table T in figure 4 after merge as described ahead. The thresholds of column and LT are respectively 1/2 and 2/3.

A		TID	A		B		TID	B
#0	A0	2	A2''	#0	B0	5	B5'
#1	A1			#1	B1		
#2	A2			#2	B2		
#3	A3			#3	B3		
#4	A4'			#4	B4		
#5	A5			#5	B5		
#6	A6'			#6	B6		
#7	A7			#7	B7		
#8	A8			#8	B8'		
#9	A9			#9	B9'		
	A11				B11		

Fig. 5. Table T after merge operation

4   Query Processing

Although the storage model has been made relatively significant changes in FBDSM, query processing is not complicated. In this section, we show how to deal with queries on FBDSM.

Many optimization on columnar storage query processing [11][12] can be applied on FBDSM, some may only by simple change. Actually, the essential difference between FBDSM and traditional columnar storage is the LT. In query processing, when need the data of an attribute, we recommend to integrate the column and its LT. At first, read LT into memory and extract those valid data with the help of BFM. Then sort these valid data by TID. With TID, it is very easy to identify the location of its old value, just iterate the column block by block and replace the old value. Then it can process queries with methods that have been proposed. For example, it can process join query with FlashJoin [12]. The extra step is to integrate column with LT before input data into Join Kernel. Because LT is relatively very small, so this step will not increase so much cost.

5   Experiment and Analysis

In this section, we present the results of experiment and analysis. Section 5.1 shows the experiment setup and update results and section 5.2 compares the query processing performance of DSM and FBDSM theoretically. In section 5.3, we illustrate that the cost of merge is so low that it doesn't affect the query performance FBDSM.

5.1   Update

We simulate the DSM and FBDSM storage models on MySQL DBMS. Here the DSM is a little different from [1], it stores only attribute column and do in-place update or insert-after-delete update. Both DSM and FBDSM can locate an attribute value by TID immediately. To simulate this, in our experiment, we add an indexed TID column to the attribute column of both DSM and FBDSM. We generate a table with 4 attributes and 530w tuples randomly. The table is stored in 3 methods: DSM on HDD, DSM on SSD and FBDSM on SSD. In the experiment, we test the performance of updating one attribute of 40000 tuples of the table randomly and compute the average cost of an update. Then, we test the performance to update 2 attributes of 40000 tuples randomly and also compute the average cost. The results is shown in figure 6. From the results we find DSM on SSD almost has no advantage over SSD on HDD, though the great advantages of SSD over HDD. This is because in-place update cause erase-before-write and erase is an expensive operation on SSD. Compared with DSM both on HDD and SSD, our approach gains more than 40% improvement. For DSM on HDD, updating a few attributes in a tuple leads to several random disk I/O, which increase the update cost greatly. To avoid this, some column-stores would rather do insert after delete than in-place update. Figure 7 shows the performance of

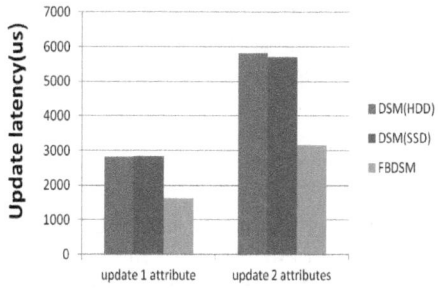
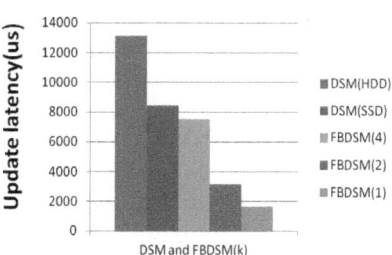

Fig. 6. The average performance comparison of DSM(HDD) and DSM(SSD) with in-place update and FBDSM to update a tuple

Fig. 7. The average performance comparison of DSM(HDD) and DSM(SSD) with insert-after-delete and FBDSM(k) to update a tuple, k is the number of updated attributes

DSM on SSD and HDD with insert-after-delete update method and FBDSM. The result in figure 7 shows that using insert-after-delete method, DSM on SSD performs much better than DSM on HDD. Compared to FBDSM, DSM(SSD) could gain equivalent performance only when all the attributes of the tuple need to be updated. Certainly this case is rare. When minority attributes need to be updated, FBDSM save much time, as FBDSM(2) and FBDSM(1) shows. So FBDSM is always better than DSM(SSD) with insert-after-delete update.

5.2   Join

Join is the most expensive operation in query processing and its performance usually dominates query processing performance, so we mainly focus on analyzing join. In [12], although FlashJoin, is proposed to optimize join on SSD with PAX layout, actually, it is a more perfect join method for column layout, no matter on SSD or on HDD. As have been mentioned in section 4, for FBDSM, it should read LT to do a merge operation in memory before join. As the cost of merge done in memory is relatively low compared to the scan cost of LT. Then, on HDD and SSD, join cost can be respectively disintegrated as follows:

$$ColumnScan_{HDD} + FlashJoin. \tag{1}$$

$$ColumnScan_{SSD} + LTScan_{SSD} + FlashJoin. \tag{2}$$

$ColumnScan_{HDD/SSD}$ is the cost to read columns from HDD/SSD, and $LTScan_{SSD}$ is the cost to read LTs from SSD. The cost difference, (1)-(2), is $(ColumnScan_{HDD} - ColumnScan_{SSD}) - LTScan_{SSD}$. For the same table, no doubt, ColumnScan on SSD is faster than HDD. While, this advantage will be counteracted by LTScan, if the LT is too large. As have been acknowledged the sequential read speed of SSD and HDD are about 160MB/s and 110MB/s respectively. If the columns on HDD and SSD are both 1G, then when the LTs is 3.2G, the cost of (1) and (2) are equal. Actually, it is impossible, because in our approach LTs are relatively small and can be integrated to columns when necessary. So, FBDSM is always better than DSM in normal workloads

6 Conclusion

This paper presented a new columnar storage model that fits for flash, named FBDSM. It not only exerts the advantages of columnar storage, but also coordinates the features of SSD. Our experiment proved that FBDSM enjoys good query and update performance.

References

1. George, G.P., Setrag, S.N., Ferreira, M.: A Decomposition Storage Model. In: SIGMOD, pp. 268–279 (1985)
2. Stonebraker, M., Abadi, D.J., et al.: C-Store: A Column-Oriented DBMS. In: VLDB, pp. 553–564 (2005)
3. MacNicol, R., French, B.: Sybase IQ multiplex-designed for analytics. In: VLDB, pp. 1227–1230 (2004)
4. Larson, P., Clinciu, C., et al.: SQL Server Column Store Indexex. In: SIGMOD (2011)
5. Abadi, D.J., Madden, S.R., Hachem, N.: Integrating Compression and Execution in Column-Oriented Database System. In: SIGMOD (2006)
6. Abadi, D.J., Madden, S.R., Hachem, N.: Column-Stores vs. Row-Stores: How Different Are They Really? In: SIGMOD, Vancouver, Canada, pp. 981–991 (2008)
7. Lee, S.-W., Moon, B.: Designe of Flash-Based DBMS: An In-Page Logging Approach. In: SIGMOD, Beijing, China, pp. 55–66 (2007)
8. Chen, F., Koufaty, D.A., Zhang, X.: Understanding Intrinsic Characteristics and System Implications of Flash Memory based Solid State Drives. In: Proc. of SIGMETRICS (2009)
9. Birrell, A., Isard, M., Thacker, C., Wobbe, T.: A Design for High-Performance Flash Disks. In: Technical Report MSR-TR-2005-176, Microsoft Research (2005)
10. Boncz, P., Zukowshi, M., Nes, N.: MonetDB/X100: Hyper-Pipelining Query Execution. In: CIDR (2007)
11. Shah, M.A., Harizopoulos, S., Wiener, J.L., Graefe, G.: Fast Scans and Joins Using Flash Drives. In: DaMoN (2008)
12. Tsirogiannis, D., Harizopoulos, S., Shah, M.A.: Query Processing Techniques for Solid State Drives. In: SIGMOD (2009)

The Contribution of Bayesian Networks to Manage Risks of Maritime Piracy against Oil Offshore Fields

Xavier Chaze[1], Amal Bouejla[1], Aldo Napoli[1], Franck Guarnieri[1],
Thibaut Eude[2], and Benjamin Alhadef[2]

[1] MINES ParisTech, CRC - Centre de recherche sur les risques et les crises,
BP 207 1 rue Claude Daunesse 06904 Sophia Antipolis Cedex, France
[2] SOFRESUD, 777 Avenue des Bruxelles, 83500 La Seyne sur Mer Cedex

Abstract. In recent years pirate attacks against shipping and oil fields have continued to increase in quantity and severity. For example, the attack against the Exxon Mobil oil rig in 2010 off the coast of Nigeria ended in the kidnap of 19 crew members and a reduction in daily oil production of 45,000 barrels, which resulted in an international rise in the price of oil. This example is a perfect illustration of current weaknesses in existing anti-piracy systems. The SARGOS project proposes an innovative system to address this problem. It takes into account the entire threat treatment process; from the detection of a potential threat to implementation of the response. The response to an attack must take into account all of the many parameters related to the threat, the potential target, the available protection resources, environmental constraints, etc. To manage these parameters, the power of Bayesian networks is harnessed to identify potential countermeasures and the means to manage them.

1 Introduction

World oil production is spread over more than 10,000 offshore fields, each of which requires tools and equipment to extract, process and temporarily store oil, and vessels that can transport the hydrocarbons between the point of production and the point of consumption.

Modern maritime piracy is currently the major threat to the security of these energy production installations and the maritime shipping of oil.

Monitoring methods, and above all protection measures are the major weaknesses in the detection of a threat on such installations. They tend to be ineffective and limited in the extent to which they can be tailored to a particular situation. Finding a system that can manage the safety of oil fields, and provide both suitable protection and effective crisis management is of primary importance.

The SARGOS project is a response to this need. It proposes a global system in the fight against acts of piracy committed against oil industry infrastructure.

H. Yu et al. (Eds.): DASFAA Workshops 2012, LNCS 7240, pp. 81–91, 2012.

This paper discusses the issues surrounding acts of oil field piracy. It describes in detail a method for the planning of countermeasures. Notably, it uses a Bayesian network to model the situation using two different inputs: the Piracy and Armed Robbery database of the International Maritime Organization (IMO) and the consolidated knowledge of experts in the domain. The results are tested using realistic and comprehensive pirate attack scenarios.

2   Problem Definition and Research Objectives

The infrastructure of the offshore oil industry is subject to a constantly rising risk of piracy. These acts have repercussions both on local operations and globally. This section describes the economic and political challenges related to these attacks. It highlights an increasingly insecure context, where the actors involved in the offshore oil industry are helpless to protect themselves, and the current tools do not effectively protect infrastructure. Finally, we outline the SARGOS project, illustrate its potential contribution to finding new ways of dealing with these issues and demonstrate their relevance.

2.1   Political and Economic Challenges

The offshore oil industry is growing rapidly. Offshore oil extraction currently accounts for about one-third of global oil production. Despite its scarcity, this source of energy is under active exploration in many parts of the world.

From an economic perspective, it is important to highlight that attacks on such infrastructure generate significant additional costs (ransom payments, insurance premiums, the installation of security equipment etc.). These additional costs directly affect the price of oil in the international market.

From the political perspective offshore oil fields are an interface between the activities of the oil industry and the maritime world. The legal status of oil rigs is complicated, although this is due more to the heterogeneity of applicable regulations than the absence of a body of law. This complexity can result in political conflicts between nations; it is often the case that the rig is located in one country, while the company operating the platform is located in another.

The importance of oil installations to the world economy and global industry, and the consequences that can arise from acts of piracy provides a strong incentive to better protect these assets.

2.2   Context and Operational Needs

Despite the fact that attacks against oil fields are infrequent and above all, receive little media coverage, they are of great cause for concern because of the serious consequences for both crew and infrastructure.

Infrastructure managers, employees and safety officers no longer want to see commercial assets become the subject of large ransoms. Nor do they want to continue to see crewmen injured, traumatized, held under extreme conditions for long periods of time, or even killed. For their part, insurers do not want to continue to insure highly expensive risks for an indefinite period of time. Finally, nations want to see an end to the situation where the price of oil is affected by such events.

The recent attacks are a perfect illustration of the weakness of existing anti-piracy tools. Currently, the safety of oil installations is provided by so-called classical tools (radio identification, radar, Automatic Identification Systems, etc.). Despite their usefulness in helping to detect threats, they cannot distinguish between different types of hazard (fishing boats, jet skis, tankers, etc.). Moreover, their effectiveness depends on many and various parameters that are related to the environment as well as technical and operational constraints.

The proposed solution is therefore to increase the degree of infrastructure protection by developing a new system (SARGOS), which is capable of generating an alarm and can set in motion an internal and external response to a confirmed intrusion.

2.3   The Contribution of the SARGOS Project

The SARGOS project (*Système d'Alerte et de Réponse Graduée OffShore*/Graduated Offshore Response Alert System) aims to meet this new need to protect vulnerable civilian infrastructure, which is exposed to acts of piracy or terrorism carried out at sea. The project aims to design and develop a comprehensive system that takes into account the whole threat treatment process, from the detection of a potential threat to the implementation of the response. It can be integrated into the infrastructures operations and respects regulatory and legal constraints.

The project is funded by the French National Research Agency (*L'Agence Nationale de la Recherche*)[1] , and is approved by other regional bodies in France. The development of a comprehensive protection system requires multi-disciplinary technical skills; challenges include the automatic detection and identification of threats, assessment of potential risks, and management of an appropriate response. The functional outline of the SARGOS system (Figure 1) shows how the threat is processed.

In the context of the current discussion, there is insufficient emphasis on the preparation of the diagnosis and the way in which parameters and constraints related to attacks should be managed. In order to address these shortcomings, we propose a new approach that can automatically draw up response plans, tailored to the type of intrusion detected.

---

[1] The SARGOS project brings together many different private sector organisations (including the French naval shipbuilder, DCNS and SOFRESUD, a supplier of high-tech equipment to the defence industry) and public research centres (including ARMINES, a French contract research organisation and TéSA, Telecommunications for Space and Aeronautics).

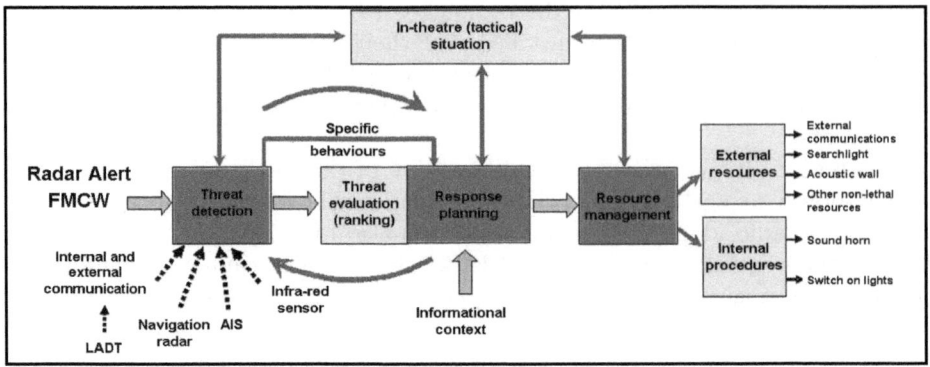

Fig. 1. Functional outline of the SARGOS system

3   Method

This paper will focus particularly on the contribution of a Bayesian inference approach, which uses input from an established database and the knowledge of experts in the maritime domain. The Bayesian network is used in the response planning process and the aim is to prepare a response that is appropriate, graduated and which can adapt as the threat evolves. Database information and the knowledge of oil industry experts are used together to address the lack of both *a priori* knowledge of the object in question and learning from experience in the applied domain.

The Bayesian network is used to model this information and knowledge; the tool is based on Thomas Bayes theorem (1), which forms the basis for probability theory.

$$\left(\frac{P(B/A) * P(A)}{P(B)}\right) = P(A/B) \tag{1}$$

A Bayesian network is a model that represents knowledge, and makes it possible to calculate conditional probabilities and provide solutions to various types of problems.

BayesiaLab software[2] was used to construct the Bayesian network. This tool for modelling Bayesian networks has many features and an intuitive graphical interface.

The SARGOS Bayesian network was developed in two stages, described below: first an initial network was constructed using data from an established, professional database, and then the final network was built using expert knowledge from the field.

---

[2] The BayesiaLab software has been developed by the French company Bayesia (http://www.bayesia.com/).

3.1   Construction of a Bayesian Network from Database Content

The first stage exploited data from the Piracy and Armed Robbery database of the International Maritime Organization (IMO). It is the only database in existence that holds records (dating from 1994) of pirate attacks in the maritime environment. On 15 th July, 2011 this database held 5,502 records and recorded the following information for each attack: the name of the asset targeted, the number of people involved, the type of weapon used, the measures taken by the crew to protect themselves, the impact on the crew and pirates, etc.

From this data the BayesiaLab software automatically generates a Bayesian network and suggests dependency relationships between the main elements found in the database. This study of the contents of the database made it possible to define the principle countermeasures adopted by the majority of entities attacked, namely: engage evasive manoeuvres, activate the Ship Security Alarm System (SSAS), contact the security vessel, move the crew to a safe location, and turn on the searchlight, etc.

These modalities and conditional probabilities were then used to construct the expert knowledge Bayesian network.

Figure 2 shows the Bayesian network created from the database content. Information such as longitude, latitude, name of the asset targeted, etc. was not used as data was not available for all attacks.

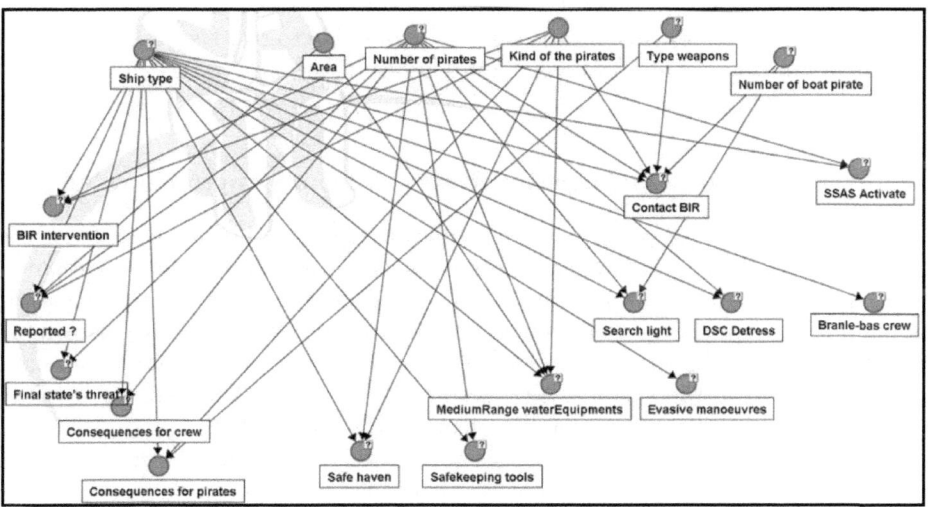

Fig. 2. The Bayesian network constructed from IMO data

In its initial form, the probability distribution of the nodes shows that most ships that come under attack are bulk carriers or tankers (this raw data does not reflect the application of constraints relating to particular attacks).

66.39 percent of attacks occur in international waters; this is due to the lack of security controls. Equally often, pirates profit from attacking in numbers: 60.39 percent of attacks are organized by pirate teams consisting of more than five people.

This network provides a very clear view of the pirates tactics, how they are armed, and above all the number of people involved.

The following example demonstrates how the network can be used to simulate a particular attack scenario. In the example, specific modalities are set for nodes that characterize an attack:

- The asset targeted: a tanker
- The location of the incident: international waters
- Type of attackers: thieves
- Type of weapons: armed personnel.

This makes it possible to identify the countermeasures used. The Bayesian network indicates that in this situation, the robbers fired shots at the potential target and that the crew, in order to protect themselves, tried to apply evasive manoeuvres and used high-pressure fire hoses on the attackers.

The Bayesian network created from the IMO data made it possible to make an initial assessment of the main tools and protection measures used by a crew under attack to protect themselves, to evaluate the effectiveness of these tools, and to determine the probability of certain types of attack.

Secondly, the use of this database made it possible to test the functionality of the BayesiaLab tool and establish the feasibility of creating a Bayesian network from existing data.

3.2   Construction of a Bayesian Network from Expert Knowledge

The first stage of the project was to construct a Bayesian network from the IMO data. In the second stage, expert knowledge from the marine community was used to construct the network. Civil and military experts in the maritime domain shared their experiences and opinions during successive brainstorming sessions to define the variables (Bayesian network nodes) and connections (links between nodes) of the system. As a result of this expert knowledge the initial conditional probabilities were set. These probabilities were then updated through an iterative process in which many scenarios were simulated in order to verify, and if necessary refine, the values which had been set for each node.

The principle is the following: when an object is detected by radar in the vicinity of an oil field, a set of variables is determined and calculated in order to identify and assess the potential danger. Such information includes, for example, the speed at which the object is moving, visibility, time of day, longitude and latitude of the object and the target etc.

This data is used to calculate the distance between the target and the moving object, and the time required for the intervention of the security vessel. This information is recorded in an alert report and each object detected is allocated a unique identifier. The SARGOS system only generates this report when the threat is identified as suspicious or hostile.

The basic architecture of the SARGOS response planning network consists of five modules and four sub-modules. The definition and the scope of each of these five modules are directly related to the meaning of the nodes that constitute them. The modules are classified as: basic parameters, the overall danger level of the situation, aggravating factors and constraints, nodes related to communication and distress calls and countermeasures. These are described in detail below.

The Basic Parameters. These are static or dynamic physical data that characterize the threat and the target. They are the direct result of, or are derived from, the intermediate calculations of the alert report. They constitute the minimum data necessary to create a model that is still sufficiently detailed to permit a full understanding of the threat and the target, and the issues involved in responding to an attack. These parameters include, for example, the identity of the threat IdentityClass (suspicious or hostile), the distance between the threat and the target DTG Threat/Asset, the criticality of the target AssetAssessment. Four modalities are defined for this node: critical, major, significant and other.

The Overall Danger Level of the Situation. The overall danger level of the situation is derived from the basic parameters. The node ShowGradationLevel is the formalization of this module in the Bayesian network. The grading system uses levels 1-4 (1 being least serious, 4 being most serious). This level, and the planning of countermeasures, is constantly adapted to each situation.

Aggravating Factors and Constraints. Aggravating factors and constraints are both internal and external elements of the system.

Aggravating factors make it possible to take into account a potential deterioration in the situation and thus to anticipate potential planning options. They represent the environment, for example the visibility (Visibility) and the time of day (PeriodOfDay).

Constraints are represented by parameters that reflect the effectiveness of the response both technically and operationally. Technical constraints are directly related to the use of countermeasures and include factors such as their availability (ImmediateReadiness) or the potential for remote control (RemoteControlled).

Communication and Distress Calls. Communication and the distress call are two indispensable resources called upon in response to a threat. Internal communication at the target can be used to notify all relevant personnel (e.g. inform the Offshore Installation Manager, InformOIM), while external communication operates at various levels to alert the various actors involved in safety at sea (for example, to request the intervention of security vessels RequestSecurityVessels, or activate the Ship Safety Alarm System, RaiseSSAS). This communication makes it possible for oil field installations and shipping to prepare their response and to ask, where possible, for outside intervention.

Countermeasures. This refers to the set of defensive measures implemented when the target is attacked, in order to protect itself against an identified threat. They are the concrete expression of the response and provide a set of means and actions to normalize the situation as quickly as possible following an attack.

The countermeasures module is divided into four sub-modules which are the core of the concept of a graduated response as they off er increasingly forceful countermeasures depending on the nature of the detected threat: they include deterrence and low-impact repulsion measures; repulsion, anti-boarding and neutralization measures; management of procedures, and ensuring the safety and security of the facility. They are described in detail below.

Deterrence and Low-Impact Repulsion Measures. These inform attackers that the target is aware of their intentions, that the target is able to follow the attackers and the target has no interest in taking action. Low-impact repulsion is the ability of the target to repel the attack using relatively low impact means such as searchlights, high-pressure fire hoses or Long Range Acoustic Devices (ActivateLRAD).

Repulsion, Anti-boarding and Neutralization Measures. These are high-impact countermeasures, whose main function is at least to mitigate if not neutralize attackers. The node EngageRepellentEquipment encompasses a growing number of resources available on the maritime piracy market that make it possible to repel an attack at a distance, while respecting the principle of non-lethal self-defence. The main function of anti-boarding measures, as with repulsion equipment, is to prevent attackers from being able to board should they approach the facility or ship.

The role of SetCrowdControlMunition is to delay the progress of the attackers to the point of exhaustion, or even neutralize them and so allow maximum time for the crew to deploy other safety actions.

Procedure Management. This consists of the following countermeasures:
The CrewMangement node refers to the sounding of crew action-stations, and for the crew to immediately report to their pre-assigned post or station on the installation.

The AssetAssaultManagement node relates to the management of safety and security at the potential target. The modalities of this node are: withdrawal to a designated safe room, evasive manoeuvres (on mobile units and ships), and activating the security station (this consists of particular procedures to be followed by individual crew members when the alarm is raised).

Ensuring the Safety and Security of the Facility. As is the case with procedure management the SARGOS system includes action plans for the management of production equipment in order to safely shut it down, and deny access to sensitive areas.

Conditional Probabilities. In the Bayesian network, each module or sub-module consists of one or many nodes that all have an effect on each other.

Each node consists of a matrix of conditional probabilities that are calculated by taking into account the various influences between nodes and the actual situation represented by the node itself.

The probabilities of the base nodes are normalized, i.e. elements that would characterize a specific attack have not been added.

4 Discussion

When the probability distribution of the different modalities has been established, and the Bayesian network has been developed, various attack scenarios can be simulated. Once the initial conditions have been determined, the network translates them into a response report. Experiments with different potential scenarios enabled the network to be finalised before it was integrated into the SARGOS system.

4.1 Attack Scenario Case Studies

The following two examples demonstrate what happens when parameters are set to simulate an attack on a Floating Production, Storage and Offloading (FPSO) unit, which is considered to be a critical asset. Figure 3 describes the first scenario where an unknown vessel creates the threat.

The parameters that were set to reproduce the scenario on an FPSO are: the identity class of the threat, the ranking between the threat and the target (Ranking Threat/Asset, which corresponds to the time in seconds required for the threat to travel the remaining distance to the target), the distance between the threat and the target (DTG Threat/Asset, in meters), the security vessel response time (TTG SecurityVessels/Asset, in seconds), the time of day and the visibility. The simulation shows that the danger level of this situation is 2, with a percentage of 64.68. In this case the countermeasures to be applied are: inform the boatswain, request the intervention of a security vessel, send a clear, strong message using a long-range loudspeaker, activate the searchlight, activate the security station, and activate repulsion equipment. Planning is tailored to the danger level of the situation and evolves in response to changes in the parameters relating to the threat and the target. In the second scenario the attacker is now hostile, armed and equipped with a highly manoeuvrable boat. This high-threat scenario is described in Figure 4.

In this scenario, the danger level is 4, with a percentage of 79.79. Figure 5 illustrates the adapted response plan. This level of danger requires internal and external communications (BroadcastField and ActivateDistressCall) but more importantly, a more vigorous response demonstrated by the following countermeasures: assemble the crew (CrewManagement), activate the security station (AssetAssaultManagement), ensure the safety and security of the production facility (EngageESDS), block access to sensitive areas (ShutLockAccesses) and delay the progress of the attackers (SetCrowdControlMunition). Finally, a low-impact repulsion measure such as the Long Range Acoustic Device (ActivateL-RAD) is put on stand-by.

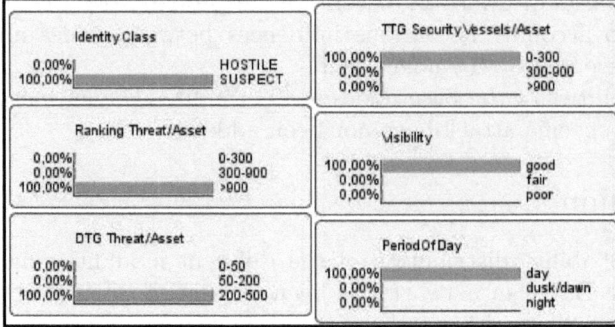

Fig. 3. Observations set for scenario1

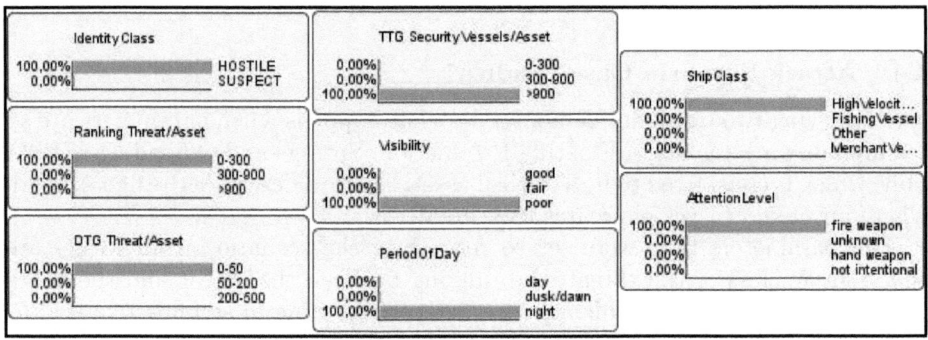

Fig. 4. Observations set for scenario2

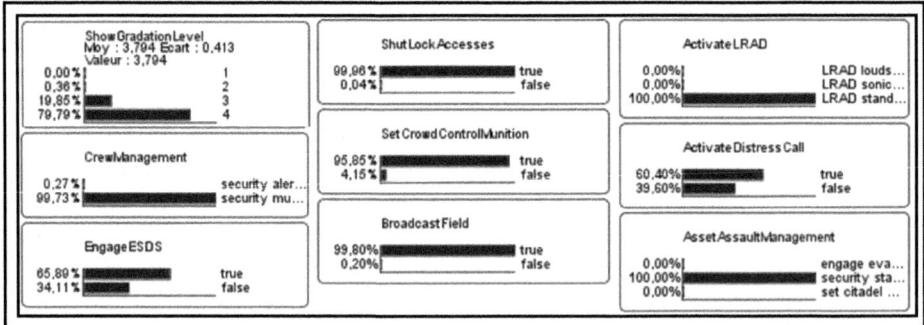

Fig. 5. Simulation of a high-threat scenario on an FPSO

Generating attack scenarios in this way helps to refine the probability distribution and tests the response of the Bayesian network to changes in parameters (the threat, the target, the environment, etc.).

4.2   Integration of the Bayesian Network into the SARGOS System

In order to integrate the Bayesian network into the SARGOS system, a prototype was developed that took as input an alert report and generated as output a response plan. The response plan contains all the countermeasures to be applied either by the crew or automatically by the system. Once the countermeasures (whose probability exceeds the activation threshold) have been selected, they are displayed in the response plan in a specific order. The main factors that determine the order are: the action mode of the countermeasure, its ease of implementation, the extent to which it is automated (or the need for a large number of people to activate it), the time needed for it to become effective, and any potential additional functions.

5   Conclusion and Future Work

The SARGOS system responds to an alert report with a response plan, which is the result of an intelligent analysis of the alert report. This response plan brings together the information necessary for the physical installation to protect itself against a threat.

An initial constraint of the project is met, as all the countermeasures are non-lethal responses.

The use of a Bayesian network for the planning of the response is a major asset of the SARGOS system as this network can handle all possible combinations of threat characteristics, the target under attack, environment, crew management and facilities. Most importantly, it adapts to changes in the danger level of the situation.

Finally, the network is able to integrate feedback from attacks that has previously been used to administer and can therefore evolve. Consequently, the planning module can be modified and improved iteratively.

References

[GA1]   Giraud, M.A., Alhadef, B., Guarnieri, F., Napoli, A., Bottala Gambetta, M., Chaumartin, D., Philips, M., Morel, M., Imbert, C., Itcia, E., Bonacci, D., Michel, P.: SARGOS: Securing Offshore Infrastructures Through a Global Alert and Graded Response. In: System Workshop MAST Europe, Juin 27-29 (2011)

[GA2]   Giraud, M.A., Alhadef, B., Guarnieri, F., Napoli, A., Bottala Gambetta, M., Chaumartin, D., Philips, M., Morel, M., Imbert, C., Itcia, E., Bonacci, D., Michel, P.: SARGOS: Système d'Alerte et Réponse Graduèe Off Shore. In: Conference WISG, Janvier 25-26 (2011)

[GV1]   Giraud, M.A., Van Gaver, A., Napoli, A., Scapel, C., Chaumartin, D., Morel, M., Itcia, E., Bonacci, D.: SARGOS: Système d'Alerte et Réponse Graduèe Off Shore. In: Conference WISG, Janvier 26-27 (2010)

[WB1]   Ware, B.S., Beverina, A.F., Gong, L., Colder, B.: A Risk-Based Decision Support System for Antiterrorism. Digital Sandbox, 8 pages (Août 14, 2002)

[NW1]   Naïm, P., Wuillemin, P.H., Leray, P., Pourret, O., Becker, A.: Les réseaux bayésiens 3, 424 pages (1999)

A Scalable Object Based Discovery Service for Global Tracing of RFID Products

Gihong Kim, Bonghee Hong, and Joonho Kwon

Department of Computer Engineering, Pusan National University,
San 30, Jangjeon-dong, Geumjeong-gu, Busan 609-735, Republic of Korea
{buglist,bhhong,jhkwon}@pusan.ac.kr

Abstract. Radio-Frequency Identification (RFID) technology can provide global visibility for each product by the Discovery Service (DS) of EPCglobal, the de facto international standard for RFID. Only the role of a DS has been introduced and no concrete standard for a DS has yet been specified. A DS should have high scalability because of the huge volume of products and therefore tracing information.

We propose a scalable distributed architecture for a DS that is Object-Based Discovery Service (OBDS). An OBDS consists of several unit-DSs and a Unit-DS Lookup Service (UDLS). A unit-DS contains all data for some objects and a UDLS enables the discovery of the desired unit-DS among distributed unit-DSs for global tracing of an object. We present an analytic comparison of several approaches for the DS and an experimental comparison of several approaches for designing a UDLS.

Keywords: RFID, Discovery Service, Global Track and Trace, Distributed System.

1 Introduction

Recently, Radio-Frequency Identification (RFID) technology has enabled automatic identification and also global tracing of tagged objects in a supply chain environment [1]. A party in a supply chain delivers a physical object such as a trade product identified by an Electronic Product Code (EPC) [2] to another party. In this environment of global data exchange, standardization is very important. EPCglobal [3] is the de facto international standard for RFID, but while EPCglobal has introduced a Discovery Service (DS) [4] that provides a means of global tracing of tagged objects, no concrete standard has yet been specified. Available literature [5][6], however, provides a high-level description of the EPCglobal DS architecture.

The EPCglobal architecture framework [7][8] defines EPC physical object exchange standards to ensure that when one company delivers a physical object to another company, the latter can share EPC data and interpret it properly.

Therefore, in this paper, we show some naive approaches for a DS. First, in the centralized approach, the data are managed by only one DS, so the main advantage of this centralized DS is its simplicity. However, its scale is limited by the capacity of

H. Yu et al. (Eds.): DASFAA Workshops 2012, LNCS 7240, pp. 92–103, 2012.

the server. A centralized DS architecture suffers from serious performance and scalability problems [10]. Next is a hierarchical approach, dividing the DS into several local DSs and a root DS. The local DS covers some region for all kinds of objects. The root DS in this approach is the bottleneck of the entire architecture and there are problems in coupling of data. We therefore propose a scalable Object-Based Discovery Service. The design concept of this architecture is that a unit-DS does not bind with a region, but with objects. This approach solves the problem of coupling of data and outperforms other approaches.

2   Naive Distributed Approach

Fig. 1 shows a typical hierarchical distributed model of a DS. In this model, a local DS is in charge of several EPCISs, and a root DS is connected to all the local DSs. The main design parameter for this distribution is the location of each EPCIS. All the objects observed in a specific EPCIS are reported to its connected local DS. As Fig. 1 shows, local DS1 is connected to EPCISs A and B. Similarly, local DS2 is in charge of C, D and E. All the local DSs are connected to a root DS. If any observation of objects occurs in EPCIS A, the observation is reported to its parent node, local DS1.

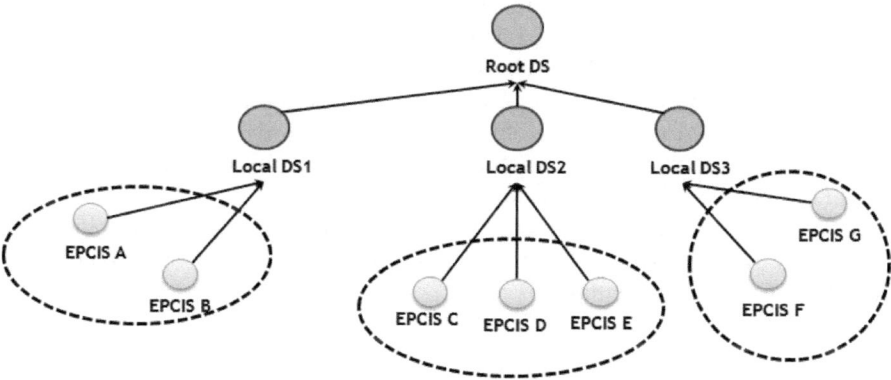

Fig. 1. Distributed Discovery Service

Fig. 2 shows an insert example for the first approach to a hierarchical distributed DS. The trajectory of object "a" is A→B→C→D and that for object "b" is F→D→C→A. Only the data of "a" is represented in the Fig. 2. Observations of an object are reported to each local DS. For example, when object "a" is observed in EPCIS A, the observed information is reported to the local DS1 and also reported to the root DS. A root DS should store all the observation information of all the objects, as indicated in Fig. 2.

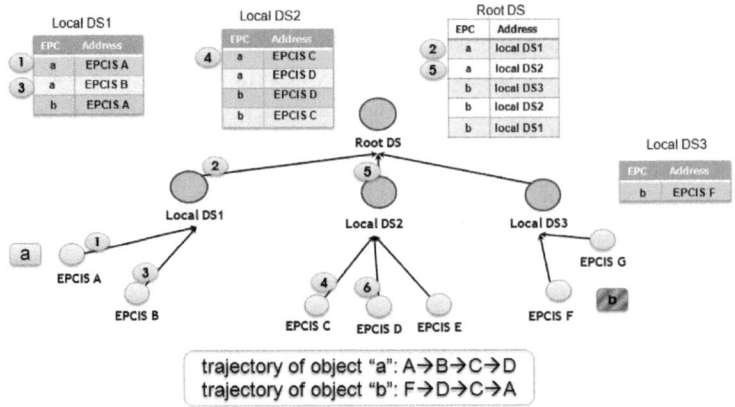

Fig. 2. Insert example

Fig. 3 shows a search example using the first approach. Users access the DS by accessing an application. The application must first access the root DS, because all the child information is stored there. If the user wants global tracing information for object "a", the root DS is first accessed, thus indirectly addressing the next-level node (the local DS), where the user can retrieve the next point information. The answer from the root DS is local DS1 and DS2. In the next step, the user queries the databases of local DS1 and DS2. The local DS also has an indirect address (the EPCIS) pointing to where the real observed information is. The EPCISs contain the actual observed information for objects. In the Fig. 3, the user can retrieve the answer that observation information for object "a" is in EPCISs A and B from the local DS1. Similarly, the information for object "a" is in EPCISs C and D through the local DS2. Finally, the user can retrieve the observation information for object "a" in the EPCISs. A user wanting all the tracing information for object "a" should query EPCISs A, B, C and D sequentially.

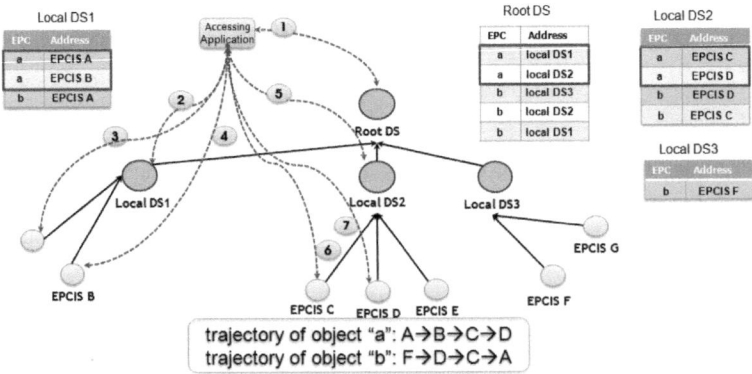

Fig. 3. Search example

3   Object-Based DS

In Section 2, we examined naive distributed approaches such as putting the local DS in charge of some EPCISs. That is, the local DS is assigned to the fixed EPCISs on the basis of locations. As in Fig. 1, if any object is observed in EPCIS A, the observed information should be reported only to its bound local DS1. For example, Korean company EPCISs are bound to only one Korean local DS. At first glance, this may seem to be an appropriate binding, but it is not. In the next section, we explain why this is not suitable for DS architecture.

3.1   Design Concept

In this section, we propose a scalable distributed architecture for a DS. The key design concept is the decoupling of observed information. This architecture proposes several unit-DSs that are independent of each other, because all the observed data for one object are stored in only one unit-DS. Therefore, a user does not need to search several unit-DSs. The design concept of this architecture is that the unit-DS does not bind with the EPCIS, but with an object, as shown in Fig. 4. The EPCIS can therefore report the data to any unit-DS, but the data for one object are stored in only one unit-DS. An additional component is the Unit-DS Lookup Service (UDLS), which finds the appropriate unit-DS address where the global tracing information for the object the user is seeking is stored. The UDLS informs the user which unit-DS contains the global tracing information for this object.

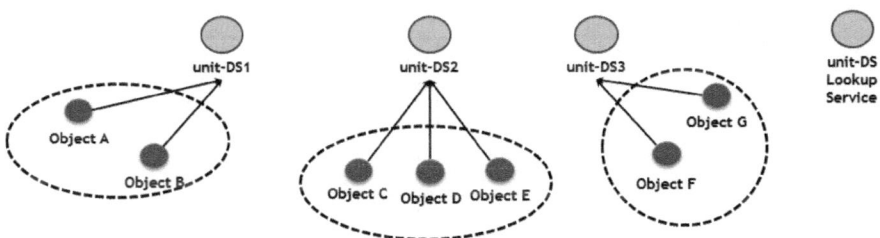

Fig. 4. Design concept of Object-Based DS

3.2   Components

Unit-DS. This component stores all the global tracing information for some objects, so there is no dependency between unit-DSs. If the user wants to find global tracing information for some object, only one unit-DS must be accessed. The information in the unit-DS is a mapping table that maps objects to a list of observed EPCISs. If an object is observed on any EPCIS, the pair of object and observed EPCIS address is inserted in the appropriate unit-DS in the inserting step. Later, the user can query the unit-DS to find the global tracing information for this object, and the unit-DS returns an address list for visited EPCISs. Table 1 shows the methods of the unit-DS.

Table 1. Methods of the unit-DS

Method name	Returned value	Parameter
getEPCISAddress (return a list of EPCIS addresses that store global tracing information for the object the user wants)	List of URLs (list of EPCIS addresses)	URI of EPC (the object the user wants to trace)
registerEPCObservation (register the EPCIS address where the object is observed)	void	URI of EPC (observed object)
		URL (what EPCIS observed the object)

Unit-DS Lookup Service (UDLS). The Unit-DS Lookup Service (UDLS) is new in our proposal. In the previous naive approaches, there is an entry point that is a root DS, but in this architecture there is no entry point. The user can find the right unit-DS through the UDLS. If the user already knows which is the right unit-DS for the object, the UDLS is unnecessary, but general users who do not know the appropriate unit-DS address can use the UDLS. The information in the UDLS is a mapping table that maps an object to a unit-DS address. This information is very static and there is very little update. Thus, once the user downloads the mapping table of a UDLS, there is no more access to a UDLS except updates of the UDLS information. Table 2 shows the methods of the UDLS.

Table 2. Methods of the Unit-DS Lookup Service (UDLS)

Method name	Return value	Parameter
getUnitDSAddress (return the unit-DS address that stores global tracing information for the object that the user wants)	URL (address of unit-DS)	URI of EPC (the object the user wants to trace)
registerUnitDSAddress (register the unit-DS address where the user can find the tracing information for an object)	void	URI of EPC (object code)
		URL (unit-DS address)

3.3   Insert and Search

We now discuss insert and search examples. In Fig. 5, the object "a" is observed in EPCIS A, and EPCIS A asks which is the right unit-DS address for object "a". The answer is unit-DS1, so EPCIS A stores the information in unit-DS1. Likewise, the information for object "b" is stored in unit-DS3.

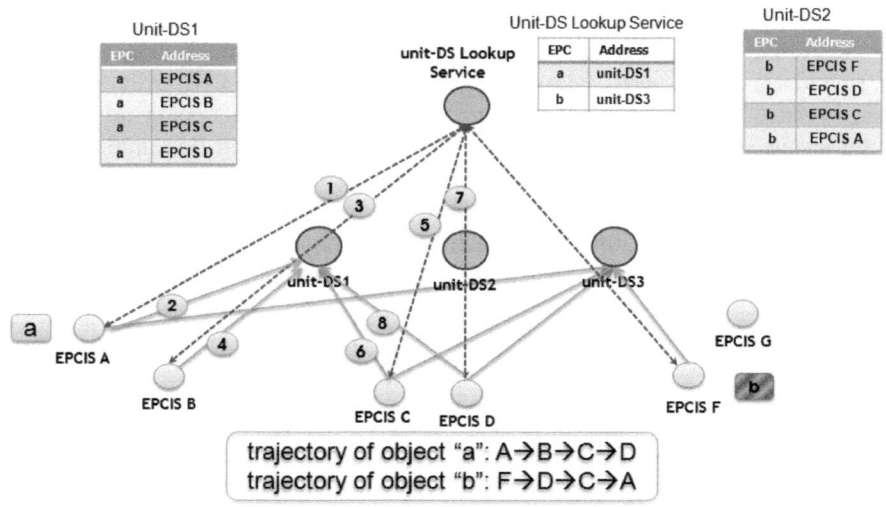

Fig. 5. Insert example of Object-Based Discovery Service

The UDLS information is a table mapping objects to appropriate addresses of unit-DSs. That information is preregistered by the system administrator. Updates of UDLS information are very rare unless the unit-DS IP is changed. The architecture in Fig. 5 therefore has room for improvement. If each EPCIS downloads the UDLS information at its first request, it need not access the UDLS every time. An example is shown in Fig. 6.

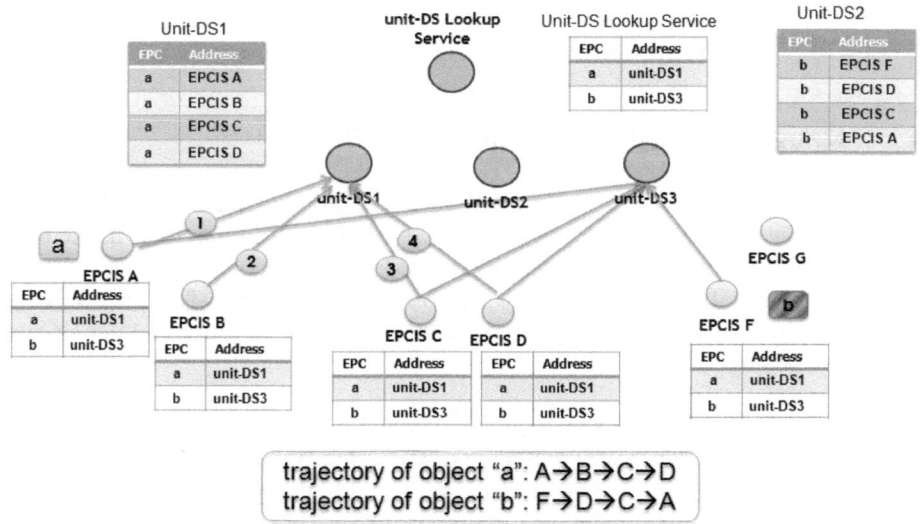

Fig. 6. Insert improvement by downloading the UDLS information in advance

We now describe a search example using the Object-Based Discovery Service in Fig. 7. A user wanting to trace object "a" first accesses the UDLS, because the information about the next access point is stored there. Unit-DS1 is the next access point to search for tracing information. In unit-DS1, all the tracing information is stored in one database, so the user accesses only the one unit-DS, unlike the naive approaches. Finally, the user can retrieve the observation information for object "a" in each EPCIS. This search mechanism also has room for improvement: users can download the UDLS information only at the first request. Later, a synchronization mechanism should be proposed when the UDLS information is updated.

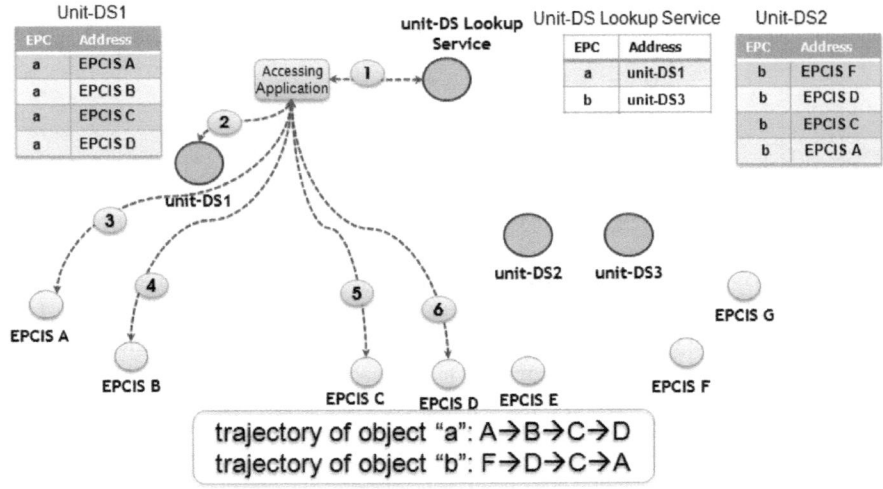

Fig. 7. Search example using the Object-Based Discovery Service

4   Design of Unit-DS Lookup Service (UDLS)

In this section, we propose two approaches for designing a Unit-DS Lookup Service (UDLS). The Electronic Product Code (EPC) uniquely identifies objects in RFID environments. It is divided into three parts: company, product and serial, as shown in Fig. 8. The serial part is not our concern. The two approaches below assign appropriate indexes to the company and product parts using a b-tree and one or more interval trees, respectively.

Horizontal Approach. This approach is horizontal because it intersects the result of each index in Fig. 8. There are two parts of the index: the first is one b-tree for the company part and the second is one interval tree for the product part, because the company data are all point data, and the product part has some point data and some range data.

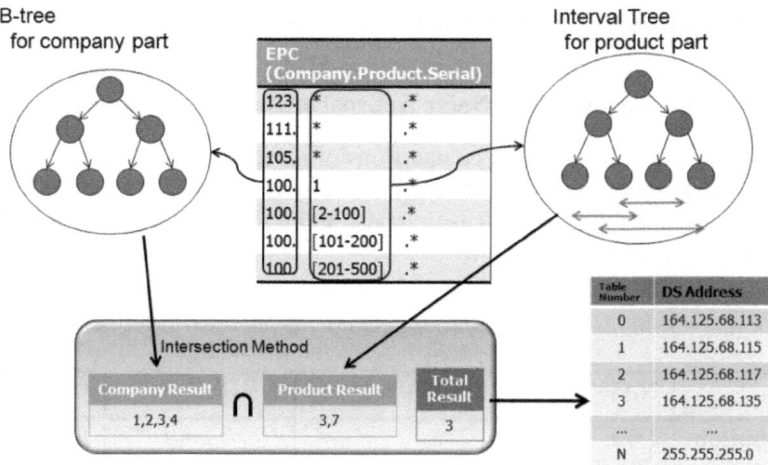

Fig. 8. Horizontal approach

Vertical Approach. In this layered architecture, the first layer is one b-tree for the company part and the second layer is several interval trees for the product part. Like Fig. 9, in the first layer, each leaf node is linked to the second layer unless the company part specifies a whole range (* means whole range). In the search flow of this design, we first access the b-tree and then access the interval tree if required.

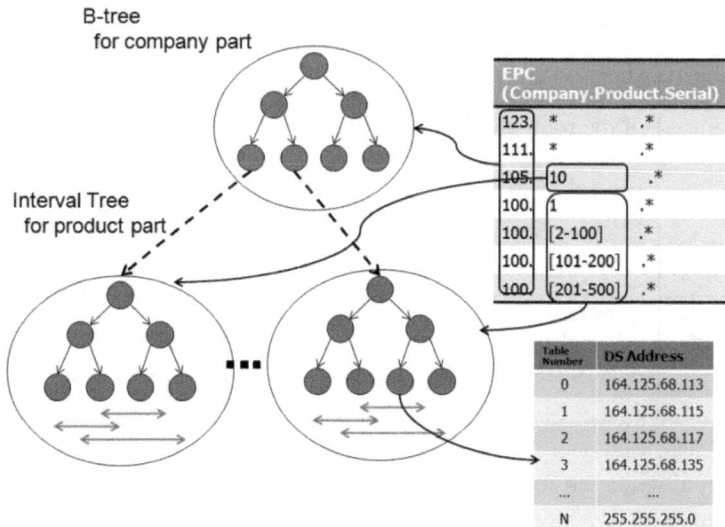

Fig. 9. Vertical approach

5   Analytic and Experimental Comparison

5.1   Analytic Comparison of Naive Approaches and New Solutions

In this section, we show analytic comparisons of naive approaches and new solutions in Table 3. The proposed solution (ODBS 1) has room for improvement in which each EPCIS downloads the UDLS information only at the first request, so no further access to the UDLS is required (ODBS 2).

Table 3. Analytic comparison of insert, search and data costs

	NAIVE 1	NAIVE 2	OBDS 1	OBDS 2
Distribution	*Tree distribution* – Coupling of data	*Tree distribution* – Coupling of data	*Horizontal distribution* – Decoupling of data	*Horizontal distribution* – Decoupling of data
Root DS, UDLS	*Root DS* – Stores the data in the root DS	*Root DS* – Does not store the data in the root DS	*UDLS* – User does not download the UDLS info	*UDLS* – User downloads the UDLS info only at first request
Insert cost	*High* – One insert to the root DS	*Low* – No access to the root DS	*Medium* – One search of the UDLS	*Low* – No access to the UDLS
Search cost	*High* – One search of the root DS – Search of several local DSs	*Very High* – One search of a root DS – Search for all local DSs – Search in several local DSs	*Medium* – One search of the UDLS – One search of a unit-DS	*Low* – No search of the UDLS – One search of a unit-DS
Data space (root DS, UDLS)	*Very High* – Serial-level data of EPC	*Low* – No data in the root DS	*Medium* – Product-level data of EPC	*Medium* – Product-level data of EPC

As Table 3 shows, in NAIVE 1, accessing the root DS on every operation is the bottleneck of the entire architecture, not only for inserts but also for searches. NAIVE 2 is very good for insert but very poor for search because of the additional search for complete local DSs all over the world. On the other hand, ODBS 1 has moderate performance for both insert and search. The performance of ODBS 2 is further improved by downloading the UDLS information in advance.

5.2   Experimental Comparison of Design Approaches

We compare the performance of the vertical and horizontal approaches on various sets of data and queries. All indexes were kept in main memory and performance was measured using a personal computer with an Intel Pentium IV 2.6 GHz processor, 2 GB of main memory and the Microsoft Windows XP operating system.

We first studied the insert cost. The dataset consisted of 10, 20 or 30 million records.

Fig. 10 shows the insertion costs of the horizontal and vertical approaches. The vertical approach outperforms the horizontal approach. Because the horizontal approach is built from one b-tree and many interval trees, the insertion cost is higher than that of the vertical approach.

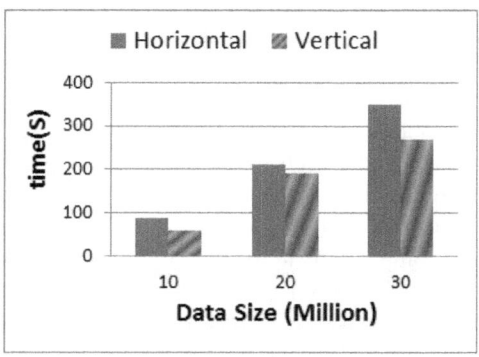

Fig. 10. Insertion cost of vertical and horizontal approaches according to data size

Next, we studied the search cost for the three types of query shown in Fig. 11. Query 1 searches for point (.) data, Query 2 searches for group ([ ]) data and Query 3 searches for asterisk (*) data.

The results show that the vertical approach has much poorer performance than the horizontal approach. The horizontal approach uses one b-tree for the company part and small interval trees for product parts, while the vertical approach uses one b-tree

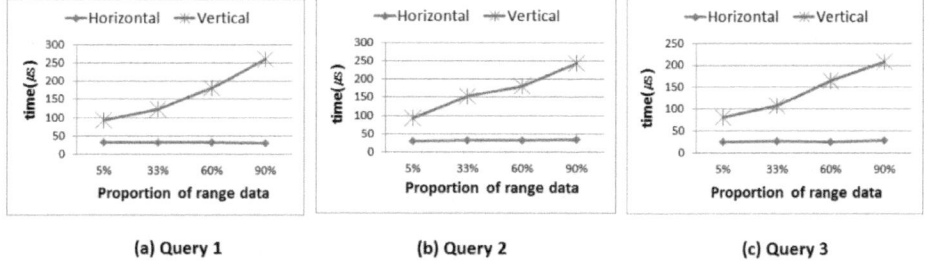

Fig. 11. Search cost of vertical and horizontal approaches according to query type

for the company part and one large interval tree for the product part. The performance of the vertical approach is poor because of the single large interval tree instead of the several small interval trees of the horizontal approach. As the proportion of range data grows, the performance of the vertical approach deteriorates severely. As the proportion of range data grows, the size of the interval tree grows, and as a result the performance declines.

6   Conclusions and Future Work

The Discovery Service (DS) is the core service of the EPCglobal Network Architecture for global tracing of RFID tagged objects. Standards for the DS have not yet been specified by EPCglobal. The DS should have high scalability because the databases are very large and the number of users is increasing rapidly.

In this paper, we have proposed a scalable architecture for DS and several approaches to the design of the Unit-DS Lookup Service (UDLS). We first established the architecture of the Object-Based Discovery Service (OBDS). We solved the problem of data coupling and showed that the OBDS outperformed the naive approaches for insert and search by an analytic comparison in the first part of Section 6. Next, we designed the UDLS using two approaches. We showed that the horizontal approach outperformed the vertical approach for various datasets and queries by experimental comparisons in the last part of Section 6.

In future work, if the information of a UDLS is updated, we should propose a synchronization mechanism for UDLS information and user-downloaded information. A more advanced solution for the UDLS is also possible: because the data of the UDLS consist of two parts, we could propose a dedicated two-dimensional index structure.

Acknowledgments. This work was supported by the grant of the Korean Ministry of Education, Science and Technology. (The Regional Core Research Program/Institute of Logistics Information Technology).

References

1. Roy, W.: An Introduction to RFID Technology. IEEE Pervasive Computing 5(1), 25–33 (2006)
2. EPCglobal Inc.: EPC Tag Data Standard (TDS), Version 1.5,
   http://www.epcglobalinc.org/standards/tds/
   tds_1_5-standard-20100818.pdf
3. EPCglobal Inc., http://www.epcglobalinc.org/
4. EPCglobal Inc.: Discovery Service Working Group Discovery Service, Version 1.0,
   http://www.epcglobalinc.org/apps/org/workgroup/subscriber/
   download.php/15448/SAG_Discovery_Services_Opt-
   In_Charter_Final.doc

5. Kürschner, C., Condea, C., Kasten, O., Thiesse, F.: Discovery Service Design in the EPCglobal Network – Towards Full Supply Chain Visibility. In: Floerkemeier, C., Langheinrich, M., Fleisch, E., Mattern, F., Sarma, S.E. (eds.) IOT 2008. LNCS, vol. 4952, pp. 19–34. Springer, Heidelberg (2008)
6. Germany, G.: Internet der Dinge. Management Information. Das EPCglobal Netzwerk, Tech. Rep.
7. EPCglobal Inc.: The EPCglobal Architecture Framework, Version 1.3,
   `http://www.epcglobalinc.org/standards/architecture/`
   `architecture_1_3-framework-20090319.pdf`
8. Verisign: The EPC Network: Enhancing the Supply Chain,
   `http://www.verisign.com/static/002109.pdf`
9. EPCglobal Inc.: EPC Information Services (EPCIS) Version 1.0.1 Specification,
   `http://www.epcglobalinc.org/standards/epcis/`
   `epcis_1_0_1-standard-20070921.pdf`
10. Vanalstyne, M., Brynjolfsson, E., Madnick, S.: Why not one big database? -Principles for data ownership. Decision Support Systems 15(4), 267–284 (1995)

On Smart and Accurate Contextual Advertising

Guandong Xu[1] and Zongda Wu[2]

[1] Centre for Applied Informatics, Victoria University
PO Box 14428, Vic 8001, Australia
Guandong.Xu@vu.edu.au
[2] Oujiang College, Wenzhou University
Wenzhou 325003, China
zongda1983@163.com

Abstract. Web advertising, a form of advertising, which uses the World Wide Web to attract customers, has become one of the most important marketing channels. As one prevalent type of Web advertising, contextual advertising refers to the placement of the most relevant commercial ads into the content of a Web page, so as to increase the number of ad-clicks. However, some problems such as homonymy and polysemy, low intersection of keywords, and context mismatch, can lead to the selection of irrelevant ads for a generic page, making that the traditional keyword matching techniques generally present a poor accuracy. Furthermore, existing contextual advertising techniques only take into consideration how to select as relevant ads for a generic page as possible, without considering the positional effect of the ad placement in the page.

In this paper, we propose a new contextual advertising framework to tackle problems, which (1) uses Wikipedia concept and category information to enrich the semantic representation of a page (or a textual ad) and (2) takes the placement position of embedded advertise into account. To accomplish these steps, we first map each page (or ad) into three feature vectors: a keyword vector, a concept vector and a category vector. Second, we determine the relevant ads for a given page based on a similarity measure which combines the above three feature vectors. In dealing with position-wise contextual advertising, the relevant ads are selected based on not only global context relevance but also local context relevance, so that the embedded ads yield contextual relevance to both the whole targeted page and the insertion positions where the ads are placed. We experimentally validate our approach by using a real ads set, a real pages set , and a set of more than 260,000 concepts and 12,000 categories from Wikipedia. The experimental results show that our approach performs better than the simple keyword matching and can improve the precision of ads-selection effectively.

Keywords: Wikipedia Knowledge, Smart Contextual Advertising, Position-wise Contextual Advertising, Global Relevance, Local Relevance.

H. Yu et al. (Eds.): DASFAA Workshops 2012, LNCS 7240, p. 104, 2012.

Review Summarization Based
on Linguistic Knowledge

Kyung-Mi Park*, Hogun Park, Hyoung-Gon Kim, and Heedong Ko

Korea Institute of Science and Technology (KIST)
Hwarangno 14-gil 5, Seongbuk-gu, Seoul 136-791, Republic of Korea
{kmpark,hogun,hgk,ko}@imrc.kist.re.kr
http://kist.re.kr/en/index.jsp

Abstract. In this paper, domain-knowledge extraction and aspect-opinion extraction are proposed in order to generate a summary from the relevant product and service review. In order to extract the word corresponding to aspect and opinion, we extract the domain-salient word and collocation information by applying statistical techniques from the bulk of the text, and construct the clue words through manual filtering. In domain knowledge extraction, in order to extract useful information, domain-salient words which occur more significantly in a given domain rather than in a public domain article are automatically extracted by using the statistical techniques. As well, collocation information has the association with high frequency words. In recognition of aspect-opinion association, words corresponding to aspects and opinions in a sentence are checked by using information of clue words, and the polarity of the sentence is determined by performing pattern-based modality analysis. Through checking the binary association based on the frequency of co-occurrence, a pair of aspect and opinion is extracted, our system can automatically acquire the scores for a review target based of the degree of positive/negative.

Keywords: Domain-knowledge extraction, aspect-opinion extraction, linguistic knowledge, review summarization.

1 Introduction

The current search method shows the results by finding a matching word between a query and the meta-data. However, it is likely to bring a different result which does not reflect the intent of the user. It is necessary to handle the query by identifying what a user's intent is. For instance, the current search way does not give the same answer for the query *Are there great restaurants near here?* or *Is*

---
* This research is supported by Korea Institute of Science and Technology under "Development of Tangible Social Media Platform" project. This research is supported by Ministry of Culture, Sports and Tourism(MCST) and Korea Creative Content Agency(KOCCA) in the Culture Technology(CT) Research & Developement Program 2011.

H. Yu et al. (Eds.): DASFAA Workshops 2012, LNCS 7240, pp. 105–114, 2012.

Med for Garlic of Gwanghwamun branch fine? Thus, it is necessary to present a summary review for evaluation factors such *dating* or *atmosphere* by analyzing the review text, just as the proposed method does. In addition, it will be able to present *recommendations* depending on the given conditions and situations.

There are various examples of the review summary. The number of positive/negative expressions for defined evaluation factors are calculated, and it can be represented as a figure. We can obtain overall rating which means the sum of the number of the positive/negative expressions. Through this, users determine the extent the users like or dislike each restaurant. Also, we can show a sentence containing positive/negative opinion regarding to evaluation factor. For each evaluation factor, sentences including opinion may be presented as a table. By providing sentences including opinion, as well as review summary scores, it is helpful for user to judge.

It is important to extract vocabulary information corresponding to the positive/negative expressions in review mining. However, it is not easy to automatically generate a positive/negative words given a particular domain. Thus, we can see the high frequency positive/negative expression words from that domain by using resources such as pre-built dictionary or through hand. Unfortunately, available resources for the Korean do not exist yet. When applied machine learning method, lexical information can be automatically obtained by utilizing useful statistical information on the training data. However, it is required to perform many manual works in order to build a large amount of training data, and it is required to build a new training data whenever the domain changes. Suggested method minimizes manual work and extracts a relatively accurate vocabulary information after the identification of people through domain-salient word and collocation extraction applied to statistical techniques.

Existing research judged positive/negative opinion on a sentence or on a whole document. However, a user can write a positive review for the specific evaluation factors even though the user was negative impression as a whole for a particular product or service. For instance, a user received overall negative impression for a particular restaurant, but the user can write good reviews for a parking lot and the amount of the food. Thus, it is important to correctly connect between the aspect and the opinion about what. Proposed method can be calculated the degree of opinion about detailed evaluation factors. To be specific, this method can present a summary review for various evaluation factors such as *taste, service, atmosphere, price, cleanliness, parking* for a restaurant review, or *lens, LCD, memory, resolution, video recording, design* for a digital camera.

This paper presents a rating system to give a score for the review target by recognizing the pair of aspect and opinion as well as domain-knowledge extraction through statistical techniques. Proposed method calculates the number of positive expressions or negative expressions for each detailed assessment element and suggests evaluation factors and statements including its opinion. When producing a review summary, in order to apply multiple domains without learning data, the proposed method recognizes domain knowledge and aspect-opinion association is analyzed and defined. For restaurants and movie reviews, as well as calculations of the system scores, it is analyzed by applying Pearson correlation coefficient for the correlation between user's star ratings and the proposed system evaluation. Experiment is performed for the restaurants or movie domains rather than a single domain in order to prove that the proposed method can be applied to various domains.

2   Related Work

Previous studies proposed to extract useful information for pattern-based method [1-2]. It proposes k-Structure method among Korean product review, which can improve accuracy by automatically extracting sentiment word from a simple sentence. This simple sentence means the maximum pattern length is 3. This means a sentence contains a sentiment word at a distance of 2 based on the attribute name of evaluation products.

Previous studies are applied the defined pattern by extracting sentences below 3 maximum word phrases to extract sentiment words expressing a positive/negative [1-2]. However, a positive/negative emotional expression can be expressed in the form of collocation like *deliberately go to find* or *meat is tender*. Existing studies extract direct expression of *good/bad*, while the proposed method can find a positive/negative expression in the form of collocation.

Using opinion mining technology, related work suggests the technique to determine the rankings in product review data depending on the intent of the user [3-4]. The related work is considered the value of the product review within the user's query, as well as the inclusion of subjective opinions in product review and entropy of sentiment polarity [3-4].

Existing studies manually built word lists that reflect positive and negative meaning [5-6]. However, proposed method automatically extracted domain-salient word likely to offer valuable information and collocation information and manually performed filtering in order to reduce the time. In addition, existing research did not analyze the association between each aspect and opinion. Suggested method identified relevance through the degree of occurrence from words corresponding to aspect and opinion in the text of the review.

Previous studies extract the product features from the product review, extract initial positive/negative predicate for each domain by utilizing average ratings present in the product review, and build the positive/negative dictionary for each domain by analyzing access information of the initial positive/negative predicate [7-8].

In order to extract keywords, existing research relies on the user's star ratings [9-10]. However, the proposed method can be calculated score of review data with no user's star ratings for each aspect. In addition the method that automatically extracts domain-salient word and collocation information is suggested by taking advantage of statistical techniques in order to minimize manual.

3   Review Summarization Based on Linguistic Knowledge

As in Figure 1, the phase of the system can be divided into domain-knowledge extraction and aspect[1]-opinion extraction. Both phases execute pre-processing of sentence segmentation and morphological analysis for the input data. Domain-knowledge extraction is a step to recognize significant information of each domain, and this extracted information is used in the aspect-opinion extraction. Domain-salient words are extracted by comparing the frequency of certain words in a given domain and the frequency occurred in newspaper articles in the public domain. Collocation information is evaluated by using likelihood ratio as a measure. The clue word that corresponds to aspects and the opinions is finally identified through the manual filtering for domain-salient word and collocation. This information is utilized to find the clue word from named-entity recognition and find associations between the most appropriate aspects and opinions from relation extraction by extracting co-occurrence information between aspect and opinion.

The aspect-opinion extraction consists of named-entity recognition and relation extraction. Named-entity recognition is the step that recognizes columns of words corresponding to aspects or opinions, and relation extraction is the step that recognizes only the association that exist relevance among aspect-opinion relationship. Named-entity recognitions identify columns of words that represent aspects or opinions by tagging the clue words through defined knowledge. In Korean verb phrase, verb phrase is recognized to supplement the utilization of complex ending morpheme and we perform modality analysis by analyzing the relationship between main verb and auxiliary verb. For modality analysis, positive or negative opinion is identified by utilizing a defined pattern. In relation extraction, the appropriate connection relationship is extracted based on the frequency of co-occurrence among possible aspect-opinion association candidates. The results of the evaluation rating system can be summarized as the number of positive and negative expressions about evaluation target. For a more detailed description of the detailed steps are as follows: section 3.1 and 3.2 presents a more detailed explanation for the domain-knowledge extraction and the aspect-opinion extraction, respectively.

3.1   Domain-Knowledge Extraction

In order to automatically extract domain knowledge, first, sentence segmentation and morphological analysis as pre-processing for a text review is performed. The sentence segmentation is done based on sentence punctuation mark such as a period. The morphological analysis is carried out using Korean Language Technology (KLT), which is a Kookmin University's morphological analyzer[2] and has a total 10 parts-of-speech.

---

[1] *Aspect* means evaluation factors.
[2] http://nlp.kookmin.ac.kr/HAM/kor/index.html

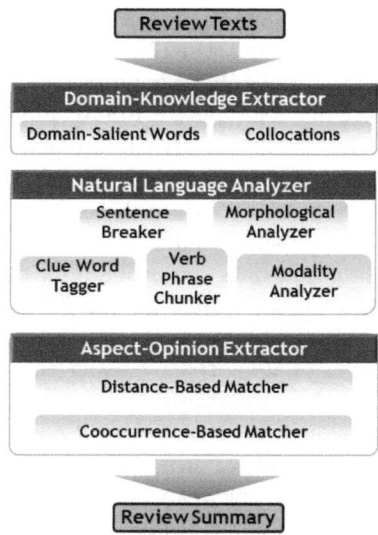

Fig. 1. System Architecture. The proposed method is divided into domain-knowledge extraction and aspect-opinion extraction. The former is the phase of extracting domain knowledge, and its extracted information is used on clue-word tagging and aspect-opinion extraction.

Domain-salient words and collocation information are automatically extracted in order to find the important words that correspond to aspects and opinion. Domain-salient words mean significantly encountered words in that domain. If restaurants reviews, they are will be applicable *taste, service, atmosphere*, and if camera reviews, they would be *lens, resolution, photos*. Collocation means the column of words which frequently occur and have cohesion. If restaurants review, they will be equivalent *food is right, there is no financial burden*, or *deliberately go find*. The proposed method finally determines the clue words that correspond to aspect and opinion from automatically extracted results through the manual filtering. The extraction of this useful information can reduce the cost of hand-work because manual filtering is performed to intend for domain-salient words and collocation information. When several aspects and opinions occurs in a sentence, co-occurrence frequency is extracted from the review data to ensure proper connection relationship.

Most clue words of aspects and opinions are domain-salient words. Domain-salient words have characteristics significantly higher incidence in a given domain compared to regular text such as news. Thus, it is not necessary to navigate all the words in order to recognize the clue words corresponding to the aspects or the opinions. It is possible for the scope of the search to qualify significantly occurred words in a given domain. The suggested method compares the frequency of a given word in news with the frequency of a given word in a review domain. If the latter occurs more frequently, that word is assumed the domain-salient word. Because domain-salient words are likely to contain a clue word that corresponds to the aspects and the opinions, they play a role to identify important words.

It is utilized Sejong data[3] in 2004 consisting of news. About the size of the data, the number of sentences and nouns is 5,018 and 61,597, respectively. Relative frequency ratio (RFR) is calculated, and it is considered a domain-salient word if the value is more than a threshold.

In the RFR formula, denominator is the probability that the word w occurred in the Sejong data, numerator is the probability that the word w occurred in the review texts of restaurants. If denominator is zero, it is regarded as the domain-salient word. The frequency is calculated by using the first morpheme of each word phrases obtained morphological analysis.

Positive/negative expression from the review domain can be various across multiple words, as well as like/dislike. For instance, in order to indicate positive, the opinions in the restaurant reviews can be represented *deliberately go find, go again, keep going, a lot of guests, always go, good quality*, as well as *like*. These words are high frequency words like collocation in a given domain, and have features that have bonding between words. The suggested method automatically extracts collocation by utilizing likelihood ratio (LR), and determines positive/negative expressions through the manual filtering.

LR is assumed as collocation, if the value is more than threshold. Frequency of consecutive two words is calculated by using the first morpheme for each word phrase obtained morphological analysis.

3.2   Aspect-Opinion Extraction

Aspect-opinion extraction consists of named-entity recognition and relation extraction. The named-entity recognition finds the words that correspond to the aspects and the opinions, and relation extraction determines the association which recognized opinion points to evaluation factors.

The named-entity recognition proceeds clue word tagging, verb phrase recognition, and modality analysis. As shown in Figure 1, clue word tagging finds the words which correspond to the aspects and the opinions in a sentence by taking advantage of the result of domain-knowledge extraction step. Positive/negative opinion from domain knowledge extraction is mostly represented by verb. To find the words expressing the opinion, the proposed method was to target the Korean main verb. The combination issue of main verb and auxiliary verb is processed through verb phrase recognition and modality analysis. For instance, in domain-knowledge extraction phase, positive expression is extracted, but negative expression is not extracted. This is handled in a later step. In this case, only the auxiliary verb which affects to determine positive/negative is

---

[3] http://www.sejong.or.kr/eindex.php

considered, rather than considering all the auxiliary verbs. The verb phrase recognition is performed by using information about part-of-speech obtained results of morphological analysis. Based on information about part-of-speech, the boundaries of verb phrase are recognized. The modality analysis is performed based on pattern. The pattern affecting in determining positive/negative was written by hand. If we encounter a word that matches the pattern, positive has been fixed as the negative, and vice versa.

Relation extraction finds the association between aspects and opinions. For instance, even though people received an overall negative impression at the restaurant, they can write a good review that the food portions are large and convenient parking. Therefore, it is necessary for the aspect-opinion extraction to recognize the exact connection by finding opinion indicated by the evaluation factors. Through the previous steps, a word that corresponds to the aspect or the opinion is recognized. The co-occurrence frequencies extracted from the review data are applied to determine which the evaluation factors correspond to recognized opinion. It is finally extracted among candidates which a word corresponding to the aspect and the opinion co-occur more than threshold. As well as domain-salient word and collocation, the first morpheme from each word phrase is used by utilizing the results of the morphological analysis.

The following steps explain detailed process that words in a sentence belonging to a specific evaluation factors find a proper opinion. First, in the case of words expressing *aspect-opinion relation* like *delicious*, we identify the association of aspect-opinion. Next, if the end morpheme of the previous word phrases based on word phrases is specific pattern, we identify the association of aspect-opinion. Next, when the number of the word phrase between the aspect and the opinion is less that 4, we identify the association of aspect-opinion. Finally, if the number of the word phrase between the aspect and the opinion is 5 to 8, and two words co-occur, we identify the association of aspect-opinion.

The followings is an example on the aspect-opinion extraction. An example of a input sentence is *interior is good, and the food price is not expensive.* First, sentence separation and morphological analysis as a pre-processing step is performed. Next, clue word tagging is performed by using defined domain knowledge. For example, [AIR], [POS], [PRI], [NEG] represent meaning class of each word. [AIR] represents that the word *interior* belongs to *atmosphere* among evaluation factors, and [PRI] represents that *food price* belongs to *price* class. [POS] and [NEG] implies respectively positive and negative opinion of each corresponding words. The modality analysis turn a negative into a positive in a given short sentence by applying specific modality pattern. Relation extraction identifys that *good* is an opinion on *interior*, and *not expensive* is an opinion on *food price*. Finally, as the result of review summary, the system score is added 1 point for evaluation factor *atmosphere* and *price*, and the related sentence including aspect-opinion association is extracted: [aspect *interior*-positive opinion *good*] [aspect *food price*-positive opinion *not expensive*].

4   Experiments

Rreviews from Wingspoon[4], Daum place[5], and Naver movie[6] are used as the
test data in order to verify the effectiveness of the proposed method. Restaurants reviews are obtaind from Wingspoon and Daum place. 450 reviews from
Wingspoon and 404 reviews from Daum place are extracted for the experiments.
The average review length from Wingspoon and Daum place is, respectively,
20.65 and 14.13 as the number of word phrase. 1,296 aspect-opinion association from Wingspoon and 862 aspect-opinion association from Daum place are
tagged. Naver movie is the reviews of two Korean movies. From Naver movie,
the average review length is 7.87, and 351 opinion are tagged.

The following experiments for the test data are carried out. The system's
overall performance for three different test data were measured. The system's
performances for the detailed evaluation factor in each test data were measured.
The performance contributions of modality analysis were analyzed. Also, the correlation is analyzed between user's star rating and system score through Pearson
correlation coefficient.

Table 1 indicates the overall performance of the system for the test data. The
precision indicates the percentage of correct answers among the aspect-opinion
association found by the system. The recall represents the fraction that found
by the system among the aspect-opinion association that corresponds to correct
answers. F-measure expresses the precision and the recall in the same weight as
a single value. The overall performance of the system, when expressed as the F-
measure, it was 83.05% from Wingspoon, 78.30% from Daum place, and 82.53%
from Naver movie.

Table 1. Overall performance for evaluating the accuracy of aspect-opinion extraction.
Each row means that the result of the system and that of the gold standard match on
the test data of WingSpoon, Daum-Place, Naver-Movie.

Data	Precision	Recall	F-measure
WingSpoon	89.70	77.31	83.05
Daum-Place	84.64	72.85	78.30
Naver-Movie	87.54	78.06	82.53

In WingSpoon data, 9 evaluation factors are determined from domain-salient
words of high frequency. The same evaluation factors were applied to Daum
place. In WingSpoon data, opinion of *taste, service,* and *price* were more analyzed based on F-measure, while *atmosphere* and *cleanliness* were relatively low
performance. In DaumPlace data, opinion analysis for *taste, price* showed a good
performance based on F-measure, but *food material,* and *cleanliness* showed a

---

[4] http://www.wingspoon.com/
[5] http://place.daum.net/
[6] http://lab.naver.com/research/

relatively low performance. The review analysis from Daum place was difficult because DaumPlace contains more reviews close to colloquial styles and shorter length of review text than WingSpoon. Thus, overall performance was lower than WingSpoon data. We analyzed movie reviews with different domains from previous results. Proposed method can be applied regardless of review data of any domain and the deviation of performance is not significant. In NaverMovie data, positive expression was relatively analyzed well than a negative expression.

We analyzed how much lower does the performance of the system when not performing modality analysis. A significant decrease happens in performance, the effect of modality analysis becomes important. In the experiment of Wingspoon data, the decline of the performance based on F-measure was approximately 3.77%.

Table 2 is the result which idenfifies suggested method predicts a score similar to the user's star rating. The columns of correct answers *Human* represent correlation between calculated scores and user's star ratings by utilizing tagging results of the aspect-opinion relationship written by hand. The columns of *System* show the correlation between system scores and user's star rating obtained by utilizing the aspect-opinion association analyzed by the system. Pearson correlation coefficients that have user's star ratings were mostly 0.6 or higher.

The following information can be derived through various experimental results. The proposed method can be applied regardless of domain of review data. In addition, the review text is analyzed by automatically extracting domain-salient word and collocation information without the cost of building the training data. Opinion on various evaluation factors can be extracted within one sentence. Our system's score can obtain like user's star rating. Through experiments utilizing Pearson correlation coefficient, we suggest that relevance of the correlation coefficient between the system scores obtained by applying the proposed method and user's star rating is 0.6 or higher.

Table 2. Pearson Correlation coefficient between star rating and human/system assessment. The second column *Human* shows the correlation coefficient between star rating and human assessment. The third column *System* represents the correlation coefficient between star rating and system assessment.

Data	Human	System
WingSpoon (Overall)	0.7257	0.6985
WingSpoon (Taste)	0.6949	0.6544
WingSpoon (Service)	0.6210	0.6036
WingSpoon (Atmosphere)	0.5586	0.4316
Daum-Place	0.6655	0.6059
Naver-Movie	0.6701	0.6068

5   Conclusion

This paper have performed recognition of domain-knowledge extraction and aspect-opinion association in order to generate review summary. We have extracted domain-salient words and collocations by applying statistical techniques from a large amount of review data. Through manual filtering, clue words corresponding to aspect and opinion have been defined. By using this information, the word that corresponds to aspect and opinion have been found in a new statement. Based on pattern-based modality analysis and frequency of co-occurrence, this paper finally have shown the system that automatically generates rating summary of review about target review by extracting a pair of aspect and opinion

References

1. Song, J.S., Lee, S.W.: Automatic Construction of Positive/Negative Feature-Predicate Dictionary for Polarity Classification of Product Reviews. Korean Institute of Information Scientists and Engineers: Software and Applications 38(3), 157–168 (2011)
2. Zhuang, L., Jing, F., Zhu, X.Y.: Movie Review Mining and Summarization. In: Proceedings of the International Conference on Information and Knowledge Management (2006)
3. Chen, H., Zimbra, D.: AI and Opinion Mining. IEEE Intelligent Systems 25(3), 74–80 (2010)
4. Titov, I., McDonald, R.: Modeling Online Reviews with Multi-Grain Topic Models. In: Proceedings of the 17th International Conference on World Wide Web, pp. 111–120 (2008)
5. Blair-Goldensohn, S., Hannan, K., McDonald, R., Neylon, T., Reis, G.A., Reynar, J.: Building a Sentiment Summarizer for Local Service Reviews. In: WWW Workshop on NLP in the Information Explosion Era (2008)
6. Ming, Z.Y., Chua, T.S., Cong, G.: Exploring Domain-Specific Term Weight in Archived Question Search. In: Proceedings of the ACM International Conference on Information and Knowledge Management (2010)
7. Esuli, A., Sebastiani, F.: Determining Term Subjectivity and Term Orientation for Opinion Mining. In: Proceedings of the European Chapter of the Association for Computational Linguistics (2006)
8. Nishikawa, H., Hasegawa, T., Matsuo, Y., Kikui, G.: Opinion Summarization with Integer Linear Programming Formulation for Sentence Extraction and Ordering. In: Proceedings of the International Conference on Computational Linguistics (2010)
9. Mei, Q., Ling, X., Wondra, M., Su, H., Zhai, C.X.: Topic Sentiment Mixture: Modeling Facets and Opinions in Weblogs. In: Proceedings of the 16th International Conference on World Wide Web, pp. 171–180 (2007)
10. Lerman, K., Blair-Goldensohn, S., McDonald, R.: Sentiment Summarization: Evaluating and Learning User Preferences. In: Proceedings of the European Chapter of the Association for Computational Linguistics (2009)

Finding Related Micro-blogs Based on WordNet*

Lin Li[1], Huifan Xiao[1], and Guandong Xu[2]

[1] School of Computer Science & Technology,
Wuhan University of Technology, China
{cathylilin,huifanxiao}@whut.edu.cn
[2] Centre for Applied Informatics, Victoria University, Australia
guandong.xu@vu.edu.au

Abstract. In the common formulation, the recommendation problem is
reduced to the problem of estimating the utilization for the items that
have not been seen by a user [1]. Micro-blog recommendation will recommend micro-blogs interest users, mostly those related to the micro-blogs
that a user had issued or trending topics. One indispensable step in realizing effective recommendation is to compute short text similarities
between micro-blogs. In this paper, we utilize two kinds of approaches,
traditional cosine-based approach and WordNet-based semantic approach,
to compute similarities between micro-blogs and recommend top related
ones to users. We conduct experimental study on the effectiveness of two
approaches using a set of evaluation measures. The results show that
semantic similarity based approach has relatively higher precision than
that of traditional cosine-based method using 548 twitters as dataset.

1   Introduction

Measuring the similarity between documents and queries has been extensively
studied in information retrieval. However, there are a growing number of tasks
that require computing the similarity between two very short texts. Micro-blog
recommendation will recommend micro-blogs interest users, mostly those related
to the micro-blogs that a user had issued or trending topics. One indispensable
step in realizing effective recommendation is to compute short text similarities
between micro-blogs. Traditional cosine-based similarity computing measure perform poorly on such tasks because of data sparseness and the lack of context,
it rely heavily on terms occurring in both two documents. If two sentences do
not have any terms in common, then they receive a very low similarity score,
regardless of how topically related they actually are. This is well-known as the
vocabulary mismatch problem. For example, UAE and United Arab Emirates
are semantically equivalent, yet they share no terms in common. This problem
is only exacerbated if we attempt to use traditional measures to compute the
similarity of two short segments of text [2]. According to conventional measures,
the more overlaps of words two sentences have, the higher similarity score they

* This research was undertaken as part of Project 61003130 funded by National Natural Science Foundation of China.

H. Yu et al. (Eds.): DASFAA Workshops 2012, LNCS 7240, pp. 115–122, 2012.
© Springer-Verlag Berlin Heidelberg 2012

will receive, which is unreasonable and inaccurate. For example, apple pie and apple phone share one word apple yet have low semantic relation.

Bearing this problem in mind, we take WordNet-based semantic approach into consideration, to see whether and how much can it improve the final accuracy than traditional word based approach. Semantic similarity is a confidence score that reflects the semantic relation between the meanings of two sentences.

We evaluate and analyze the two methods on tweet-tweet similarity task using 548 English messages (tweets) containing 1184 tokens sampled from twitter API(from August to October, 2010). All of the 548 tweets had been normalized before using as our dataset.

The remainder of this paper is laid out as follows. First, we provide an overview of related work and describe the two approaches in Section 2. The experimental results are then reported in section 3. In section 4 we discussed some related works. Finally, conclusions are given in section 5 with directions of future work.

2   Finding Related Micro-blogs

2.1   Overview

Many methods have been proposed to measure the similarity between short text, including purely lexical measures, stemming, language modeling-based measures, and hybrid measures, as studied in [2]. In this paper, we utilize two approaches, traditional cosine-based approach and WordNet-based semantic approach, to evaluate similarities between micro-blogs. We also conduct experimental study on the effectiveness of two approaches using a set of evaluation measures to see which one is better.

2.2   Cosine-Based Approach

We assign to each term in a tweet a weight for that term, that depends on the number of occurrences of the term in the tweet. The way we used in this paper is to assign the weight to be equal to the number of occurrences of term t in a tweet, which is referred to as term frequency and is denoted tf. For one document (tweet here), the set of weights determined by the tf weights above(or indeed any weighting function that maps the number of occurrences of t in a document to a positive real value) may be viewed as a quantitative digest of that document. In this view of a document, known in the literature as the bag of words model, the exact ordering of the terms in a document is ignored but the number of occurrences of each term is material (in contrast to Boolean retrieval). We only retain information on the number of occurrences of each term. Thus, the document "Mary is quicker than John" is, in this view, identical to the document "John is quicker than Mary". Nevertheless, it seems intuitive that two documents with similar bag of words representations are similar in content.

There's a problem that are all words in a document equally important? Clearly not. We looked at the idea of stop words that we decide not to index at all, and therefore do not contribute in any way to retrieval and scoring. At this point, we may view each tweet as a vector with one component corresponding to each term in the dictionary, together with a weight for each component that is given by *tf*. For dictionary terms that do not occur in tweet, this weight is zero. This vector form will prove to be crucial to scoring and ranking. The representation of a set of documents as vectors in a common vector space is known as the vector space model and is fundamental to a host of information retrieval operations ranging from scoring documents on a query, document classification and document clustering.

The standard way of quantifying the similarity between two tweets τ_1 and τ_2 is to compute the cosine similarity of their vector representations, $\boldsymbol{V}(d_1)$ and $\boldsymbol{V}(d_2)$.

$$sim(d_1, d_2) = \frac{\boldsymbol{V}(d_1) \cdot \boldsymbol{V}(d_2)}{|\boldsymbol{V}(d_1)||\boldsymbol{V}(d_2)|} \tag{1}$$

$$sim(d_1, d_2) = \boldsymbol{V}(d_1) \cdot \boldsymbol{V}(d_2) \tag{2}$$

The effect of the denominator of equation 1 is thus to length-normalize the vectors $\boldsymbol{V}(d_1)$ and $\boldsymbol{V}(d_2)$ to unit vectors $\boldsymbol{V}(d_1) = \frac{\boldsymbol{V}(d_1)}{|\boldsymbol{V}(d_1)|}$ and Thus, equation 2 can be viewed as the dot product of the normalized versions of the two document vectors. This measure is the cosine of the angle θ between the two vectors.

Viewing a collection of N documents as a collection of vectors leads to a natural view of a collection as a term-document matrix: this is an $M \times N$ matrix whose rows represent the M terms (dimensions) of the N columns, each of which corresponds to a document. As always, the terms being indexed could be stemmed before indexing; for instance, jealous and jealousy would under stemming be considered as a single dimension.

The following is a basic introduction to the overall process of cosine-based approach that used to measure tweets similarity. The first step we are supposed to do is to split one tweet (short text message) into several terms. Then, we need to normalize ill-formed terms and pick out all the stop words, punctuation marks and signs, say "_", "#", "@", and numbers. Third, find distinct terms and establish continuous and unique index for them. After indexing, the term-document matrix can be created using distinct term as horizontal dimension, the tweets as vertical dimension, and each matrix unit represents how many times the specific row term appeared in the correspondent column tweet. Finally, we compute Euclidean length for each tweet vector (the column of term-document matrix) and utilize equation 2 to get similarity score between two tweets.

2.3   WordNet-Based Approach

WordNet-based approach utilizes dictionary-based algorithms to capture the semantic similarity between two sentences, which is heavily based on the WordNet semantic dictionary.

WordNet is a lexical database which is available online and provides a large repository of English lexical items. It was designed to establish the connections between four types of Parts of Speech (POS) - noun, verb, adjective, and adverb.

The smallest unit in a WordNet is synset, which represents a specific meaning of a word. It includes the word, its explanation, and its synonyms. The specific meaning of one word under one type of POS is called a sense. Each sense of a word is in a different synset. Synsets are equivalent to senses structures containing sets of terms with synonymous meanings. Each synset has a gloss that defines the concept it represents. For example, the words night, nighttime and dark constitute a single synset that has the following gloss: the time after sunset and before sunrise while it is dark outside. Synsets are connected to one another through the explicit semantic relations. Some of these relations (hypernym, hyponym for nouns and hypernym and troponym for verbs) constitute is-a-kind-of (holonymy) and is-a-part-of (meronymy for nouns) hierarchies. For example, tree is a kind of plant, tree is a hyponym of plant and plant is a hypernym of tree. Analogously, trunk is a part of a tree and we have that trunk as a meronym of tree and tree is a holonym of trunk. Therefore, the whole dictionary can be treated as a large graph with each node being a synset and the edges representing the semantic relations.

Malcolm Crowe and Troy Simpson have developed an open-source .NET Framework library for WordNet called WordNet.Net. The codes we used in our experiment are downloaded from Google Code repository and are licensed under The Code Project Open License (CPOL). Given two sentences, it determines how similar the meaning of two sentences is. The higher the similarity score is, the more similar the meaning of the two sentences. The following is steps for computing semantic similarity between two sentences:

1. First each sentence is partitioned into a list of tokens. Each sentence is partitioned into a list of words and we remove the stop words. Stop words are frequently occurring, insignificant words that appear in a database record, article or a web page, etc.
2. Part-of-speech disambiguation (or tagging). This task is to identify the correct part of speech (POS - like noun, verb, pronoun, adverb. . .) of each word in the sentence.
3. Stemming words. We use the Porter stemming algorithm. Porter stemming is a process of removing the common morphological and inflexional endings of words.
4. Find the most appropriate sense for every word in a sentence (Word Sense Disambiguation)
5. Finally, compute the similarity of the sentences based on the similarity of the pairs of words. And we capture semantic similarity between two word senses based on the path length similarity, in which we treat taxonomy as an undirected graph and measure the distance between them in WordNet.

2.4 Summary

The essence of cosine-based method lies in counting the number of overlap terms between two vectors, while WordNet-based method is more concerned with the meaning of words. Generally speaking, the performance of WordNet-based method would outperform the traditional method as discussed by Zhao et al. in [14].

Instead of computing similarity, other scholars have chosen other ways helped in realizing microblog recommendation and related works. Ramage et al. argues that latent variable topic models like Labeled LDA provide a promising avenue toward solving two categories of unmet information needs: improving methods for following new users and topics, and for filtering feeds in [3], and it effectively models important similarity information in posts, improving performance on two concrete tasks modeled after information needs: personalized feed re-ranking and user suggestion. It will be a promising topic for our future study.

3 Experiment

3.1 Dataset

We demonstrate the working of two approaches on the dataset extracted from [15]. It contains 548 English messages sampled from Twitter API (from August to October, 2010) and contains 1184 normalized tokens. All ill-formed words had been detected, and generates correction candidates based on morphophonemic similarity. Both word similarity and context are then exploited to select the most probable correction candidate for the word. We performed experiments to compare the performance of WordNet-based measure with that of the classic cosine-based similarity measure on it.

3.2 Evaluation Measure and Methodology

We use precision and kendall tau distance [16] as evaluation metrics to assess the performance of two approaches.

Precision

For each of the 548 tweets, we compute its similarity with all the 548 tweets(including itself) using the two approaches respectively. For the 548 similarity scores of each tweet, we sort them decreasingly, then pick up top ten scores and capture the corresponding tweet number. After doing this, we can get two 548 × 10 matrix with each row represents top ten related tweets of the specific tweet that the row stands for. Then, we manually judge how many of these relative high related tweets are really related. The precision metric is defined as the ratio of the number of correctly selected relative tweets to the number of pairs of tweets that are really related.

Kendall tau distance

The Kendall tau distance is a metric that counts the number of pairwise disagreements between two lists. The larger the distance, the more dissimilar the two lists are. Kendall tau distance is also called bubble-sort distance since

Table 1. Precision score of two approaches for top 5 and top 10 relative tweets

	Cosine-based	WordNet-based
Top5	0.8276	0.8306
Top10	0.8350	0.8366

it is equivalent to the number of swaps that the bubble sort algorithm would make to place one list in the same order as the other list. The Kendall tau distance between two lists τ_1 and τ_2 is:

$$K(\tau_1, \tau_2) = |(i,j) : i < j, (\tau_1(i) < \tau_1(j) \wedge \tau_2(i) > \tau_2(j)) \vee (\tau_1(i) > \tau_1(j) \wedge \tau_2(i) < \tau_2(j))|$$

$K(\tau_1, \tau_2)$ will be equal to 0 if the two lists are identical and $n(n-1)/2$ (where n is the list size) if one list is the reverse of the other. Often Kendall tau distance is normalized by dividing by $n(n-1)/2$ so a value of 1 indicates maximum disagreement. The normalized Kendall tau distance therefore lies in the interval [0,1]. From the Kendall tau distance equation we can see that it can also be defined as the total number of discordant pairs.

3.3   Experimental Results

Before showing our experimental results, what should be mentioned here is that, because of lacking of clear and specific criteria, and because of that tweet is short English messages without context, it is difficult to judge accurately whether two tweets are within one topic. Furthermore, since the quantity of our dataset is small, the total number of related tweets is also scant, which means the denominator of metric precision is in small number.

Precision

The precision metric is defined as the ratio of the number of correctly selected relative tweets to the number of pairs of manually judged related tweets. The accuracy of precision may be affected by personal understanding toward tweets, which contain large amounts of informal abbreviations and expressions. We can see the results from 1 that the precision scores for two approaches are similar, while the WordNet-based approach is slightly higher than the other even the total amounts of related tweets(119 pairs of related tweets by one human judger) is small. According to previous experience, we believe when the experimental dataset become much larger, WordNet-based approach would get higher precision as studied in [14].

Kendall tau distance

Kendull tau distance is used to measure the agreement of the two recommendation lists produced by our cosine based and WordNet based approaches. The average tau distance for 548 pairs of list of similarity scores is approximately 0.0397922914, which shows general high agreement in the order of each pair of list.

3.4   Discussion

As we can see from the experimental results, the two approaches achieve similar precision scores that we cannot determine which of them is better simply from this experiment. There are several reasons contribute to the phenomenon. Firstly, twitters 140 character limit on tweets presents a challenge to the two methods because short messages would cause data sparseness and the lack of context. Secondly, since the informal expressions and abbreviations being frequently used have not been included in Wordnet, which is a traditional dictionary without trending words and terms, the similarity score between tweets would no doubt be affected, in most cases, reduced. Thirdly, the lacking of abundant dataset together with tweets about daily lives without clear topic may lead to inaccurate human judging of related tweets, thus affecting the value of precision. In order to better evaluate related tweets, we are thinking of expanding tweets with replies or web search results to make it longer and normalized.

4   Related Work

Many twitter related work and problems have been investigated in the literature. Kwak et al. have made the first quantitative study on the entire Twitter sphere and information diffusion on it. They studied the topological characteristics of Twitter and its power as a new medium of information sharing and have found a non-power-law follower distribution, a short effective diameter, and low reciprocity, which all mark a deviation from known characteristics of human social networks [4].

Yin et al. analysis link formation in micro-blogs in [5]. They found that 90 percent of new links are to people just two hops away and the dynamics of new link creation are affected by the users account age. Their experimental results showed that in the very beginning (within 100 days), the users add many friends and then for the older users (100-400 days), their friends seem more stable, while for much older users (more than 500 days), their number of new friends is larger and larger. Results also showed that the older the user, the larger the increase in followers.

On the other hand, computation of short text similarity has also been studied in many literatures through various angles. Many techniques have been proposed to overcome the vocabulary mismatch problem, including stemming [6,7], LSI [8], translation models [9], and query expansion [10,11]. Query expansion is a common technique that used to convert an initial, typically short, query into a richer representation of the information need [10,11,12]. This is accomplished by adding terms that are likely to appear in relevant or pseudo-relevant documents to the original query representation. Sahami and Heilman proposed a method of enriching short text representations that can be construed as a form of query expansion [13]. Their proposed method expands short segments of text using web search results. The similarity between two short segments of text can then computed in the expanded representation space.

5 Conclusion

In this paper we studied the problem of measuring related micro-blogs, which first step is to compute similarity between short texts. We looked at two types of similarity computing approaches, traditional cosine-based approach and WordNet-based semantic approach. We showed that WordNet-based approach works slightly better under a small number dataset containing only 548 tweets. We will conduct experiment on larger dataset in the future to see whether WordNet-based approach would get much better result than the cosine-based approach as we thought.

References

1. Adomavicius, G., Tuzhilin, A.: Toward the next generation of recommender systems: A survey of the state-of-the-art and possible extensions. IEEE Transactions on Knowledge and Data Engineering 17(6), 734–749 (2005)
2. Metzler, D., Dumais, S., Meek, C.: Similarity Measures for Short Segments of Text. In: Amati, G., Carpineto, C., Romano, G. (eds.) ECIR 2007. LNCS, vol. 4425, pp. 16–27. Springer, Heidelberg (2007)
3. Ramage, D., Dumais, S., Liebling, D.: Characterizing Microblogs with Topic Models. Association for the Advancement of Artificial Intelligence (2010)
4. Newman, M.E.J., Park, J.: Why social networks are different from other types of networks. Phys. Rev. E 68(3), 036122 (2003)
5. Yin, D., Hong, L., Xiong, X., Davison, B.D.: Link Formation Analysis in Microblogs. In: Proceedings of SIGIR 2001, Beijing, China, pp. 24–28 (2001)
6. Krovetz, R.: Viewing morphology as an inference process. In: Proceedings of SIGIR 1993, pp. 191–202 (1993)
7. Porter, M.F.: An algorithm for suffix stripping. Program 14(3), 130–137 (1980)
8. Deerwester, S., Dumais, S., Landauer, T., Furnas, G., Harshman, R.: Indexing by latent semantic analysis. JASIST 41(6), 391–407 (1990)
9. Berger, A., Lafferty, J.: Information retrieval as statistical translation. In: Proceedings of SIGIR 1999, pp. 222–229 (1999)
10. Zhai, C., Lafferty, J.: Model-based feedback in the language modeling approach to information retrieval. In: Proceedings of CIKM 2001, pp. 403–410 (2001)
11. Lavrenko, V., Croft, W.B.: Relevance based language models. In: Proceedings of SIGIR 2001, pp. 120–127 (2001)
12. Rocchio, J.J.: Relevance Feedback in Information Retrieval, pp. 313–323. Prentice-Hall (1971)
13. Sahami, M., Heilman, T.: A web-based kernel function for measuring the similarity of short text snippets. In: Proceedings of WWW 2006, pp. 377–386 (2006)
14. Zhao, S., Du, N., Nauerz, A.: Improved recommendation based on collaborative tagging behaviors. In: Proceedings of the 13th International Conference on Intelligent User Interfaces, New York, NY, USA
15. Han, B., Baldwin, T.: Lexical normalisation of short text messages: Makn sens a #twitter. In: Proceedings of the 49th Annual Meeting of the Association for Computational Linguistics, Portland, USA
16. Kendall tau distance, http://en.wikipedia.org/wiki/Kendall_tau_distance

An Efficient Path Nearest Neighbor Query Processing Scheme for Location Based Services

Yonghun Park, Kyoungsoo Bok, and Jaesoo Yoo[*]

Dept. Information and Communication Engineering
Chungbuk National University, Cheongju, Chungbuk, Korea
{yhpark1119,ksbok,yjs}@chungbuk.ac.kr

Abstract. A path nearest neighbor (PNN) query finds the closest objects from paths consisting of line segments. The existing method has redundant searches as a problem. In this paper, we propose PNN methods that avoid redundant searches in an index. In order to avoid redundant searches, the proposed methods find the closest objects from a path sequentially and determine whether the next closest object is found. To prove the superiority of the proposed methods, we evaluate their performance.

Keywords: Nearest neighbor, Query processing, Path nearest neighbor, Location based services, Location based social networks.

1   Introduction

According to the development of mobile devices and location aware technologies, the interests of the location based services (LBSs) have been increased. LBSs are to provide information related to the locations that a user finds. Finding nearest friends, taxies and buses, and finding the gas stations in certain area are the examples of LBS. The quality of LBS is currently insufficient but its growth potential is unlimited in the future. It is because the production and demand of mobile devices including the GPS modules, such as iphones and smart phones, are increased and various contents using locations of clients are developed these days.

Location-based social networks have been studied, recently. Location-based social networks provide an important new dimension in understanding human mobility [8, 9, 10, 11]. The networks depend on the similarity of the users' activities [12]. The users who have similar patterns of the activities, will be interested in same objects, possibly. It means that the many of the interesting place which they will stop by, will be overlapped. To recommend their interesting objects on their path is very useful application. We have researched on the method to find the recommended objects on the path efficiently.

A nearest neighbor (NN) query is one of the representative query types for LBSs. The NN query finds the closest object from the point that a user indicates. The studies

---

[*] Corresponding author.

H. Yu et al. (Eds.): DASFAA Workshops 2012, LNCS 7240, pp. 123–129, 2012.

on NN query processing have been progressed for long time. However, they focus on NN from a point. The path nearest neighbor (PNN) query finds the closest object from paths consisting of line segments. PNN is applicable to continuous queries in the moving queries and static objects environments.

The purpose of this paper is to propose the efficient PNN query processing methods to find the interesting objects on my path. [1] proposed a continuous query processing method in the moving queries and static objects environments. The method is applicable to PNN query processing method. However, it processes a NN algorithm to find each neighbor object of the path. The method performs NN algorithm as many as the number of objects in the result of the query and it leads to the redundant searches in the index.

In this paper, we propose PNN methods that avoid redundant searches in the index. We fix the previous algorithm to be efficient. The previous method finds the objects vertically but the proposed method finds the objects horizontally. In another words, to avoid the redundant searches, the proposed methods find the closest objects from a path sequentially and determine whether the next closest object is found. To prove the superiority of the proposed methods, we evaluate their performance.

The rest of this paper is organized as follows. Section 2 presents the related work and section 3 describes the proposed methods and the cost model. Section 4 shows the performance evaluation results. Finally, we conclude this paper in section 5.

2   Related Work

[1] is a method to process a continuous NN query efficiently on R-tree according to the movement of a client. The method estimates the future path of the client and provides the continuous query result by pre-computing the result on the estimated path. The result of the query defined in [1] is the same with PNN. Figure 1 shows the query processing method for Continuous Nearest Neighbor Searches (CNNS). To calculate the k-NN objects on the path from point s to point e, CNNS searches point a as the nearest object of s. And then, CNNS searches another object that exists within the circle area that the center is e and the radius is the distance between a and e. It is shown in figure 1(a), where $|e, a|$ denotes the distance between points e and a. When they search the nearest object c of the point e in the circle, it divides the path by the point s_i on the path that $|s_i, a|$ is equal to $|s_i, c|$. It is shown in Figure 1(b). This process is repeated again for the two paths made by the previous process until nearest objects from the paths are not found any more.

There are other researches to process continuous queries in the static objects and moving queries environments. To process continuous k-NN queries efficiently, a number of methods were also proposed [3, 4, 5]. In [3], they pre-compute the Voronoi diagram of the data and store it in R-tree. When a nearest neighbor query arrives at the server, the Voronoi diagram is used to efficiently compute the nearest neighbor. In [4], they pre-compute the future results on the predefined path and provide them to clients. In [5], they use the safe region which guarantees the same results. [6, 7]

propose the efficient index structure to process kNN queries efficiently. In [6], they fix R-tree structure like B$^+$-tree by adding the link information to the leaf nodes between neighbor leaf nodes. In [7], they use grid structure instead of the tree structure so save the cost travel the tree structure.

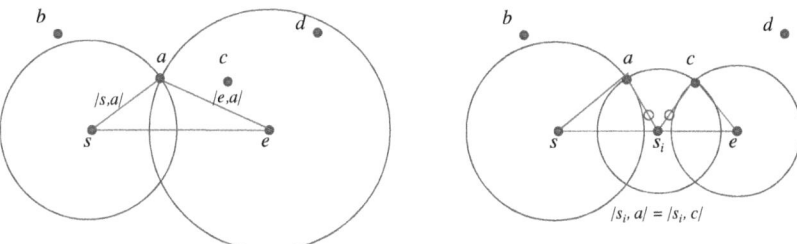

(a) Predefined path and the nearest point a from s (b) Nearest point c from e and division of the path

Fig. 1. CNNS query processing method

3   The Proposed Method

We propose three methods processing a PNN query. The method uses circles and the principals in [1]. The previous method finds the objects vertically but the proposed method finds the objects horizontally. In the method, we find the closest objects to a path sequentially and draw the circles that the center is on the path and passes two neighboring objects. The objects inside the circles are only evaluated. While finding the closest objects sequentially, the circles are re-drawn and become smaller. While finding the closest objects, if the object is the closest, it is put in the result set. The result set is sorted in the order of the horizontal distance to the left end point of the path. For the first point of the result set, the circle is drawn with the left point of the path as its center, and passes the first point. Similarly, for the last point of the result set, the circle is drawn with the right point of the path as its center and passes the last point. And then, we draw the all circles that the center is on the path and passes two neighboring objects in the result set. If the next closest object is overlapped with the circle drawn by the result set, the object is put into the result set and the circles are redrawn. The method is processed until objects do not exist in the circles.

Figure 2 shows the process of the proposed method. O_1 is first found and the two circles are drawn. The circles and pass O_1, and the centers of the circles are start point and end point of a path P, respectively, as shown in Figure 2(a). After O_2 is found, the circle containing O_2 is split up into two smaller circles as shown in Figure 2(b). When drawing the circles, if the circle is most right or left circle of the path, the center of the circle is the end point or the start point. Otherwise, the center of the circle is a point on the path and the circle passes two horizontally neighboring objects. It is continued until all objects in the circles are found. After O_3 and O_4 are found, the algorithm is finished in Figure 2. Figure 2(c) and Figure 2(d) show the process.

Figure 3 shows the pseudo code of the algorithm. The algorithm first finds the nearest neighbor object of a path P (line 05). After that draw two circles the centers are the start point and end point of P, respectively (line 06~11). $DIST(p_i, p_j)$ returns the distance between points p_i and p_j. Next, it finds the next nearest neighbor object of P and splits the circle containing the object (line 12~34). This step is continued until there are no objects in the all circle areas (line 15).

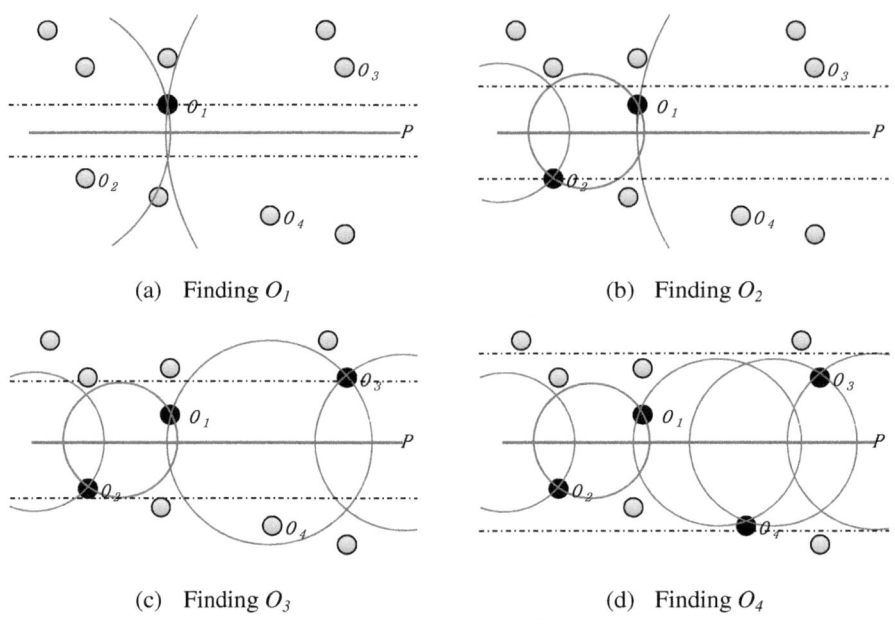

(a) Finding O_1

(b) Finding O_2

(c) Finding O_3

(d) Finding O_4

Fig. 2. Proposed PNN query processing method using circles

We analyze the algorithms to compare the computation efficiency. We formulate the cost model in terms of the relation between the size of the searched area and search cost of the index. Equation 1 presents the cost of previous method on tree index structure. A_{nnq} denotes the average size of the searched area for processing NN query. $N_{result\_objects}$ and $N_{circles}$ denote the number of result objects and circles maid by the query processing, respectively. $logA_{nnq}$ means the cost of processing a NN query that the size of the searched area is A_{nnq}. The NN algorithm is processed as many as $N_{result\_objects} + N_{circles}$.

$$C_{tree\_previous} = (N_{result\_objects} + N_{circles}) \times log\,A_{nnq} \qquad (1)$$

Equation 2 presents the cost of proposed method on tree index structure. A_{pnnq} denotes the size of the total searched area for a PNN query. As shown in equation 3, A_{pnnq} is much smaller than $A_{nnq} \times N_{circles}$ because of the overlapped areas between the circles. If we substitute $A_{nnq} \times N_{circles}$ for A_{pnnq} in equation 2, $C_{tree\_proposed}$ is $log(A_{nnq} \times N_{circles})$. Therefore, it is obvious that $C_{tree\_proposed}$ is smaller than

$C_{tree\_previous}$. To tell the truth, log in the equations can be changed and it depends on the index structure for the data. Equation 4 and 5 are the example.

$$C_{tree\_proposed} = log\, A_{pnnq} \qquad (2)$$

$$A_{pnnq} \ll A_{nnq} \times N_{circles} \qquad (3)$$

Equation 4 and 5 present the cost of processing previous and proposed method on grid index structure. The cost depends on the resolution of the grid and the size of searched area. α denotes the size of a cell. We estimate the number of accessed cells by dividing the size of searched area by α. As following equation 3, $C_{gird\_proposed}$ is also smaller than $C_{grid\_previous}$.

$$C_{grid\_previous} = (N_{result\_objects} + N_{circles}) \times \lceil A_{nnq}/\alpha \rceil \qquad (4)$$

$$C_{gird\_proposed} = \lceil A_{pnnq}/\alpha \rceil \qquad (5)$$

PNNQueryProcessing(*path P*)

```
01   Circle c, c_right, c_left;
02   Object o, o_right, o_left;
03   ResultSet result;
04   List circleList;
05   o = FindNearestObject(P);
06   c_right.center = StartPoint(P);
07   c_right.radius = DIST(o, c.center);
08   InsertCircleIntoList(circleList, c_right);
09   c_left.center = EndPoint(P);
10   c_left.radius = DIST(o, c.center);
11   InsertCircleIntoList(circleList, c_left);
12   While () {
13       o = FindNextNearestObject(P);
14       c = {c_i | o ⊂ c_i, c_i ⊂ circleList};          // find a cell containing o
15       if (c is NULL) then break;
16       DeleteCircleFromList(circleList, c);
17       o_right = GetRightNearestObject(result);
18       o_left = GetLeftNearestObject(result);
19       if (o_right is NULL) {                          // draw right circle
20           c_right.center = StartPoint(p);
21       } else {
22           c_right.center = { point p_i | DIST(o, p_i) = DIST(o_right.center, p_i), p_i ⊂ P }
23       }
24       c_right.radius = DIST(o, c_right.center);
25       InsertCircleIntoList(circleList, c_right);
26       if (oright is NULL) {                           // draw left circle
27           c_left.center = StartPoint(p);
28       } else {
29           c_left.center = { point p_i | DIST(o, p_i) = DIST(o_left.center, p_i), p_i ⊂ P }
30       }
31       c_left.radius = DIST(o, c_left.center);
32       InsertCircleIntoList(circleList, c_left);
33       PutObjectIntoResultSet(result, o);
34   }
35   return result;
```

Fig. 3. PNN query processing algorithm

4   Performance Evaluation

We evaluate the PNN method using circles as a representative algorithm in this paper and compare the algorithm with the existing method [1]. The PNN method using circles and the existing method are denoted by PNN-circle PNN-previous, respectively. We index objects by the R-tree index structure [2]. The number of objects is set to 100K and the number of queries is set to 10. We measure the number of disk I/Os according to the path length from 0.1% to 1% of the side length of the index space. Figure 4 shows the performance comparison of two methods in terms of the number of disk I/Os. We can see easily from figure 4 that the number of I/Os in PNN-circle is slowly increased while the number of I/Os in PNN-previous is highly increased according to the increase of the path length. The gap between the algorithms presents the number of the redundant disk I/Os. Figure 5 shows the performance comparison of two methods using grid structure in terms of the number of disk I/Os. We set the resolution of the grid structure to 100×100. It also haves the same aspects as shown in figure 4.

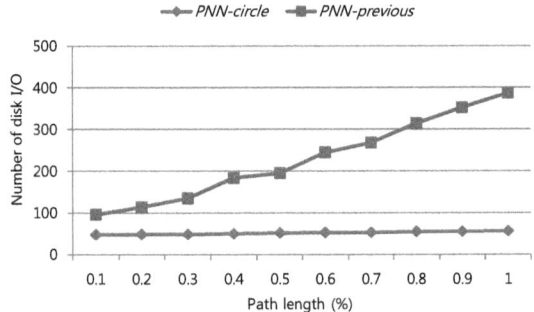

Fig. 4. Performance comparison using r-tree structure according to the length of path

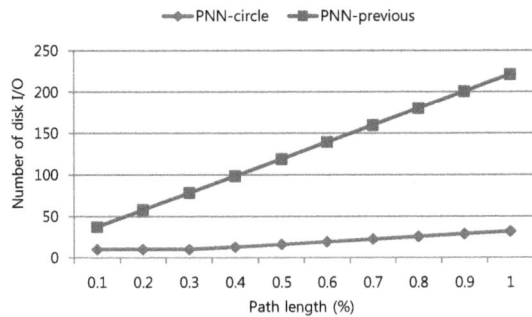

Fig. 5. Performance comparison using grid structure according to the length of path

5   Conclusion

In this paper, we proposed PNN methods avoiding redundant searches. To avoid the redundant searches, the proposed methods basically find the closest objects from a path sequentially and determine whether the next closest object is found. To prove the superiority of the proposed method using circles, we evaluated its performance. Compared with PNN-previous, PNN-circle performs about 32% disk I/Os on average to process PNN queries when the path lengths of the queries are 0.1 ~ 1.0% of the side length of index space in the simulation environment. For the further work, we will adapt the method to find the relationship between nodes (Minimum Bounding Rectangle: MBR) in the R-tree index structure.

Acknowledgment. This work was supported by the Korea Institute of Science and Technology Information (K-12-L06-C02-S03) and Basic Science Research Program through the National Research Foundation of Korea (NRF) grant funded by the Korea government (MEST)(No. 2009-0089128).

References

[1]   Tao, Y., Papadias, D., Shen, Q.: Continuous Nearest Neighbor Search. In: Proc. Intl. Conf. Very Large Data Bases, pp. 287–298 (2002)
[2]   Sellis, T., Roussopoulos, N., Faloutsos, C.: The R+-Tree: A dynamic index for multi-dimensional objects. In: Proc. Intl. Conf. Very Large Data Bases, pp. 507–518 (1987)
[3]   Zheng, B., Lee, D.L.: Sementic Caching in Location Dependent Query Processing. In: Proc. Intl. Symp. Advanced in Spatial and Temporal database, pp. 97–116 (2001)
[4]   Tao, Y., Papadias, D., Shen, Q.: Continuous Nearest Neighbor Search. In: Proc. Intl. Conf. Very Large Data Bases, pp. 287–298 (2002)
[5]   Zhang, J., Zhu, M., Papadias, D., Tao, Y., Lee, D.: Location-based spatial queries. In: Proc. ACM SIGMOD Intl. Conf. Management of Data, New York, pp. 443–454 (2003)
[6]   Park, Y., Seo, D., Park, H., Yoo, J.: An Index Structure for Efficient k-NN Query Processing in Location Based Services. In: Proc. Intl. Conf. Ubiquitous Information Technologies & Applications, pp. 165–170 (2009)
[7]   Park, Y., Seo, D., Lim, J., Lee, J., Kim, M., Bao, W., Ryu, C., Yoo, J.: A New Spatial Index Structure for Efficient Query Processing in Location Based Services. In: Proc. Intl. Workshop on Ubiquitous and Mobile Computing (2010)
[8]   Humphreys, L.: Mobile social networks and social practice: A case study of dodge ball. Journal of Computer-Mediated Communication (2008)
[9]   Noulas, A., Scellato, S., Mascolo, C., Pontil, M.: An empirical study of geographic user activity patterns in foursquare. In: Proc. Intl. AAAI Conf. Weblogs and Social Media (2011)
[10] Scellato, S., Noulas, A., Lambiotte, R., Mascolo, C.: Socio-spatial properties of online location-based social networks. In: Proc. Intl. AAAI Conf. Weblogs and Social Media (2011)
[11] Cho, E., Myers, S.A., Leskovec, J.: Friendship and mobility: user movement in location-based social networks. In: Proc. ACM SIGKDD Intl. Conf. Knowledge Discovery and Data Mining (2011)
[12] Scellato, S., Mascolo, C.: Measuring User Activity on an Online Location-based Social Network. In: Proc. Intl. Workshop on Network Science for Communication Networks (2011)

Social Community Based Blog Search Framework

Ok-Ran Jeong[1] and Jehwan Oh[2]

[1] Department of Software Design and Management,
Gachon University, Republic of Korea
orjeong@kyungwon.ac.kr
[2] Department of Computer Science and Engineering,
University of Minnesota, Minneapolis, MN, USA
ohxxx245@umn.edu

Abstract. This study proposes a blog search framework which enables a more in-depth search on a given topic by extracting the collective intelligence features in social community sites and through the query extension using these features. The characteristics of blog contents is that it has a lot of information made up of user experience and trusted more by most users than the contents gained by general search. The proposed framework extends the query using the answer information related to the query which the user wishes to search and gets applied to the blog search on this basis. The information gained from various types of social community sites could be considered as one form of collective intelligence while this has been applied to the blog search. The framework proposed in this paper utilizes the important Q&A information of social community to let the user gain more reliable and useful search results.

Keywords: Social Community, Collective Intelligence, Q&A information, Blog Search Engine.

1 Introduction

As the web-based digital industry plays an important role of domestic and foreign economy, the social network based technology is rapidly growing under the environment of next generation technology. This paper uses the community based knowledge search information of asking and answering questions to each other under a new paradigm called the social network environment while trying to propose a framework for applying this information on the blog search.

Recently, the blog is becoming a new place for the user to produce and share information on the web. The success factor of blog is because the writing method is simple and it is easy to share information among users. It is also able to provide one's daily routine, opinion, commentary, emotions or articulating ideas, etc. [26]. The blog search sets the purpose as finding the blog which is focused on and repeats dealing with the topic given by the query entered by the user from the various blogs provided this way. Blog search is distinguished from the normal search of information from the fact that it sets the blog which is a set of posts or from the fact that it handles the blog

H. Yu et al. (Eds.): DASFAA Workshops 2012, LNCS 7240, pp. 130–141, 2012.

documents with unrestrained subjective thoughts, expressions and styles of the user who has created the blog. For the typical search of information, most of the knowledge and information are searched from digital library or online search websites. As the social network based online Q&A websites [7] have become active, the user demand on the search is getting more diverse each day to expect accurate and satisfactory answers. Due to the rapid development of general search, a mass quantity of search results (regular documents or blog, etc...) on the information one tries to find can be gained. But the reality is that whether the searched result provides in-depth knowledge on the topic or whether it only provides the general or superficial knowledge is not being considered. In order to make improvement on such disadvantage, the search on a blog which deals with the topic in-depth has been proposed as a new subject [24].

Let's think about this simple question. If you wish to search the blog on "Alzheimer's Disease", the user would try to find out the cause of this disease, various answers of the people who have experienced this diseases or even the in-depth knowledge for its treatment and management. The situation is that only the general and superficial answers can be obtained on this question if you simply search the blog. The user will use the method of getting answers by uploading the question on, Yahoo! Answers [23], Twitter [18] or Naver Knowledge iN [1], etc. The answer gained from the question above is the search method using the Q&A based *collective intelligence* that has focused on the interaction between users based on user participation. The method above is the one that has used Collective Intelligence in order to find the valuable and in-depth information desired by the user among many search targeted contents.

In this study, the Q&A information in social community will be analyzed as the first step for using collective intelligence. The Q&A information gathers and analyzes the answers of other users on the questions written by the user to select the most appropriate answers among them. While using this social network, if we limit the social community domains containing the applicable Q&A information, it will be effective for finding the in-depth blog search results containing various social factors. This study will try to propose a framework for providing the in-depth blog knowledge through the query extension using the Q&A information of these social communities. The proposed framework was organized into the following three parts.

- First, when a search term or query is given by the user for the search, a keyword analysis algorithm for deciding the meaning of this sentence and the domain is applied. Here, the Preprocessing is accomplished for the analysis.
- Second, after deciding which social community domain must be searched through the keyword analysis, the Q&A information of the users in the social community domain of that field is analyzed. To extend the useful related information, the corresponding category is extended using the set of Q&A's to find the keywords according to this extended category.
- Third, the blog search is performed targeting the social community domain using the keyword followed by the extended category. The matched blog search engine performs the blog search according to the corresponding field.

The remainder of this paper is organized as follows. In Section2, explains the related studies on Q&A system and blog search. In Section 3, explains on the overall structure of the proposed framework and actually implements the proposed framework in stages to describe the process of applying data in each stage. In Section 4, performs experiment and evaluation. In Section 5, describes the conclusion.

2   Related Works

The social communities for the search of knowledge such as the social network based Yahoo! Answers [23] uses a shared service in a form of bulletin where other users answer the questions written by the user. Especially, the Q&A information inside the blog is getting attention as the knowledge search service for the last several years as a new form of window to obtain information with the advantage of being able to get answers from many people if you write the questions on even the information in a form that cannot be easily found with regular search engines such as opinions, advices and know-how, etc.

The form of knowledge search service is made up of a similar type as online bulletin or Usenet.   Zhongbao [9] has performed the social network analysis using online bulletin to show that the behavior pattern of users vary depending on their area of interest. Jeon [8] has improved the performance of search by extracting the keywords. Agichtein [5] has proposed an improved quality measurement method of documents using non-text information such as rating, no. of recommendations and no. of inquiries as well as the text information such as the length of Q&A document or number of punctuation marks. Whittaker [13] has found a statistical pattern including the number of users and the length of characters targeting Usenet.   Adamic [10] has clustered as three types of classifications according to the characteristics such as length of document, no. of answers per question and interaction pattern between users using K-Means algorithm on the categories of Yahoo! Answer.

Aardvark Company has developed a social network based search engine called 'Aardvark' [4]. It is in a form of finding the people who can answer the corresponding question within the extended social network of the user and looking for the answer to that question if the user asks a questions through instant message, e-mail, web input, text message, voice input using iPhone or Twitter)[18]. This engine not only finds the answer from the library, but also was based on the intimacy of user using the social network based village paradigm. Aardvark analyzes the question once the question comes in through the gateway so that the conversation manager looks for the association of user. It provides the information of that person along with the general search result to the inquirer by searching for the people who can answer the corresponding question using the social network extended afterwards.

In the TREC of 2009, various methods for in-depth blog feedback search was proposed. Li [25] has proposed a search method that had used the length of blog post as the qualification for decision on the in-depth blog. This method has assumed that a lengthy post shows an in-depth topic. Keikha [27]   has assumed that a blog including the words that are not shown in regular blogs as in-depth blog and reflected on the ranking by calculating the Cross Entropy between the set of blog post and entire document to make this possible. Jiang [28] has assumed that an in-depth mostly includes the words describing objective facts and proposed the method of applying the language model estimated through the query extension that has used subjective lexicon to make this possible. McCreadie [29] has proposed the classification based method using various qualifications such as length of document, length of sentences and number of personal pronouns, etc. However, these methods have shown lower search performance than the ordinary methods that had been used in the existing blog feedback search in the in-depth blog feedback search.

3   Proposed Framework

The social websites created based on social network have the services providing Q&A information. The advantage of more useful collective intelligence can be utilized through the use of this Q&A information. This paper is able to use the essential features of the social websites proposed in the existing studies [22] [30].

In this study, a Q&A based social network is used in order to extend the meaning of the keywords extracted from the search term or query. Through this, the information on domains and keywords that can be extended with the corresponding keyword is acquired to perform blog search.

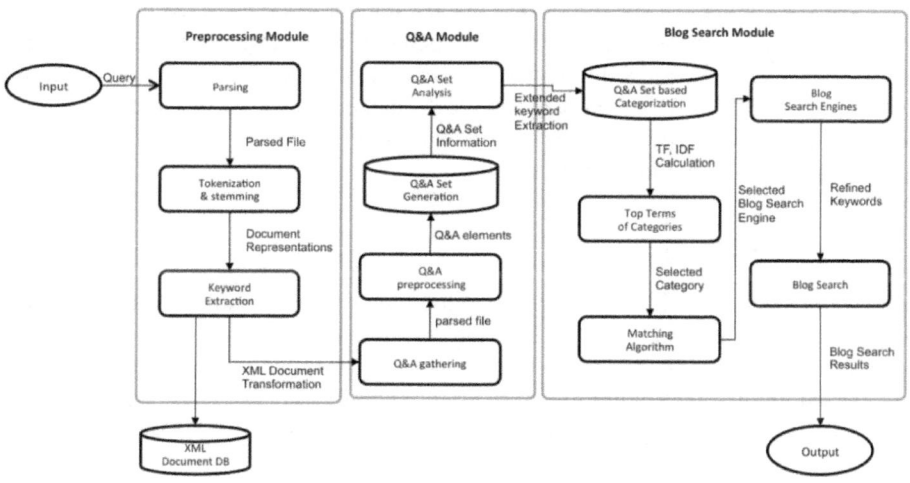

Fig. 1. The Overall Structure of Proposed Framework

Fig.1. shows the overall structure map of the blog search framework proposed in this paper. The proposed framework is composed of three modules. It is composed of the Preprocessing module for the basic task, the Q&A module for extending the Q&A contents and the Blog Search module.

3.1   Three Modules

(1) Preprocessing Module

The data preprocessing algorithm is applied on the feature extraction and Q&A module. In case the natural language gets included in the research stage, the preprocessing algorithm of data must be applied out of necessity. Especially, the information existing on the social network websites does not get written according to an official format such as user created contents or web bulletin. The preprocessing must also consider processing on the abbreviations or jargons frequently used by users as well as the stemming or elimination of stopwords. For the sentence analysis, the task of parsing and tokenization on the document of primary result searched after entering the query first. Only the core keywords are extracted through elimination of stopwords and stemming.

(2) Q&A Module

The related Q&A information is found from the Q&A community websites using the extracted core keywords. Through this, the information of useful users could be gained in relation to this other than the related search keywords. Among the features for utilizing the acquired information, the Term Frequency indicating how often a specific term appears within the document and the Document Frequency indicating in how many documents a specific term appears among all targeted documents. This is for giving weight on the extracted terms depending on their priority. For the detailed stages, the Q&A data collecting Q&A information is gathered first to perform preprocessing on this data. The next stage is Q&A analysis stage so that the TF and IDF (inverse document frequency) are calculated using the Q&A set searched with the query keyword to find out the priority.

(3) Blog Search Module

In the blog search module, the blog search engines corresponding to the Q&A keyword related to the query produced as the result of previous stage are matched. The blog search engine can be crawled real-time based on the keyword, or the existing specialized field search engine can be used according to the ODP (open directory project) [31] category. In this study, the existing specialized field search engine is matched by ODP based list to show the results.

3.2   Implementation of Proposed Framework

In this study, a framework which uses the information on Q&A to the existing search engine and applying this information on the specialized blog search engine. In this section, the proposed framework will be explained in stages by dividing the process of applying by designing and implementing into 6 stages.

(1) Query Preprocessing

While parsing is performed first to analyze the user query, the user query is divided into tokens in the parsing stage. The next stage is the keyword extraction through the elimination of stopword and stemming. It is necessary to identify the property of each term in order to eliminate the stopword. We have used the term tagging method that was used in the 'Bracketing Guidelines for Treebank II Style Penn Treebank Project [3]' performed by Carnegie Mellon University. In this project, the library on about 70,000 terms is provided. If this is used, it becomes possible to identify the properties of the tokens appearing in the user inquiry. Stem is the part which remains after eliminating the prefix and suffix from a term. Although the term inquired by the user and the term included in the related documents during the search usually correspond with each other in the actual content, it can become a cause of lowering the performance of search result due to the structural modification. This problem can be solved by replacing each term with a stem. In this study, the stem for each token was found using 'The Porter Stemming Algorithm [17]' for the stemming. The Porter Stemming Algorithm is a typical algorithm which applies 37 rules by going through 6~7 stages to find the stem of a term. In the final stage, the token for search is selected after going through the stopword elimination and stemming. We have selected a keyword that has noun as a morpheme for the search.

Fig. 2. Extraction of Keyword through Preprocessing Stage: Extracting a Keyword Called 4G and LTE Using the Query "4G LTE"

(2) Gathering Q&A Information

In this stage, Yahoo! Answers website was used to gather the Q&A information. Using the API provided by Yahoo! Answers website, the search result on each keyword can be obtained in XML format through the preprocessing of user query. Fig.3 shows the Q&A document gathered as an XML format. The Q&A document includes the information such as question id, Subject, content, chosen answer and category id.

Fig. 3. Storage of Gathered XML Document into the Database: The XML document includes the information such as question id, subject, content, chosen answer and category id, etc.

(3) Generating the Q&A Set

The Q&A document basically exists as one pair of question and answer unlike ordinary documents while including various types of information in addition. We have organized the Q&A as one set in order to express such characteristics of Q&A. A set of Q&A is composed of Question Title (Q1), Question Contents (Q2), Best Answer (A1) and Other Answers (A2). A set of Q&A is expressed as shown in the Equation (1).

$$Q \& A\_Set = \{Q_1, Q_2, A_1, A_2\}$$

Q$_1$: Question Title   (1)
Q$_2$: Question Contents
A$_1$: Best Answer
A$_2$: Answers other than A$_1$

Also, the elements forming up the set of Q&A are expressed as shown in the Equation (2).

$$Q_1 = \left\{ t_{q1_1}, t_{q1_2}, t_{q1_3}, \dots, t_{q1_{n'}} \right\}$$
$$Q_2 = \left\{ t_{q2_1}, t_{q2_2}, t_{q2_3}, \dots, t_{q2_{n''}} \right\}$$
$$A_1 = \left\{ t_{a1_1}, t_{a1_2}, t_{a1_3}, \dots, t_{a1_{n'''}} \right\}$$
$$A_2 = \left\{ t_{a2_1}, t_{a2_2}, t_{a2_3}, \dots, t_{a2_{n''''}} \right\}$$

$$(2)$$

Fig. 4. shows the composition of Q&A provided by Yahoo! Answers. This website also basically provides 'question title', 'question contents', 'answer contents' and 'best answer' while providing supplementary information such as 'asker', 'user's rating' or 'date' in addition.

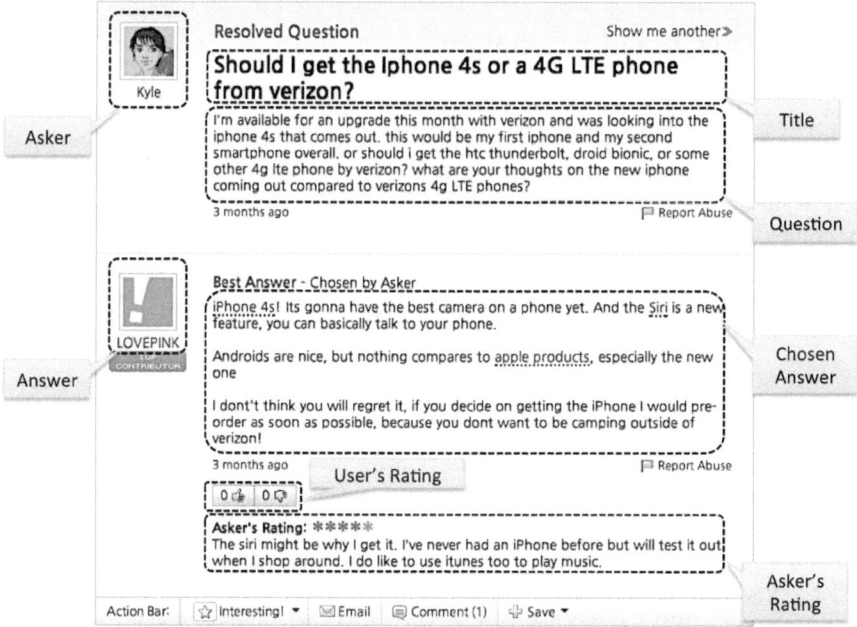

Fig. 4. The Q&A Interface Provided by Yahoo! Answer

(4) Q&A Preprocessing

The preprocessing is required in order to analyze the meaning of the generated Q&A set. First, the preprocessing of Q&A selects the core keywords by performing the same process as query preprocessing (parsing, stopword elimination and stemming).

(5) Q&A Analysis

Q&A analysis is the stage of calculating TF and IDF using the Q&A set searched with the query keyword. First, the Q&A sets having same category are classified to calculate TF and IDF to show how important a certain term is in a specific Q&A set.

(6) Blog Search Module

In the blog search module, the corresponding specialized blog search engine is matched using the core Q&A keywords related to the query to search again using the matched search engine. For the match of blog search engine, the blog search engine of the each applicable field gets selected by analyzing the core Q&A keywords. The search result by each corresponding category is shown using the selected blog search engine. After extending the keyword inquired by the user using the Q&A set, the search result gets shown by matching with the blog search engine.

4   Experiment and Evaluation

For the experiment, the core keywords extracted through the proposed system on a specific query, the keywords with high TF-IDF value extracted from the search result page provided by general search websites and the keywords high TF-IDF value extracted from the search result page provided by specialized blog search websites have been compared. It has experimented on how much the core keywords extracted through the proposed system are matching with the core keywords extracted from the general search websites and the specialized blog search websites. We will be able to determine that the proposed framework is useful as the consistency gets higher.

4.1   Experiment Data

The data used for extracting the core keywords in the proposed system have been gathered on the questions that have already answered from Yahoo! Answers. The 1,074 Q&A sets have been gathered by searching for '4G LTE' query in Yahoo! Answers and 12 categories have been found from this data. We have set the subject as the 312 Q&A sets having best answers in 3 out of 12 categories ('Cell Phones & Plans', 'Internet', 'PDAs & Handhels'). We have also gathered the documents on each of 10 upper pages among the search result provided by Google and Yahoo! on the corresponding query for the general search. And for the specialized blog search websites, the blog search engine provided by Google [32] and the blog search engine provided by Yahoo [33] have been used to gather the documents on each of 10 upper pages among the search result on the corresponding query. The data gathered for experiment are shown in Table 1. We have extracted the terms that can be used for the experiment by performing the preprocessing stage described in section 4 among the terms found for the experiment. The ratio of terms used for the experiment among the terms that have been found is 9.2%.

Table 1. The Data Gathered for Experiment

Category	Gathered Website	No. of Gathered Doc.	No. of Terms Found	No. of Terms Used for experiment	Rate of Usage (%)
Proposed System	Yahoo! Answer	312	8,643	932	10.78
General Search	Google	10	10,243	853	8.33
	Yahoo	10	8,335	911	10.93
Specialized Blog Search	Google Blog	10	4,236	353	8.33
	Yahoo Blog	10	3,930	301	7.66

4.2   Experiment Results

We have extracted the core keywords from the proposed system, general search and specialized blog search using the data mentioned in section 5.1. For the core keyword extraction, TF-IDF has been used. The consistency of the core keywords appeared in general search and specialized blog search have been calculated depending on the number of keywords extracted from the proposed system. Table 2. shows the number of core keywords appeared in general search and specialized blog search depending on the number of keywords extracted from the proposed system. Table 2. shows the consistency of general search and specialized blog search followed by the increase of core keywords. The specialized blog search showed a consistency of about 29%. But the general search showed a consistency of about 13%. We can see the consistency being increased as the number of core keywords extracted from the general search and specialized field search are increased. This is because the probability of including the core keyword extracted through the proposed system gets higher as the number of core keywords gets increased. If the number of core keywords exceeds a specific value, the consistency is expected to be shown as 100%. As shown in Table 2, we could verify that the core keyword extracted through the proposed system appeared more in the result of specialized field search than the result of general search. The core keyword extracted from the proposed system could be considered as the keyword extracted from the specialized field. This means that the user is able to gain the search result of specialized field through the proposed system. It seems that a method to raise the consistency with the core keyword extracted from the proposed system from the same amount of core keywords would be necessary in the future studies.

Table 2. The Consistency of Core Keywords In General Search and Specialized Field Search followed by the Increase of Core Keywords

No. of Core Keywords / Category		10	20	30	40	50
General Search	Google	1	3	7	12	18
	Yahoo	0	2	3	7	13
	Average	0.5	2.5	5	9.5	15.5
Specialized Blog Search	Google Blog	2	6	13	18	21
	Yahoo Blog	2	5	9	15	17
	Average	2	5.5	11	16.5	19

5   Conclusion

In this study, a framework that can gain the information of collective intelligence and the user information within the blogs based on social community has been proposed and implemented. This study has designed and developed focused on the framework to be applied in stages with the real data. The characteristics of blog contents is that there is a lot of information made up of user experience and the information is trusted more by most users than the contents acquired by general search. Therefore, the proposed framework will become very useful when the blog page made up of active social community unlike the conventional web page is set as the search target. In the future researches, the experiment for performance analysis considering specialty and practicality will be performed a little further. The search result of the time when the search engine that has applied the proposed technology was used would improve the reliability and satisfaction of users as the information containing the experience and knowledge of users can be searched.

Acknowledgments. This work was supported by Gachon University Research Fund in 2012, and by Basic Science Research Program through the National Research Foundation of Korea (NRF) funded by the Ministry of Education, Science and Technology (2011-0014747).

References

1. NaverKnowledge iN, http://kin.naver.com/
2. Lee, J.-W., Park, S., Lee, S., Park, J., Lee, S.-G.: Semantic Search and Recommendation of e-Catalog Documents through Concept Network. KIPS 15(3), 131–145 (2010)
3. Bracketing Guide lines for Treebank II Style Penn Tree bank Project, http://bulba.sdsu.edu/jeanette/thesis/PennTags.html/
4. Horowitz, D., Kamvar, S.D.: The Anatomy of a Large-Scale Social Search Engine. In: Proc. of the WWW 2010, pp. 4331–4440 (2010)
5. Agichtein, E., Castillo, C., Donato, D., Gionis, A., Mishne, G.: High-quality content in social media. In: Proc. of the International Conference on Web Search and Web Data Mining, pp. 183–194 (2008)

6. eurekster, http://www.eurekster.com/
7. Maxwell Harper, F., et al.: Predictors of Answer Quality in Online Q&A Sites. In: Proc. of the Twenty-Sixth Annual SIGCHI Conference on Human Factors in Computing Systems, pp. 865–874 (2008)
8. Jeon, J., Corft, W.B., Lee, J.H.: Finding similar questions in large question and answer archives. In: Proc. of the 14th ACM International Conference on Information and Knowledge Management, pp. 84–90 (2005)
9. Zhongbao, K., Changshui, Z.: Reply networks on a bulletin board system. Physical Review E 67(3) (2003)
10. Adamic, L.A., Zhang, J., Bakshy, E., Ackerman, M.S.: Knowledge sharing and yahoo answers: everyone knows something. In: Proc. of the 17th International Conference on World Wide Web, pp. 665–674 (2008)
11. Alessandro, M., et al.: Exploiting Syntactic and Shallow Semantic Kernels for Question/Answer Classification. In: Proc. of ACL 2007, pp. 776–783 (2007)
12. ODP, http://www.dmoz.org
13. Whittaker, S., Terveen, L., Hill, W., Cherny, L.: The dynamics of mass interaction. In: Proc. of the 1998 ACM Conference on Computer Supported Cooperative Work, pp. 257–264 (1998)
14. Search engine guide, http://www.searchengineguide.com/
15. Search-engine-index, http://www.searchenginesindex.com/
16. SearchMedica, http://www.searchmedica.com/
17. The Porter Stemming Algorithm, http://tartarus.org/~martin/PorterStemmer/
18. Twitter, http://twitter.com/
19. Verticalsearch, http://www.verticalsearch.com/
20. Virtual Sites, http://www.virtualfreesites.com/
21. WebMD, http://www.webmd.com/
22. Kim, W., Jeong, O.-R., Lee, S.-W.: On social Web sites. Information Systems 35(2), 215–236 (2010)
23. Yahoo! Answers, http://answers.yahoo.com/
24. Macdonald, C., Ounis, I., Soboroff, I.: Overview of TREC-2009 Blog track. In: Proc. of TREC 2009 (2010)
25. Li, S., Gao, H., Sun, H., Chen, F., Feng, O., Gao, S.: A study of faceted blog distillation - PRIS at the TREC 2009 Blog Track. In: Proceedings of the 18th Text Retrieval Conference (TREC 2009), Gaithersburg, US (2009)
26. Nardi, B.A., Schiano, D.J., Gumbrecht, M., Swartz, L.: Why we blog. Communications of the ACM 47(12), 41–46 (2004)
27. Keikha, M., Carman, M., Gwadera, R., Gerani, S., Markov, I., Inches, G., Alidin, A.A., Crestani, F.: University of Lugano at TREC 2009 Blog Track. In: Proc. of TREC 2009 (2010)
28. Jiang, P., Yang, Q., Zhang, C., Niu, Z.: BIT at TREC 2009 Faceted Blog Distillation Task. In: Proc. of TREC 2009 (2010)
29. McCreadie, R., Macdonald, C., Ounis, I., Peng, J., Santos, R.L.T.: University of Glasgow at TREC 2009: Experiments with Terrier. In: Proc. of TREC 2009 (2010)
30. Oh, J., Jeong, O.-R., Lee, E., Kim, W.: A framework for collective intelligence from internet Q&A documents. IJWGS 7(2), 134–146 (2011)
31. ODP, http://www.dmoz.org
32. https://www.google.com/blogsearch?hl=en&tab=wb
33. http://blog.search.yahoo.com/

LSA as Ground Truth for Recommending "Flickr-Aware" Representative Tags[*]

Xian Chen[1], Hyoseop Shin[1], and Minsoo Lee[2]

[1] Web Intelligence Laboratory, Konkuk University
{chenxian,hsshin}@konkuk.ac.kr
[2] Dept. Computer Science and Engineering,
Ewha Womans University
mlee@ewha.ac.kr

Abstract. Most item recommendation systems nowadays are implemented by applying machine learning algorithms with user surveys as ground truth. In order to get satisfactory results from machine learning, massive amounts of user surveys are required. But in reality obtaining a large number of user surveys is not easy. Additionally, in many cases the opinions are subjective and personal. Hence user surveys cannot tell all the aspects of the truth. However, in this paper, we try to generate ground truth automatically instead of doing user surveys. To prove that our approach is useful, we build our experiment using Flickr to recommend tags that can represent the users' interested topics. First, when we build training and testing models by user surveys, we note that the extracted tags are inclined to be too ordinary to be recommended as "Flickr-aware" terms that are more photographic or Flickr-friendly. To capture real representative tags for users, we apply LSA in a novel way to build ground truth for our training model. In order to verify our scheme, we define Flickr-aware terms to measure the extracted representative tags. Our experiments show that our proposed scheme with the automatically generated ground truth and measurements visibly improve the recommendation results.

Keywords: LSA, User Survey, Tag Recommendation, Flickr.

1 Introduction

Nowadays, most item recommendation systems are implemented by applying machine learning algorithms to automatically generate personal recommendable items for users. However, when applying machine learning algorithms, it requires training data with ground truth. In order to get believable ground truth, it requires a mass of user surveys. Most researches perform user surveys to use as their ground truth, such as [1][2]. However, to get enough user surveys is not an easy task. Additionally, sometimes human's opinions are subjective and personal. Hence user surveys cannot

[*] This work (No. 2011-0029729) was supported by Mid-career Researcher Program through NRF grant funded by the MEST.

tell all aspects of the truth. In the image detection or image recognition field, ground truth may be generated by algorithms, instead of human effort, such as SIFT, SURF, semi-automatic ground truth, clustering, etc. In the field of text-based technology, however, there are few researches that consider the use of algorithms for generating ground truth. In this paper, we try to generate ground truth automatically instead of using user surveys. We built our experiment with Flickr to recommend tags that can represent users' interested topics. In the previous work of [3], they extracted textual and social features of tags for each Flickr user and applied machine learning algorithms with user surveys to suggest representative tags for Flickr users. From their user survey-based approach, it seems that the evaluators choose common and ordinary tags as users' representative tags (e.g. "cloud", "sun", "nature"). However, we noticed that in Flickr, power users[4] frequently use more photographic or Flickr-friendly terms (terms that are frequently used in Flickr) (e.g. "Bokeh", "HDR", "AnAwesomeShot") than normal terms. Those more photographic and Flickr-friendly terms are too difficult for evaluators to choose. But in a specialized photo community like Flickr, photographic terms can be more important than normal terms, because photographic terms can express the photo in a more professional way. To solve the limitation of user surveys, we provide a novel scheme; we apply LSA to each user with each tag, and choose the top K tags for each user as the ground truth for the training model. Furthermore, in order to verify our scheme and compare the differences between the user survey results and results from our proposed scheme, we define a new measurement, Flickr-aware terms, which are terms widely used among power users, but less used by non-power users in Flickr.

The rest of this paper is organized as follows. In Section 2 we introduce the related work. In Section 3, we briefly describe our research background. In section 4, we provide our ground truth generation. In Section 5, we describe our data set and experiments. In Section 6, we discuss the experimental results. Finally, we summarize the conclusions of this paper in Section 7.

2   Related Work

Recently, there have been many researches about item recommendation systems for online community users. For instance, one can recommend books, movies, music, photos, tags to users in amazon, movieLens, IMDB, Flickr, delicious. These kinds of recommendation systems are based on analysis of relationships between users and items in online communities. Many researches study this problem by applying machine learning algorithms and obtaining training data through user surveys. Even though they use heuristic algorithms, user surveys are frequently used as ground truth to evaluate the results. [5] provided an approach to leverage online collaborative tagging in product design through user study.[2] proposed a scheme to learn semantic relationships between entities in Twitter through user survey. [1] proposed a method to recognize the quality tags for users in movieLens by SVM with ground truth of user survey. [6] proposed a method for content-based tag recommendation that can be applied to Web pages through user study. [7] provided a visualization scheme for tag recommendation via user study.

These researches on item recommendation or analysis of relationships between users and items are based on user survey as ground truth. However, to obtain enough information through user surveys is not an easy task. It would take a lot of time and human effort. However, even if you have enough user surveys, the result cannot tell all the aspects of the truth. Therefore, we provide a new scheme to generate ground truth for the training model to extract representative tags for Flickr users instead of using user surveys. We use LSA in a novel way to generate ground truth: (i) build matrix between users and tags; (ii) apply LSA to matrix; (iii) train and test extracted features of tags for users with generated ground truth; (iv) define a new measurement as Flickr-aware terms to verify the extracted tags.

3   Research Background

To validate that our assumption is feasible and valuable, we build our experimental model using Flickr. We would like to extract tags for Flickr users that can be representative of the users' personally interested and favorite topics, by applying LSA to generate the ground truth. To compare the results with the previous work of [3] which applied user surveys as ground truth, we set the same environment and used the same features which are described in detail in [3]. In this section, we will briefly describe the whole process of our extraction scheme.

3.1   Tags Features

To extract representative tags, we use the following features of tags for Flickr users.

Textual Features:

- *tf*: the number of one tag used by one user;
- *tf_ratio*: is derived from dividing *tf* by the sum of frequencies of all tags for one user;
- *iuf*: the logarithm of the quotient obtained from dividing the total number of users by the number of users who used one tag;
- *in_title*: the times of one tag appeared in the titles of one user's posts;
- *in_content*: the times of one tag appeared in the contents of one user's posts;
- *in_comment*: the times of one tag appeared in the comments of one user's posts.

Social features:

- *fav_coocc_freq*: in Flickr, user can maintain the list of favorite photos that are uploaded by other users. This feature is described as the times of one tag appeared in one user's favorite photos;
- *fav_coocc_ratio*: an alternative representation of *fav_coocc_freq* for considering the ratio against all the favorite photos; among total favorite photos, the ratio of which included the tag;

- *fav_giver_freq*: in Flickr, we defined the users who mark one user's photo as their favorite as this user's favorite givers. This feature is the number of one user's favorite givers who also use one tag;
- *fav_giver_ratio*: it is obtained from dividing *fav_giver_freq* by the sum of one user's favorite givers.

3.2   Algorithm

With the tags' features, we applied the Naïve Bayes for the training and testing model. We performed binary classification, which only marked whether a tag is representative or not. The formula is as follows:

$$P(Y|F_1, ... F_n) = P(Y) \prod_{i=1}^{n} p(F_i|Y) \qquad (1)$$

Where P(Y) is the probability of positive instances in training set. F_i is one of the feature of tags.

4   Ground Truth

For preparing the ground truth of the training and testing model, we apply LSA to generate the ground truth. To validate our scheme, we also use the user survey for comparison.

4.1   User Survey

We make an interface that can show Flickr users, their photos and tags. We ask several evaluators to watch the photos and read tags for each Flickr user. By viewing photos and tags for each Flickr user, evaluators can conclude the interested topics for the corresponding Flickr user, and then they can select the representative tags which are related to the users' interested topics.

4.2   LSA

LSA Latent Semantic Analysis (LSA) is a fully automatic statistical technique for extracting and inferring relations of expected contextual usage of words in passages of discourse[8].

The reason why we apply LSA as ground truth is as follows; user surveys cannot tell all the aspects of the truth among limited evaluators. It is difficult to collect massive amounts of user information. Therefore, several evaluators' opinions cannot represent most of the human's. In the user survey, we noticed that evaluators mostly chose ordinary tags for each user, such as "landscape", "sun" and "cloud". However, in Flickr, users used more photographic terms (e.g. "Bokeh", "HDR" and "exposal") as tags, but the evaluators miss this part of tags, because they may not know enough about photographic or Flickr-friendly terms. However these terms also describe users' professional interests. If users use "HDR" several times, we can infer that they may be interested in the "HDR" photos. Therefore, photographic tags are also important which can show users' professional interests.

In Flickr, there are a huge number of users, and they use all kinds of tags, therefore, for one user, his/her own tags are relatively small and sparse, when compared to the entire tags. To reduce the sparsity and find the latent relationship between users and tags, LSA is one choice to solve these problems.

In our environment, we build a matrix between users and tags, where each row represents one tag and each column represents one user, so each cell contains the frequency of one tag used by one user. Before SVD is computed, in the original LSA, each cell should be divided by the row entropy value. However, in Flickr, tags are arbitrarily used by users, if we apply entropy in our matrix, which means that the higher entropy of one tag is, the fewer users use it. In this case, rare tags which may be only used by one user and has non-sense to other users would have high values, but normal tags which are used by most of users and have useful meanings would have low entropy values. Hence, we only use the frequency of each tag for each user and apply SVD. The SVD formula is as follows:

$$\{X\} = \{T\}\{E\}\{U\}' \tag{2}$$

$\{X\}$ means original matrix, and $\{T\}$, $\{E\}$, $\{U\}$ represent the decomposed three matrices. Matrix $\{T\}$ describes tag entities as vectors of derived orthogonal factor values; matrix $\{U\}$ describes the user entities as vectors of derived orthogonal factor values; matrix $\{E\}$ is composed of eigen values. The reduction for dimensions depends on the number of eigen values which are chosen for reconstructing a new matrix.

5   Experiment and Measurement

5.1   Data Set

We implemented our experiments by using the data set of Flickr. We collected 813, 353 users who registered in Flickr and downloaded their photos with tags for the whole month of Dec, 2009. Among these Flickr users, there are 262,722 users who used tags to denote their or others' photos. Hence, we built our data set with these 262,722 users, their photos and tags. Table 1 shows the basic information about our data set. We have 10,591,157 photos, 1,600,349 tags and 4,828,926 favorite feedbacks for these 262,722 users. The maximum number of tags assigned by a user is 3,702 and the average number of tags assigned by a user is 52. The whole data set is used for calculating the features of the tags.

Table 1. Experiment data set from Flickr in Dec. 2009

Experiment Data	Number
Total number of users	262,722
Total number of photos	10,591,157
Total number of tags	1,600,349
Total number of favorite feedbacks	4,828,926
Maximum number of tags per user	3,702
Average number of tags per user	52

Table 2. User survey results

User survey	Number
Total number of candidate users	177
Total number of photos for candidate users	1,770
Total number of tags of candidate users	13,464
Total number of evaluators	7
Total number of representative tags chosen	1,355
Average number of representative tags per user	7.7

5.2   Training and Testing Model

We apply Naïve Bayes Classifier to our training and testing data with two different ground truths. One is from the user survey result, the other one is from calculating the LSA values. Considering the user survey, it is impossible for evaluators to choose tags for the whole 262,722 users. And From Table 1, we can see that users do not give tags to all of the photos. That is why we want to choose users who upload photos and add tags consistently. At the same time, we need users who are in the social network to generate their social features, so we filter the users under conditions: the number of post was more than 10; for each post, the number of tags was more than 10 but less than 15; the favorite number of each post was more than 10. However, for our features calculation, we use the whole data set.

• User survey results as ground truth

For the user survey, we asked seven evaluators to choose representative tags for each Flickr user. Although, we do not have many evaluators, these seven evaluators' results are fine to use. We use Kappa statistic [9] to measure agreement between every two evaluators. And then we used the average Kappa coefficient value 60.3% which means the agreement of our evaluators is fine to do the experiment. The user survey results are shown in Table 2.

• LSA calculation as ground truth

We build the matrix of our data set where each row represents one tag and each column represents one user, then each cell contains the frequency of one tag which is used by one user. We apply LSA to the same data set as the 177 users' data set which is chosen for the user survey. Finally, according to the LSA calculation, we choose the top K tags for each user.

5.3   Define Flickr-Aware Term as Measurement

We define Flickr-aware terms as the tags which are only widely used by power users, but not often used by non-power users in Flickr. Power users (or high-ranked users) are named for the users who get the high scores based on the user ranking scheme[10].

We define the formula to extract Flickr-aware terms as follows:

$$power\_value(t) = \frac{N_{p,t}}{N_p} \tag{3}$$

$$non\_power\_value(t) = \frac{N_{np,t}}{|U| - N_p} \tag{4}$$

Equation (3), the power_value(t) means among top N power users, the number of users who use tag t. N_p denotes the top N power users and the numerator is the number of power users who used tag t. Equation (4), the non_power_value(t) means except top N power users, the number of non-power users who use tag t. |U| represents all the users in data set.

To extract the Flickr-aware terms, we use α and β as the thresholds for power_value(t) and non_power_value(t). The tag should satisfy the following expression (5). Then this tag can be chosen for the Flickr-aware terms.

$$power\_value(t) > \alpha \ \&\& \ non\_power\_value(t) < \beta \tag{5}$$

5.4   Experiment Results

For user the survey result, we tried the Naïve Bayes and got the best result with the feature combination (tf_ratio, iuf, in_title and fav_giver_freq) which obtained a result of precision 49%, recall 54.5% and F-score 51.6%.

For using LSA values as ground truth, we choose 10 eigen values for the reduction of the dimensions. Figure 1 shows the results when the number of eigen values equals to 10 and top K LSA values. Hidden tags are defined as the tags which are in the top K LSA values for one user, but not used by this user. The representative tags mean the tags which are not only in the top K LSA values for one user, but also used by this user. In our environment, when we apply LSA for users and tags, if one user uses several tags about "landscape", and also uses several tags about "girls", but another user is only interested in "landscape", then, from LSA calculation, the LSA values of tags about "girls" would be also higher for the latter user, because the tags about "girls" are used by the previous user who also used tags about "landscape". In this case, it is not reasonable to choose many hidden tags for users. On the contrary, it is better to choose only the tags which are used by users themselves. From Figure 1, we can also infer that with less reduction (that means the number of eigen value increases), the hidden tags are increasing, which means dimension reduction can also reduce noisy tags. But the chosen tags trend is stable. That is why using chosen tags as representative tags for users are more reasonable.

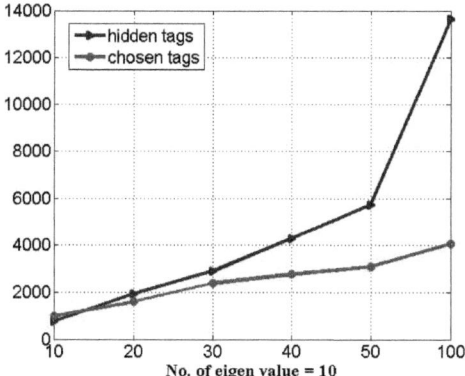

Fig. 1. The hidden tags and representative tags from top K

Finally, we chose the top 30 tags and removed hidden tags for each user based on the LSA values as ground truth. By applying the Naïve Bayes, we obtained the accuracies as shown in Table 3. For combining textual features and social features of LSA, the precision was 74.1%, recall 83.4% and f-score 78.5%, which outperformed the user survey by almost 30%.

Table 3. The accuracies for LSA as Ground Truth

Features	Precision	Recall	F-score
TF+TF	79.8%	73.5%	76.5%
SF+SF	69.6%	78.6%	73.8%
SF+TF	74.1%	83.4%	78.5%

6   Experimental Results Evaluation

6.1   Comparison for Extracted Tags

In this section, first we show some different tags which are extracted from two different ground truths in Table 4. It is easy to see that the user survey extract normal tags which describe the photos in a normal way, while LSA not only extract normal tags, but also photographic or Flickr-friendly terms (e.g. "bokeh" is the blur in out-of-focus areas of an image, "HDR" is a greater dynamic range between the lightest and darkest areas of an image). However, in photo-specific online communities, the photographic terms are more significant than normal terms.

Table 4. Tags extracted by different schemes

User survey	Sun, cloud, nature, landscape, flower, boats, fishing, car, leaf, winter
LSA	HDR, Bokeh, exposure, HBW, macro, long exposure, an awesome shot, canon, explore

Table 5. Top 9 Flickr-aware tags based on $\alpha = 0.1$ and $\beta = 0.02$

Top 9 tags when $\alpha = 0.1$ && $\beta = 0.02$		
TheUnforgettablePictures	AnAwesomeShot	explore
texture	supershot	APlusPhoto
hbw	Bokeh	long exposure

6.2 Evaluation Based on Flickr-Aware Terms

From the user survey result, we notice that evaluators mostly choose the normal tags as users' representative tags, but the photographic or Flickr-friendly terms are missing.

To evaluate the result of our proposed scheme which applied LSA to generate ground truth, we define Flickr-aware terms. We use power_value and non_power_value to control the Flickr-aware terms scope. Table 5 shows the top 9 tags when $\alpha = 0.1$ and $\beta = 0.02$. There are fewer normal terms in the list. Figure 2 shows the numbers of representative tags generated from the two ground truth, based on different power values. We set the parameter $\beta = 0.02$ and the top 1000 power users. With low α, the normal tags are easily in the scope of Flickr-aware terms, such as when $\alpha = 0.02$ or 0.04, there are lots of tags which are generated from both LSA and user survey can be Flickr-aware terms. While increasing α, the scope of Flickr-aware terms are getting smaller. There are fewer tags which were extracted from user surveys for Flickr-aware terms. When $\alpha = 0.1$, there are no Flickr-aware terms extracted from user surveys. But for LSA, even though α is as high as 0.1, it still can generate Flickr-aware terms, which proves using LSA as ground truth can generate much better results compared with user surveys.

6.3 F-Score Comparison between Two Ground Truth

In this section, we compare the F-scores between two ground truths for each feature. We applied Naïve Bayes to the training and testing models with two ground truths. Figure 3 shows the F-score for both of the two schemes. When we apply LSA to generate the ground truth, we use the tag frequency, to avoid over-fitting, we do not use *tf* as a feature in training and testing. However, when only using *tf_ratio*, the performances of LSA and user survey are similar, both are quite low. By using *iuf* for LSA, a higher F-score is obtained, but the *iuf* used for user surveys has no effect for extracting tags, which not only tells us LSA can extract latent tags for users, but also infers that evaluators are not familiar with Flickr-aware terms, but most Flickr-aware terms can be representative for users' interested topics. For social features, they work better both in user survey and LSA than textual features, which tell us features of social relationship would be better to filter representative tags. However, when combined textual and social features, both of the methods obtain the best F-score. It also shows that LSA outperforms the user survey.

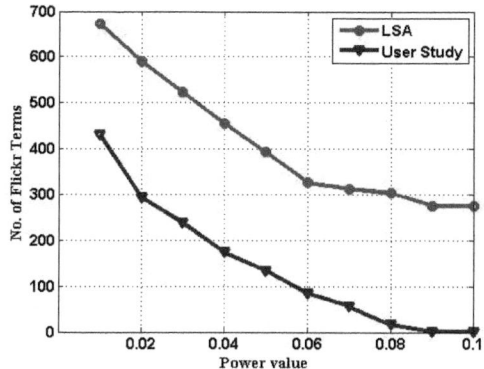

Fig. 2. Representative tags as Flickr-aware terms for different power values

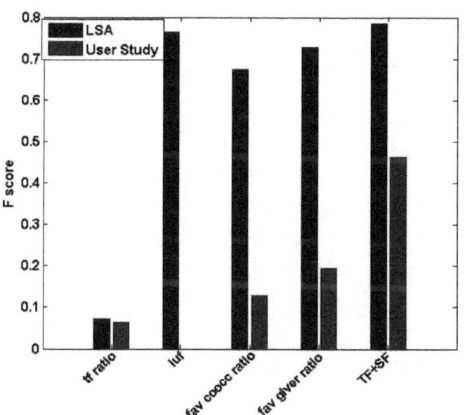

Fig. 3. F-score for both user survey and LSA results

6.4   Comparison Based on Social Features

Finally, Figure 4 and Figure 5 show the distributions of positive instances for different social features' values. The tags which have higher values of social features are frequently used by Flickr users. And Flickr users who have higher values of social features are more socially active. We strongly consider that most of these users are power users. The higher the social features values are, the more representative tags can be extracted from LSA, which indicates power users have more activities than non-power users and they use Flickr-aware terms for their social activities more frequently. That is why LSA can detect more representative tags. This also infers that our method can generate Flickr-aware terms well and Flickr-aware terms evaluate our experiments appropriately.

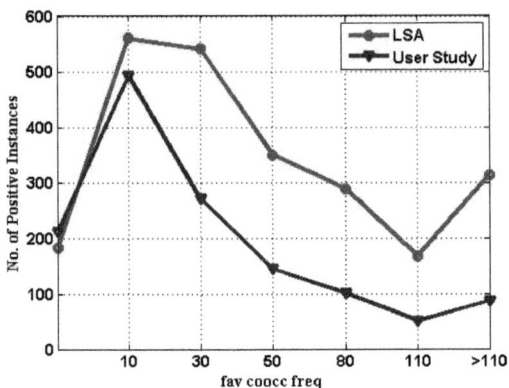

Fig. 4. Positive instances for different *fav_coocc_freq*

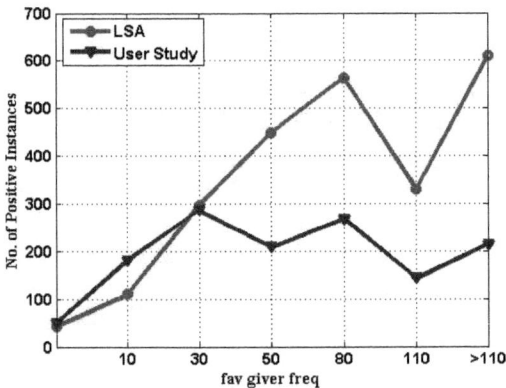

Fig. 5. Positive instances for different *fav_giver_freq*

7   Conclusion

In this paper, we suggested to use an algorithm to generate the ground truth for the training model instead of using user surveys. We built our experiment with Flickr to extract tags which can represent users' interested topics for Flickr users. In the process, we proposed a novel way to use LSA for generating ground truth. To prove our method is feasible and valuable, we defined Flickr-aware terms to evaluate our scheme. Comparing with the result of user surveys, our proposed method extracts more photographic tags which professionally describe photos or Flickr-friendly tags. On the other hand, user surveys extracted normal tags. The accuracy of LSA outperformed the user survey by 30%. Finally, we proved that our method to generate Flickr-aware terms is possible and Flickr-aware terms evaluate our experiment very well.

Each of the schemes has its own drawbacks: normal evaluators cannot recognize the photographic or Flickr-friendly terms in Flickr; LSA may extract too many latent tags which may not be related to users' interested topics. Therefore, a combined scheme of the user survey and LSA would be necessary to produce a better result for users' representative tags.

In the future work, we would like to apply PLSA for generating ground truth. However, compared with LSA, PLSA adds a probability model which can control the dimension reduction, and we expect that it would get similar results with the LSA.

References

1. Sen, S., Vig, J., Riedl, J.: Learning to Recognize Valuable Tags. In: Proceedings of International Conference on Intelligent User Interfaces. ACM Press, Florida (2009)
2. Celik, I., Abel, F., Houben, G.: Learning Semantic Relationships Between Entities in Twitter. In: In Proceedings of the 11th International Conference on Web Engineering, Cyprus, pp. 167–181 (2011)
3. Chen, X., Shin, H.: Extracting Representative Tags for Flickr Users. In: In Proceedings of the 10th IEEE International Conference on Data Mining Workshops, pp. 312–317. IEEE Press, Sydney (2010)
4. Shin, H., Xu, Z., Kim, E.: Discovering and Browsing of Power Users by Social Relationship Analysis in Large-scale Online Communities. In: Proceedings of 8th IEEE/WIC/ACM International Conference on Intelligent Agent Technology, pp. 105–111. IEEE Press, Sydney (2008)
5. Das, M., Das, G., Hristidis, V.: Leveraging Collaborative Tagging for Web Item Design. In: Proceedings ofthe 17th ACM SIGKDD International Conference on Knowledge Discovery and Data Mining, pp. 538–546. ACM Press, San Diego (2011)
6. Lu, Y., Yu, S., Chang, T., Hsu, Y.: A Content-based Method to Enhance Tag Recommendation. In: Proceedings of the 21st International Joint Conference on Artificial Intelligence, Pasadena, pp. 2064–2069 (2009)
7. Song, Y., Qiu, B., Farooq, U.: Hierarchical Tag Visualization and Application for TagRecommendations. In: Proceedings ofthe 20th ACM International Conference on Information and Knowledge Management, pp. 1331–1340. ACM Press, Glasgow (2011)
8. Landauer, T., Foltz, P., Laham, D.: Introduction to Latent Semantic Analysis. Discourse 25, 259–284 (1998)
9. Phan, N., Hoang, V., Shin, H.: Adaptive Combination of Tag and Link-based User Similarity in Flickr. In: Proceedings ofthe 10th International Conference on Multimedia, pp. 675–678. ACM Press, Firenze (2010)
10. Shin, H., Lee, J., Hwang, K.: Separating the Reputation and the Sociability of Online Community Users. In: Proceedings of 25th Symposium on Applied Computing, pp. 1807–1814. ACM Press, Lausanne (2010)
11. Singular value decomposition website,
   http://en.wikipedia.org/wiki/Singular_value_decomposition

Adaptive Access Control Enforcement in Social Network Using Aspect Weaving

Frédéric Cuppens, Nora Cuppens-Boulahia, and Eduardo Pena Viña

Télécom Bretagne
{frederic.cuppens,nora.cuppens,eduardo.penavina}@telecom-bretagne.eu
http://www.telecom-bretagne.eu

Abstract. Current social network systems support a large range of applications with very different security requirements. Even if available social network solutions provide some security functionalities, users do not control these functionalities and cannot customize them to handle their specific security needs. In this paper, we suggest a new approach to handle these issues. This approach is based on Aspect Oriented Programming (AOP) which enables the enforcement of an independent, reusable access control policy through the modification of the program at runtime. This makes possible to externalize the security concerns and weave them into an existing social network. Using this approach, it is possible to customize security of social network at different levels. First, one can specify the global security policy of the particular social network application and then, each member of this social network can further refine this global policy to specify their specific security requirements. This approach is illustrated on the open source social network system Elgg.

Keywords: Security, Social Network, OrBAC, Access Control, AOP.

1 Introduction

Security is a major concern in social network applications. Even if recent social network implementations provide further security functionalities, they inherently suffer from the following limitations:

- Users do not control which functionalities are provided by these social networks and how they are implemented
- Generally, security is viewed as an a posteriori concern in social network implementations. The developers often manually scan their code and insert security controls as an afterthought. This approach may lead to major security flaws simply because developers miss to insert appropriate security controls.
- When access control is manually hard coded into the code, it becomes very difficult to maintain the resulting code. Changing the functionalities or the security controls can create additional flaws.

H. Yu et al. (Eds.): DASFAA Workshops 2012, LNCS 7240, pp. 154–167, 2012.
© Springer-Verlag Berlin Heidelberg 2012

The solution suggested in this paper is to control access at running time using aspect oriented programming. Thus security requirements are represented as aspects that are dynamically woven into the code implementing the social network functionalities.

This approach provides the following advantages:

- Users can control and adapt their own access control policy (including the security model used to express this policy) and weave the corresponding security aspects they wish into the code.
- Risks of flaws into the code due to bad enforcement of security requirements into some parts of the code are mitigated.

The objectives of this paper are the definition and deployment of a complete methodology for users to easily and efficiently secure their social network application. This methodology is illustrated through a use case based on the OrBAC model [1] for security requirements and Elgg as a social network application [2]. Elgg is used because it is an open source social network application used in many companies and universities[1]. OrBAC is used because it provides a complete model to specify contextual and dynamic access control requirements. The advantages of using OrBAC to specify security policies in the context of social network applications are further explained in section 3. However other simpler access control models like RBAC [3] could be used as well. Since Elgg is coded in PHP, an AOP framework for PHP will be used[2] [4].

This paper is structured as follows. Section 2 illustrates through a scenario the necessity to enforce adaptive access control policies in social network applications. Section 3 presents a distributed approach for security policy specification in a social network where every member of a social network can easily customize the global security policy of this social network with their own security requirements. Section 4 introduces Aspect Oriented Programming, the concept of *intercept* the execution flow of a program and its consequences. Section 5 defines a complete methodology to secure the program, from the requirements identification to the final result. Section 6 is the use case of a social network in which a remote access control policy is enforced. This use case is publicly reachable at **http://seres.mine.nu**. Finally, section 7 concludes the paper.

2   Motivating Example

Initially, Mike manages a small social network of closely related friends. The social network manages confidential information but its members decided to share this sensitive information. So the social network initially manages a very simple confidentiality policy:

- *Everyone* inside the network is *permitted* to access everything in the network.
- *Everyone* outside the network is *prohibited* to access the network.

---

[1] **http://elgg.org/powering.php**
[2] **http://www.cmsdevelopment.com**

Mike decided not to use widely distributed social network applications like Facebook but prefers to consider an open source solution based on Elgg. That allows him to keep control of the information exchanged and to add custom functionalities. In addition, it will allow him to enforce more customized access control rules than current commercial solutions.

However, as the community grow bigger, the initial simple security policy appears to be no longer appropriate. Users start demanding different, more complex access rules. For example:

1. *Alice* shares everything with *Bob.*
2. *Trudy* is the indiscreet boss of *Bob* and *Charlie.* Since they cannot ignore him and neither they can grant him total access – if so they may lose their jobs – they add him as a *limited friend.* This means that they show a limited profile from 9 a.m. to 5 p.m, which is slightly less restricted than the public profile. The rest of the time they show their public profile.
3. *Users* want to hide sensitive information from *everyone* but *close friends.*
4. *Users* want to be able to edit their security policy at any time.

Elgg provides some functions to implement security requirements. For each item of content created the user can set its level of privacy among:

- Private (visible only for oneself)
- Friend (visible only for friends)
- Logged in users (visible for everyone who has an account)
- Public (visible for everyone)

Users can be grouped into collections: for example, close friends, work colleagues, relatives, etc. When an item is uploaded, the user is presented with a pulldown menu containing the default options (Public, Logged in users, Friend and Private), as well as any collections he has created. In this way, the user can grant permissions to all the users belonging to a custom collection.

However, it turns out that these functions were not appropriate to handle the security policy expressed by the social network members. Elgg's access control policy lacks two major features:

- Although it supports groups of users, it does not support groups of objects depending on their confidentiality (for example, *family_pictures* or *public_wallposts*).
- Activation/desactivation of rules depending on external conditions (current time, current location, etc.).

Since Elgg is an open source application, Mike could try to modify this code to handle the policy. But this would be an error prone solution. Previous analysis showed that Elgg has more than 100 places which need access control when accessing the database. Mike concluded that he had not sufficient skill and time to undertake this task. He carried out some research in the field and realized that many universities have come up with solutions to epxress and manage security requirements [1,5,6,3,7,8,9]. So he considers a solution based on the OrBAC model to express security requirements and AOP to weave these security requirements into Elgg. In order to do so he needs to *intercept* the execution of Elgg program and somehow call the access control policy to ask if some pieces of code – i.e. methods – can be executed or not.

3   A Security Policy for a Social Network

Most of existing social network systems provides some functionalities to enforce security requirements. However, definition of these functionalities is generally not based on formal security models. Instead, they use ad-hoc principles like "friend" or "friend of a friend" which are not free of ambiguities. As a consequence, it may be complex for members of these social networks to determine how other users can access their data and what the consequences of security policy updates are.

In this context, using a formal security model has many advantages. For instance, the RBAC model [3] is now widely used in many applications to define access control requirements. This model is based on the concept of role: Permissions are assigned to roles and then roles are assigned to users. However, in a social network, where each user may want to define its own security requirements, the concept of role defined in RBAC has some limitations. To illustrate this issue, let us consider the two following roles, Member and Friend and the two following permissions:

- R1: Member is permitted to read public news in the social network,
- R2: Friend is permitted to read friends-related news in the social network.

Rule R1 must be likely interpreted as "a member of the social network must be permitted to read *every* public news". By contrast, it is less likely that rule R2 must be similarly interpreted as "a friend of the social network must be permitted to read *every* friends-related news" (except if one considers that someone may be a friend of everyone in the social network). Instead, we would like to consider that rule R2 means that "a friend of the social network must be permitted to read the friends-related news of *his or her friends*". However, in RBAC, if a particular *user* is empowered in the role *Friend*, it would be a *friend* for everyone. So, in RBAC, it is not easy to make such a difference of interpretation between rules R1 and R2.

By contrast, it is straightforward to specify this distinction in the OrBAC model [1]. The advantage of OrBAC over RBAC is that OrBAC handles the concept of *organization*. An organization may be viewed as any entity that has to manage a security policy. Thus, in a social network, the social network itself corresponds to an organization. But, we can also consider that each user member of this social network is also an organization since these members must be able to manage their own security policy.[3] Actually, the social network members will

---

[3] Notice that there is nothing wrong in OrBAC to consider that a user is also an organization.

be considered sub organizations of the social network, so that every member will inherit from the social network.

In OrBAC, roles are assigned to users locally to an organization. For example, the fact $Empower(SN, Alice, Member)$ means that the role $Member$ is assigned to the user $Alice$ in the (organization) social network SN. Similarly, considering that $Mike$ is the administrator of the SN (and thus the administrator of $everyone$), we can specify the following fact: $Empower(SN, Mike, Admin)$.

By contrast, the fact $Empower(Alice, Bob, Close\_Friend)$ means that the role $Close\_Friend$ is assigned to the user Bob in the organization $Alice$. Or, what is the same, Alice is a close friend of Bob –which is not necessarily reciprocal.

We can then specify the access control rule which corresponds to R1:

- $Permission(SN, Member, Consult, Public\_News, Default)$

This rule is an abstract organizational permission which says that in the organization SN, members are permitted to consult public news. Abstract entities $Member$, $Consult$ and $Public\_News$ respectively represent a role, an activity and a view. The concept of role was already explained before and is assigned to users with similar permissions. Similarly, the concepts of activity and view are used in the OrBAC model to respectively group actions and objects which similar permissions apply to. $Default$ represents a context. In OrBAC, every access control rule is associated with a context that must be satisfied to activate the corresponding rule. A taxonomy of context and formal method to manage them is defined in [10]. Examples of context are temporal, spatial, provisional and application-dependent contexts. In permission R1, the $Default$ context represents a condition which is always true.

It is possible to respectively assign concrete entities to abstract entities role, activity and view. For example, the facts $Empower(SN, Alice, Member)$, $Use(SN, news1, Public\_News)$ and $Consider(SN, read, Consult)$ respectively means that (1) Alice is a member of organization SN, (2) The object $new1$ is a public news in organization SN and (3) the action $read$ implements the activity $Consult$ in organization SN.

Using these three facts, it is then possible to automatically derive that the user $Alice$ is concretely permitted to read the object $news1$, which is represented by the following derived fact: $Is\_permitted(Alice, read, news1)$.

The SN organization can define a global set of roles, activities and views. These roles, activities and views will be inherited by the different sub organizations (members) of SN. In the following, we shall consider the role $Admin$, $Member$ and other roles corresponding to different degrees of friendliness: Limited Friend, Actual Friend, Close Friend and Myself.

These different roles are organized hierarchically. For example, we shall specify $subrole(SN, Actual\_Friend, Limited\_Friend)$. As a consequence, the role $Actual\_Friend$ inherits the permissions from the role $Limited\_Friend$.

These roles are defined in the organization SN, and thus all sub-orgnizations (members) will inherit the same structure of roles. However, as explained later in this section, the members can easily customize this structure of roles by adding their own roles locally.

Regarding the structure of views, we shall assume that there are different categories according to their level of confidentiality. Thus, objects can be used in the following views: Public news, Limited news, Normal news, Close news, and Private news. These different views are also organized hierarchically. For example, we shall specify *subview(SN, Limited_news, Public_News)*, so that the view *Limited_news* inherits the permissions assigned to the *Public_News*. Thus someone authorized to access *limited* news is also authorized to access *public* news. The same happens with the rest of views (*limited* and *normal*, *normal* and *close*, *close* and *private*). Similarly to roles, this structure of views is defined for the organization *SN*. Members can create their own views locally.

We can then define the organizational access control rules that apply to the whole social network *SN*. For example, the following rule corresponds to the above permission R2:

– *Permission(SN, Actual_Friend, Consult, Normal_news, default)*

If there is no user empowered in the role *Actual_Friend* in the organization *SN*, we cannot directly derive any concrete permission from this organizational permission. However, due to organizational hierarchy, this organizational permission is inherited in sub organizations of *SN*, for instance *Alice*. So if we consider the following facts: *Empower(Alice, Bob, Actual_Friend)* and *Use(Alice, telephone_number_alice, Normal_news)* then, using the inherited permission:

– *Permission(Alice, Actual_Friend, Consult, Normal_news, default)*,

we can derive the following concrete permission:

– *Is_permitted(Bob, read, telephone_number_alice)*.

Notice that the problem mentioned above for the RBAC model is solved: Only Alice's friend can get an access to Alice's normal news.

In OrBAC, it also possible to specify access control rules that apply in specific context, for example the following permission which only aplies in the temporal context *working_hours*:

– *Permission(SN, Limited_Friend, read, Limited_news, working_hours)*

At this point users have a major capacity to customize their own access control policy. They can assign the objects they publish into the different *views* according to their confidentiality level. Then, they just need to *empower* the other users in the suitable *role* to limit their access. In addition, OrBAC can provide further customization. Users can create their own roles and assign other users to them. Then, they can create access control rules that only apply to these news roles. For example, let us consider that Alice wants to create the role *Office_Friends*. Then she creates the view *Office_news* and assigns the object (for example, *mobile_phone_number*) she wants them to grant an access: *use(Alice, mobile_phone_number, Office_news)*.

The permission to enforce this access control rule will be:

- $Permission(Alice, Office\_Friends, read, Office\_new, default)$

The access control policy acquisition in OrBAC can be done using a support tool called MotOrBAC [11]. This tool provides a friendly user interface so that the members of a social network can easily define their own security requirements and customize the global security policy of the social network applications. MotOrBAC also provides means to define the administration policy using the AdOrBAC model [12]. This would be useful to define the administrative permissions assigned to the different users of a social network. However, due to space limitation, we do not further develop this functionality in this paper.

Now that we have explained how to specify the security policy of a social network application using the OrBAC model, let us show how to implement this policy using Aspect Oriented Progamming. Before, let us first briefly present what is AOP.

4   What Is AOP?

In software development there is always the question of how much effort should be invested in software design. Of course a good design allows an efficient comprehension of the code by others and thus improves the teamwork, but too much design will result in a not always justified spent of time addressing scenarios that are unlikely to happen and thus result in bloated software. The AOP paradigm addresses this issue, which was called by Laddad the architects dilemma [13].

Aspect Oriented Programming (AOP) has almost endless applications, from software conception to privacy injection. In the same way as OOP optimizes software conception by modularizing the overall program in more-or-less independent objects that interrelate among them, AOP addresses the modularization of concerns, in particular the cross-cutting concerns.

Pointcuts are marks in the program that trigger the execution of an external piece of code, called *Advice*. It is this *Advice* who requests the security policy whether the user has the right to carry out that action on a particular object or not [14]. The procedure of interleaving the *advices* with the actual program is called *weaving*.

The interception of the program can be carried out in the source code but also in the bytecode. There has been attempts to implement AOP in machine code, but so far these approaches have major limitations.

One of the primary goals of introducing aspects and considering aspect languages is to abstract away the aspects from the components. Aspect code should be independent from components and from other aspects [15]. In particular, AOP can separate functional specification from security requirements. That allows to easily update the functional specification of the security requirements without redesigning the overall application. Thus the security concerns can be managed by an expert team who does not need to have a *low level* understanding of the main program.

5   Expression of Security Requirements Using AOP

Aspect Oriented Programming provides an efficient way to enforce a security policy by modifying the execution flow of a program. Nevertheless, a reusable approach which can deal with a generic program has not been addressed yet. In order to overcome this difficulty, the *injected* code – i.e. the *Advice* – must be split in two functional parts, as described in Figure 1:

- The *Adaptor*, which is a custom-made part in charge of gathering all the information required to request for permission. Thus, this part is *adapted* to each case, since the way to obtain this information may vary among programs and even methods. This information does not depend on the access control policy employed. Functionally it is composed by these three methods:

  - *getSubject*
  - *getAction*
  - *getObject*

  The number of *Adaptors* depends on how this information is retrieved. In some cases it is passed as arguments, but it might be, for example, in a database. As a consequence, the number of *Adaptors* cannot be predicted before an analysis of the program.
- The *Middleware*, the program-independent part, who asks the access control policy for permission and gives the answer back to the *adaptor*. In order to encapsulate the access control policy and make it reusable, it can be deployed as a web service. In this paper it has been deployed as a SOAP web service, though it can also work with any other web service architecture, REST for example[4].

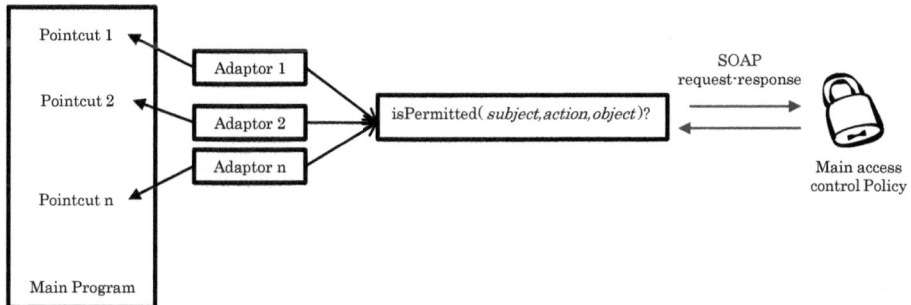

Fig. 1. *High-level* schema of the interception of the program

This structure make this approach language independent – the only condition is that a framework for Aspect Oriented Programming exists.

---

[4] In this case the Middleware will exchange HTTP requests instead of SOAP messages.

For scenarios where the access control policy needs to be edited by different entities –for example, each organization handles its own policy– new rules can be added. In the schema of Figure 2 it is called *Administrator access control*. This second access control policy says *who* can modify *which permission*. It is also composed by an *Adaptor* and a *Middleware*. The *Adaptor* is the responsible of gathering the information regarding the *Subject*, the *Action* and the *Object*. The *Middleware* asks the administrator access control policy for permission and if cleared to proceed it modifies the main security policy.

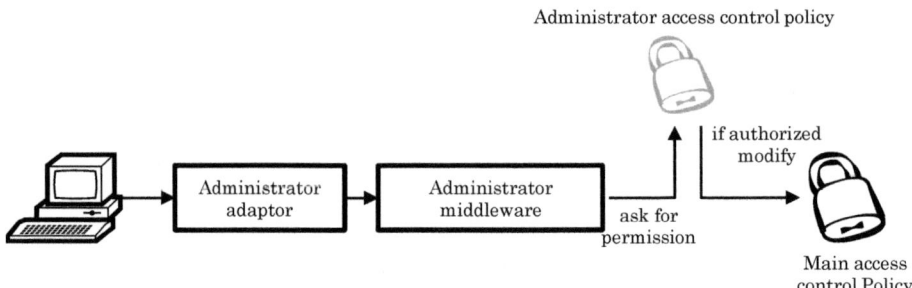

Fig. 2. Administrator access control

The procedure of securing a program from the start to end is composed by four main steps:

1. Formal definition of the security concerns. This step is independent of the security policy. It is the formal definition of access rules.
2. These three steps are independent and can be carried out simultaneously.
   - Definition of the overall security policy. For this paper the OrBAC model has been chosen since it supports context-dependent rules.
   - Identification of relevant methods to the security policy (for example, if the access control policy defines who can write a file, the relevant methods are those involved in opening, writing and closing the file). This part is specially sensitive. All the methods which carry out the same action must be intercepted. Many security flaws come from this step.
   - Coding of the Middleware. The Middleware queries the access control policy with the information provided by the adaptors. It will not vary as long as the deployment of the policy's web service does not change, so it is a good practice to store them for future reutilization.
3. Coding of the adaptor. Once the relevant methods have been identified, the adaptors will be coded and woven into the program.
4. Validation.

Listing 1.1 shows the pseudocode of an example of Adaptor. In this case the tasks to retrieve the *action* and the *object* cannot be encapsulated since they require information only available to the *advice*. Consequently they have been included in the body of the Adaptor. Listing 1.2 shows the pseudocode of the MiddleWare.

```
new Aspect ();
var pointcut=adaptor−>pointcut ("call*")
pointcut −>_before ("adviceCode()")

function adviceCode(){
  var subject=Get_subject_from_session ()
  var action = get_method_name ()
  var object = get_object_from_args ()
  if (MiddleWare(subject , action , object ))
  { continue execution}
else
  {prevent execution}
}
```

Listing 1.1. Pseudocode of the Adaptor of the schema proposed in Figure 1

```
global url=[address to the wsdl]

function MiddleWare(subject , action , object ){
  var raw_response=callSOAPService(
    url , isPermitted ,
    subject , action , object
    )

var response=parseResponse(raw_response)

return response
}
```

Listing 1.2. Pseudocode of the MiddleWare of the schema proposed in Figure 1

As a practical remark it should be noted that the pointcuts should be placed in the last method involved in the execution of a given action. Consider the example of a social network in Figure 3. In order to output the list of friends the system needs to ask a database. If the access control policy prevents the database to be read, the output method will inevitably fail. As a consequence, it is important to ensure that blocking a method will not raise future problems. In addition, this will also reduce the risk of information leakage. Following with the example of Figure 3, if only the method *getFriendList* is intercepted, Mike must intercept the methods *getFriendsListAlphabetically*, *getFriendsByEmail* and similar ones. And he is likely to forget some. This issue will be handled in the use case in Section 6.

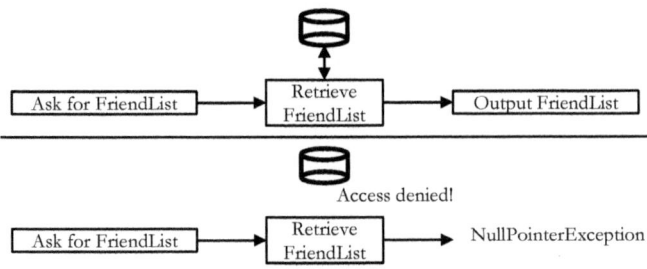

Fig. 3. Workflow interrupted by policy

6   Application to Social Network Security

Let us now show how this approach is used to enforce security in a complex, real world application, a social network. In order to do so, the Open Source social network Elgg has been considered. It has been deployed at `http://seres.mine.nu` following a LAMP architecture[5].

First, the built-in access control functions provided by Elgg are deactivated by assigning to every object the Public level. As a consequence, from the point of view of Elgg, every object can be freely accessed and there is no conflict with the security requirements that are woven in Elgg source using aspects.

According to the procedure explained in Section 5 the pointcuts have been placed in the last link of the execution chain. In Elgg's case, everything that is sent back to the client pass through some of these three methods:

- *elgg_view*: this method returns the HTML code of the profile's basic information. It receives information about *subject*, *action*, *object* and *organization*.
- *elgg_echo*: this method is used to intercept the messages belonging to the Wall. Unfortunately, it is not passed as many arguments as *elgg_view*, and thus it must be used in conjunction with it.
- *page_draw*: this is the last method called. It sends back the HTML output to the client. This method does not provide much versatility, since preventing its execution will result in a blank page, so it will not be intercepted.

This approach provides the additional advantage of automatically securing new functionalities *plugged* to Elgg. Indeed, one of the major features of Elgg is its great number of *third party* applications. In this case study, Mike allows his users to keep messages in a semi-public Wall and to upload their pictures in albums. In order to secure this two new features, he just needs to modify the Adaptor to make it able to intercept the object *wallpost* or *picture* and eventually the action, *read* in this case. Nevertheless, Mike must not forget that although this approach is fast and powerful, he still needs to be extremely careful with security flaws. These flaws may arouse when finding the methods to intercept (the step of

---

[5] LAMP stands for Linux as operating system, Apache as web server, MySQL as the database and PHP as the server side language.

Identification of relevant methods in the procedure explained in Section 5). That is why Mike must be sure that he intercepts *all* the methods that carry out a particular action. In order to minimize this risk Section 5 encourages to intercept the last one of the methods involved in the action. Additionally, Mike must make sure that there are no unsecured ways of calling a method. For example, if only the PHP methods which communicate with the database are intercepted, Mike must be sure that there are no other public ways of requesting information, as a passwordless telnet connection. Refer to Appendix A for a description about security bugs.

As explained in Section 3, Mike created a global security policy. In this policy users can easily classify their objects according to their level of confidentiality. Then, they can add other members to authorized roles. For example, if Alice wants to make public her website address, she will simply do:

$$use(Alice, website\_address, Public\_news)$$

Once these rules are enforced, the different users will have different views of the other users' profiles.

7   Related Works

Several other papers have already addressed the issue of managing security requirements using AOP.

The approach proposed by Piessens et al. [16] is the protection of a FTP server. They proved that actual, complex applications can be properly secured with AOP.

Huang et al. [17,18] conceived a reusable JAVA aspect API called Java Security Aspect Library. It provides the following frameworks: *JAAS* – Java Authentication and Authorization Service – for authentication and authorization and *JCE* – Java Cryptography Extension – for encryption/decryption among others. Chen et al. [19] proposed an approach to enforce security in web applications based on the works by De Win et al. [20,21]

Although one limitation of this approach is that the language of the main program must support AOP, there is a growing interest in creating AOP frameworks for many languages, even for compiled languages as C [22].

8   Conclusion

This paper presented a realistic example about the owner of a small social network who needs to enforce complex access control rules. Then, a generic methodology to enforce an access control policy in an existing program using AOP has been proposed. Finally, this approach has been followed to secure the social network presented in the motivating example[6]. In Appendix A a security bug serves as an example on how the aspects must be applied with care.

---

[6] And which is publicly reachable at `http://seres.mine.nu`

References

1. Kalam, A.A.E., Baida, R.E., Balbiani, P., Benferhat, S., Cuppens, F., Deswarte, Y., Miege, A., Saurel, C., Trouessin, G.: Organization based access control. In: Proceedings of the IEEE 4th International Workshop on Policies for Distributed Systems and Networks, POLICY 2003, pp. 120–131. IEEE (2003)
2. Sharma, M.: Elgg social networking. Packt Publishing, Birmingham (2008)
3. Sandhu, R., Coyne, E., Feinstein, H., Youman, C.: Role-based access control models. Computer 29, 38–47 (1996)
4. Sheiko, D.: Aspect-oriented software development and php (May 22, 2010), http://dsheiko.com/weblog/aspect-oriented-software-development-and-php
5. Gao-Feng, J., Yong, T., Yun-Cheng, J., Hong-Yi, Y.: A description logic approach to represent and extend rbac model. In: 1st International Symposium on Pervasive Computing and Applications, pp. 151–156 (2006)
6. Ferraiolo, D., Kuhn, D.: Role-based access controls. Arxiv preprint arXiv:0903.2171 (2009)
7. Roos Lindgreen, E., Herschberg, I.: On the validity of the bell-la padula model. Computers & Security 13, 317–333 (1994)
8. Saltzer, J., Schroeder, M.: The protection of information in computer systems. Proceedings of the IEEE 63, 1278–1308 (1975)
9. Thomas, R.K.: Team-based access control (tmac): a primitive for applying role-based access controls in collaborative environments. In: Proceedings of the Second ACM Workshop on Role-Based Access Control, pp. 13–19. ACM (1997)
10. Cuppens, F., Cuppens-Boulahia, N.: Modeling contextual security policies. International Journal of Information Security 7, 285–305 (2008)
11. Autrel, F., Cuppens, F., Cuppens-Boulahia, N., Coma, C.: Motorbac 2: a security policy tool. In: 3rd Conference on Security in Network Architectures and Information Systems (SAR-SSI 2008), Loctudy, France, pp. 273–288 (2008)
12. Cuppens, F., Miege, A.: Adorbac: an administration model for or-bac. International Journal of Computer Systems Science & Engineering 19, 151–162 (2004)
13. Laddad, R.: Aspectj in action. Practical Aspect Oriented Programming (2009)
14. Wand, M., Kiczales, G., Dutchyn, C.: A semantics for advice and dynamic join points in aspect-oriented programming. ACM Transactions on Programming Languages and Systems (TOPLAS) 26, 890–910 (2004)
15. Lämmel, R.: Declarative aspect-oriented programming. In: ACM SIGPLAN Workshop on Partial Evaluation and Semantics-Based Program Manipulation, pp. 131–146 (1999)
16. De Win, B., Joosen, W., Piessens, F.: Aosd & security: a practical assessment. In: Workshop on Software engineering Properties of Languages for Aspect Technologies (SPLAT 2003), Citeseer, pp. 1–6 (2003)
17. Huang, M., Wang, C., Zhang, L.: Toward a reusable and generic security aspect library. AOSD: AOSDSEC 4 (2004)
18. Parnas, D.: On the criteria to be used in decomposing systems into modules. Communications of the ACM 15, 1053–1058 (1972)
19. Chen, K., Lin, C.: An Aspect-Oriented Approach to Declarative Access Control for Web Applications. In: Zhou, X., Li, J., Shen, H.T., Kitsuregawa, M., Zhang, Y. (eds.) APWeb 2006. LNCS, vol. 3841, pp. 176–188. Springer, Heidelberg (2006)
20. De Win, B., Vanhaute, B., De Decker, B.: Security through aspect-oriented programming. Advances in Network and Distributed Systems Security, 125–138 (2002)

21. De Win, B., Piessens, F., Joosen, W., Verhanneman, T.: On the importance of the separation-of-concerns principle in secure software engineering. In: Workshop on the Application of Engineering Principles to System Security Design, WAEPSSD, Boston, MA, USA (2002)
22. Viega, J., Bloch, J., Chandra, P.: Applying aspect-oriented programming to security. Cutter IT Journal 14, 31–39 (2001)

A   Security Bugs

In order to efficiently and securely enforce an access control policy, one must be extremely careful with the *Validation* part. This is particularly true when securing a software not programmed by oneself. On November 26, 2011 a security flaw concerning the pictures has been detected. It has been deliberately kept to be studied in this paper.

This flaw concerns the plugin for Elgg that allows the user to upload pictures. It is, thus, a *third party* program added time after the security policy has been implemented. As explained in the case study in Section 6 Mike has taken advantage of the fact that all the communication with the client passes through three methods. Consequently, it should be enough to intercept and enforce the access rules in these three methods. And under normal, non threatening circumstances this approach works fine.

In the example proposed in this paper, Eliane is the only user who has uploaded pictures. She has two pictures *dancing-cards* – available for everyone – and *bread-and-butterflies* – available only for friends. Since the verification of privileges is carried out in the methods who send the HTML back to the client, a user will not be able to find a picture he is not authorized to open.

But what happens if the user directly writes the URL of the picture *without* previously visiting any other page, and thus without performing any security check? In this scenario the only security that pictures have is the unreliable *security through obscurity*, with the additional danger that in Elgg the direct URL for pictures is very easy to guess. In any case, an unauthorized user – Trudy in this case – could ask an authorized user – Alice for example – for the URL of the protected picture[7] and open it.

Once the bug has been spotted it is almost trivial to fix it. The solution is to intercept the action to show the picture and check permission with the same data which the method to output the previous page received. In this way, if the user tries to directly request the picture, its access will be refused.

---

[7] Which is `http://seres.mine.nu/pg/photos/thumbnail/20/large/`

Exploring Reflection of Urban Society through Cyber-Physical Crowd Behavior on Location-Based Social Network

Shoko Wakamiya[1], Ryong Lee[2], and Kazutoshi Sumiya[1]

[1] Graduate School of Human Science and Environment, University of Hyogo
ne11n002@stshse.u-hyogo.ac.jp, sumiya@shse.u-hyogo.ac.jp
[2] National Institute of Information and Communications Technology
lee.ryong@gmail.com

Abstract. Due to the explosive growth of cyber-real space and crowds over the recent social networking sites, the real space in which we are making daily lives is also storongly tied with the imaginary cyber-social space. In this respect, it would be more and more crucial to understand the relationships and the interactions of crowds, physical-real space and cyber-social space. In this paper, we derive images of city in terms of cyber-physical crowds who are nowadays quite common people sharing their lifelogs over social networking sites with location-aware information. Especially, we attempt to explore urban characteristics which are reflected on the social networks through crowd behavior. For this purpose, we model crowd behavior in urban areas by utilizing crowd's daily lifelogs. Then, we explore latent relationships between urban regions and the crowd behavior by means of NMF (Non-negative Matrix Factorization). In the experiment, we show the urban characteristics based on significant crowd behavioral patterns.

1 Introduction

The unexpected growth of today's social networking sites has lots of critical implications in our society. Especially, on behalf of the large population on the social networks usually via mobile devices, we are witnessing a radical change of our society, where people's lives are being extended with the cyber-social space based on the physical-real space. In fact, crowds are publishing their daily activities and thoughts with the detailed location information to social networks such as Twitter[1] and Facebook[2] using recent high-functional devices like smartphones. Such social and technological trend enables crowds to virtually exist in both physical-real and cyber-social spaces; crowds living in the real space try to get various information and share up-to-date lifelogs with their friends by way of cyber space. This seemingly trivial change of crowd's common habits represents

---

[1] Twitter: http://twitter.com/
[2] Facebook: http://www.facebook.com/

H. Yu et al. (Eds.): DASFAA Workshops 2012, LNCS 7240, pp. 168–179, 2012.
© Springer-Verlag Berlin Heidelberg 2012

the popular term, 'crowd-sourcing' invoking a new possibility to take advantages of utilizing massive crowd's life experiences.

In this paper, we will explore such crowd behavior delivering real-world information and situations to the cyber-social space of location-based social network sites (LBSN). In particular, we are looking into the images of city by observing and analyzing the reflection through crowd behavior. Particularly, we attempt to extract urban characteristics in terms of crowd behavior based on the representative LBSN, Twitter, where generally numerous people are participating intermittently or frequently by publishing their up-to-dates with geographic location and time information. Based on the vast amount of geographically distributed and temporally different crowd's lifelogs, we measure the crowd behavior on how they are activated or moving. Specifically, we model crowd behavior using their microblogs over LBSN and generate a crowd behavioral vector in each urban area. Then, we extract latent crowd behavioral patterns by exploring the relationships between urban areas and crowd behavior using NMF (Non-negative Matrix Factorization). In the experiment with massive dataset from Twitter, we show some examples of urban characteristics reasoned by geographical facilities in the urban areas.

The remainder of this paper is organized as follows: Section 2 introduces our research model and surveys related work. Section 3 describes an overall process of urban characteristics extraction based on LBSN. Section 4 illustrates the experiment with massive Twitter data for extracting urban characteristics based on crowd behavior. Finally, Section 5 concludes this paper with a brief description of future work.

2   Exploring Cyber-Physical Crowd's Lives

In this section, we introduce our research model to explore the relationships and interactions between physical-real space, cyber-social space and crowds. Next, we explain our goal to extract urban characteristics utilizing the reflection of urban areas on the cyber-social space through the crowds.

2.1   Deep into the Cyber-Physical Crowd's Lives

We first explain our research model to study the relationships and interactions of physical-real space, cyber-social space and crowds. As illustrated in Fig. 1, the crowds are actively bridging the two spaces by emitting huge amounts of information and lifelogs about the real happenings and experiences from the physical-real space to the cyber space which can be any type of information space from the Web to narrowly the SNS. In this paper, we especially focus on the LBSN significantly tied with the physical-real space. Among these three parts, active crowds between the two spaces can be consequently regarded as a tightly integrated space, so-called, cyber-physical space. Thus, we refer to the crowds who are living in the both spaces by "Cyber-Physical Crowds" or simply crowds in this paper.

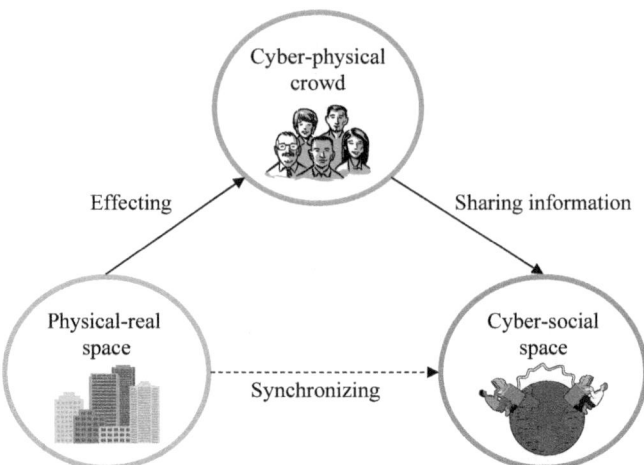

Fig. 1. Research Model of Cyber-Physical Crowd's Relationships and Interactions

In particular, we will explore the reflection of cyber-physical crowd's experiences on the cyber-social space to look into the real-urban space's characteristics. Indeed, crowds are nowadays reporting daily lifelogs which are probably connected with the characteristic features of urban spaces.

2.2   Reflection of Urban Characteristics on Cyber-Social Space

We purpose a method to extract "urban characteristics" by means of crowd behavior over cyber-physical space. For the better understanding to urban characteristics which we are targeting for, we explain several approaches to study this interesting and useful topic about our living space.

In fact, urban characteristics have been researched so far from various perspectives by physical geographic shape or diverse objects such as streets or landmarks, or cultural and structural aspects such as residential, commercial, and industrial districts. These views have been well studied in many research fields. Especially, Kevin A. Lynch's seminal contribution in his book titled "The Image of the City" [8] defined five fundamental elements of a city; paths, edges, districts, nodes, and landmarks. Based on these elements, Lynch thought that we could characterize our living space within the appearance of a city to imagine ourselves living and working there as shown on the left side in Fig. 2. In another remarkable work, Tezuka et al. [11] extracted geographic objects and their roles frequently mentioned on the Web contents which can be regarded as a mirror of the crowds' minds to the real world as shown in the middle of Fig. 2.

These two different types of urban characteristics seem to be obviously useful to grasp the image of the city. However, the former approach is only focusing on static elements in the physical-real space, and the latter one is focusing on extracting geographic information from the cyber-social space. On the other

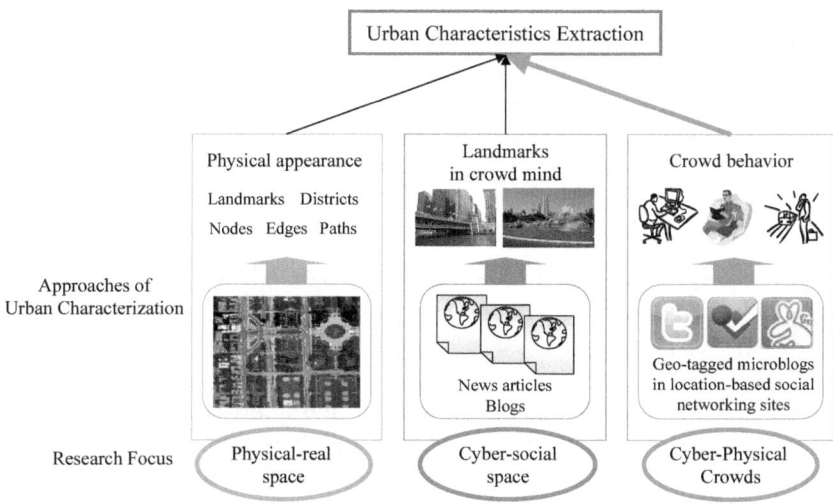

Fig. 2. Urban Characterization Approaches and their Focuses

hand, our approach relies on "crowds" living in both spaces which would be considered a critical factor for observing urban characteristics. Therefore, we attempt to derive a new kind of geo-socail characteristics based on cyber-physical crowd, especially in terms of their behavior, utilizing the LBSN such as Twitter, Foursquare[3], Gowalla[4], etc. as shown in the right side of Fig. 2. Needless to say, because plenty of crowd-sourced data quite involving crowd behavior in local areas are dynamically reflecting the physical-real world on such sites, we can enough utilize the remarkable data to extract the image of cities.

2.3   Related Work

Recently, social networking sites are regarded as a novel source which enables us to acquire lots of daily crowd life logs almost freely. However, in terms of social and individual benefits from the new open sharing space, it must be a critical issue to be researched and explored in the academic field as well as the business world.

Interestingly, some researches focusing on cooperation with Twitter for analyzing some natural incidents in the real world have been introduced. De Longueville et al. [4] analyzed the temporal, spatial and social dynamics of Twitter activity during a major forest fire incident in the South of France in July 2009. Sakaki et al. [10] constructed an earthquake reporting system in Japan using tweets which are posted from each Twitter user regarded as a sensor. In this method, tweets which are reporting the occurrence of earthquakes are extracted by using textual information. Cheng et al. [3] studied human mobility patterns revealed

---

[3] Foursquare: `https://foursquare.com/`
[4] Gowalla: `http://gowalla.com/`

by the check-ins over location sharing services and explored the corresponding factors that influence mobility patterns, in terms of social status, sentiment, and geographic constraints. This study focused on analysis of massive personal trace data based on characterizing geographic facilities. In our previous work [6], we also proposed a method to discover local social and natural events by monitoring unusual statuses of local users' activities utilizing geo-tagged crowd lifelogs over Twitter.

3   Urban Characteristics Extraction Based on Cyber-Physical Crowd Behavior over Location-Based Social Networks

In this section, we describe a process of urban characteristics extraction based on crowd behavior observed using crowds' daily lifelogs. Fig. 3 shows an overview of the process including preprocessing steps; 1) collecting crowd lifelogs from location-based social networks, 2) setting urban regions, 3) measuring crowd behavior in urban regions (Section 3.1), and 4) exploring latent urban characteristics based on crowd behavior (Section 3.2). We will briefly describe the preprocessing steps 1) and 2) in Section 4. As for the core steps 3) and 4) of this method, we will give the details in this section.

3.1   Measuring Cyber-Physical Crowd Behavior

Crowd behavior in the physical-real space must be reflected on the cyber-social space through their daily lifelogs posted over LBSN. Therefore, we consider that it could be possible to easily observe crowd behavior in urban regions using geo-tagged tweets obtained from Twitter representing LBSN. In this section, in order to derive crowd behavior, we describe a method to measure a crowd behavior in each urban area based on geo-tagged tweet's metadata such as user ID, time stamp, and location information. In this work, we define three primitive measures as follows:

$\#Tweets|_{r_i,p_j}$: The total number of tweets occurring inside of an urban region r_i in a time period p_j. This measure indicates the activeness of behavior of cyber-physical crowds to the cyber-social space.

$\#Crowd|_{r_i,p_j}$: The number of distinct crowds in an urban region r_i during a specific period of time p_j. This measure means the scale of the cyber-physical crowds.

$\#MovCrowd|_{r_i,p_j}$: The number of mobile crowds in an urban region r_i in a time period p_j.

These values of measures are calculated for urban areas. In order to grasp crowd's activities and lifestyles, we need to monitor daily crowd behavior periodically. In the experiment, we calculated the values of measures in urban regions for eight time periods from p_1 to p_8 by dividing a day by three-hours respectively as $p_k = [3(k-1):00, 3k:00)$ $(1 \le k \le 8)$.

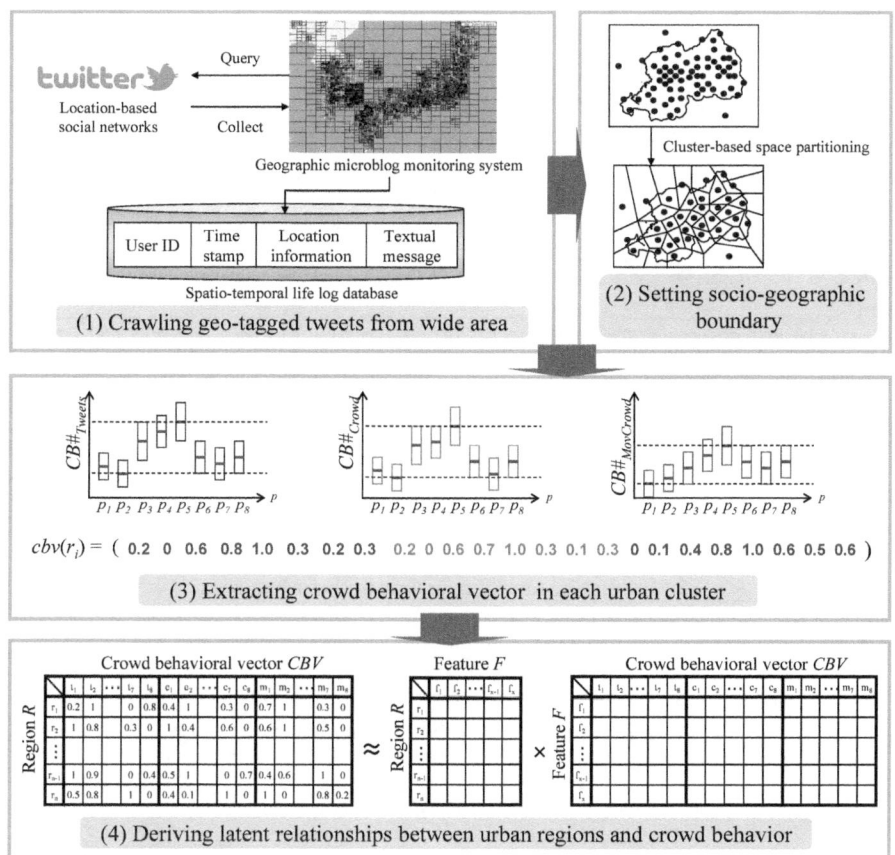

Fig. 3. Process of urban characteristics extraction in terms of cyber-physical crowd behavior

In order to derive crowd behavior using the measures, we generate a crowd behavioral vector $cbv(r_i)$ in an urban area r_i. The vector consists of a sequence of relative temporal change of crowd behavior based on the measures. Specifically, we first aggregate the crowd monitored data based on the parameter $x \in \{\#Tweets, \#Crowd, \#MovCrowd\}$ periodically as shown in Fig. 4 (1). Next, we extract crowd behavior $CB_x(r_i)$ by summarizing values of the parameter of the same time period from p_1 to p_8 through abstracting with boxplot [9] as shown in Fig. 4 (2). Finally, we represent the crowd behavior as a vector consisted of the normalized medians from 0 to 1.0 as shown in Fig. 4 (3).

3.2 Extracting Crowd Behavioral Patterns as Urban Characteristics

In order to extract urban characteristics based on crowd behavior, we need to examine characteristic clusters of crowd behavior. For the purpose of

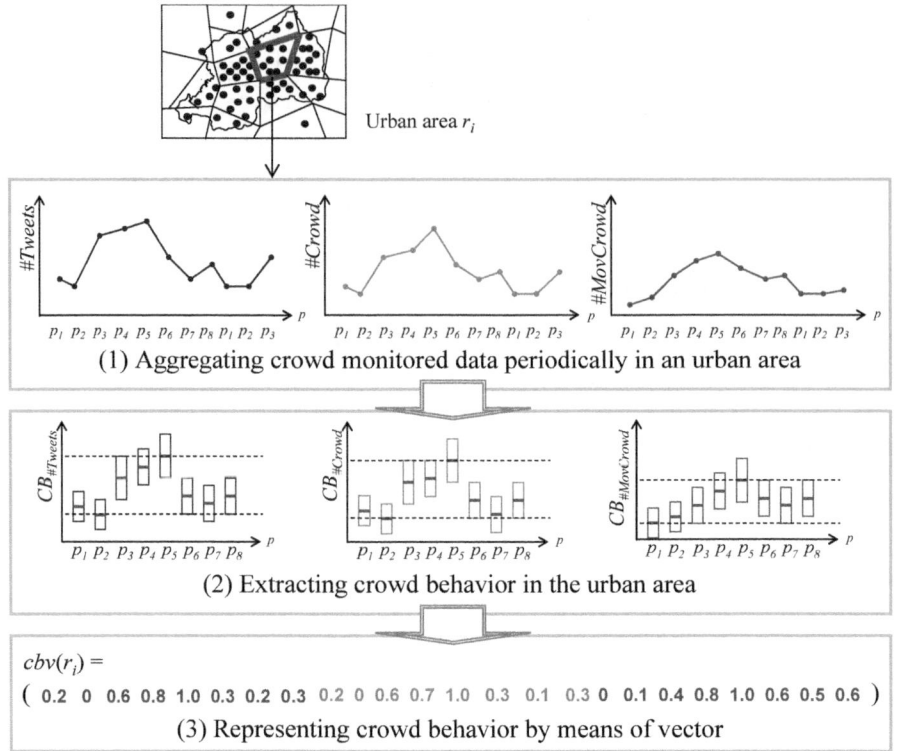

Fig. 4. Process of generation of crowd behavioral vectors

investigating the latent and characteristic relationships between urban clusters
(R) and crowd behavior features (CBV), we adopted NMF (Nonnegative Matrix Factorization) [5] which is based on localized features by factorizing a data
matrix with non-negativity constraint; Specifically, the algorithm decomposes
a matrix X to two matrixes W and H as follows; $X = W \cdot H + \alpha$, where the
dimensions of the matrix factors W and H are $n \times f$ and $f \times m$, respectively. The
feature f of the factorization is generally chosen so that $(n + m)f < n \times m$, and
the product WH can be regarded as a compressed form of the data in X. The
non-negativity constraint permits the combination of multiple basis features to
represent a crowd behavior, though only additive combinations are allowed since
the non-zero elements of W and H become all positive.

In this work, we first construct a matrix X made of urban clusters (R) and
crowd behavior vectors (CBV) and then decompose it into a basic matrix W
($R \cdot F$) consisting of clusters R and latent feature classes F and a coefficient
matrix H ($F \cdot CBV$) consisting of latent feature classes F and crowd behavioral
vectors CBV.

4 Experiment

4.1 Dataset

In order to conduct the experiment, we first need to collect lots of geo-tagged Twitter messages. For this, we set a region surrounding Osaka in Japan, so-called Kinki region, as a target region for exploring latent urban characteristics. In order to obtain lifelogs from crowd in the Kinki region, we monitor Twitter from Oct. 9th, 2011 to Oct. 22nd, 2011 in the range of [33.384555: 35.839419], [134.126551: 136.58890] using a geographical microblog monitoring system developed in our previous work [6]. As a result, we could obtain 75,279 geo-tagged tweets posted from 8,272 distinct users.

Next, in order to configure urban regions for investigating latent relationships with crowd behavior, we partitioned the target region to 145 clusters based on crowd behavior be means of a clustering algorithm. Specifically, we first split the data into two groups of high-frequency and low-frequency parts for reducing the data size of massive lifelogs to much smaller and computable size by means of the Nearest-neighbor cleaning algorithm [2]. Then, we generated clusters by the EM algorithm [1] and depicted a Voronoi diagram [7] using the center points of the clusters. As a result, we obtained the formed polygonal urban clusters as shown in the top of Fig. 5.

4.2 Mining Crowd Behavioral Patterns

By using the dataset described in Section 4.1, we calculate the values of three measures every 3-hours in 145 urban regions respectively, and generate crowd behavioral vectors based on the values of the measures as shown in the bottom of Fig. 5.

Next, we generated the matrix X consisting of 145 regions and 24-dimentional crowd behavioral vectors ($R \cdot CBV$). By applying NMF to this matrix, we decomposed it to the basic matrix W consisting of 145 regions and 13 latent feature classes as shown in Fig. 6 (a) and the coefficient matrix H consisting of 13 latent feature classes and 24-dimentional crowd behavioral vectors as shown in Fig. 6 (b). Subsequently, 13 crowd behavioral patterns were extracted and urban regions were clustered based on the 13 patterns. Fig. 6 shows the result of the matrix decomposition expressed by means of heat map. In this figure, (a) shows the basic matrix W and (b) shows the coefficient matrix H. Both the horizontal axis of the heat map in Fig. 6 (a) and the vertical axis of the heat map in Fig. 6 (b) show latent feature classes from f_1 to f_13.

The latent feature classes have similar patterns based on t_k reflecting the crowd behavior towards the cyber-social space and c_k reflecting the crowds themselves in each time period p_k were observed; f_2, f_4, f_7, f_9, f_{10}, f_{12}, and f_{13}.

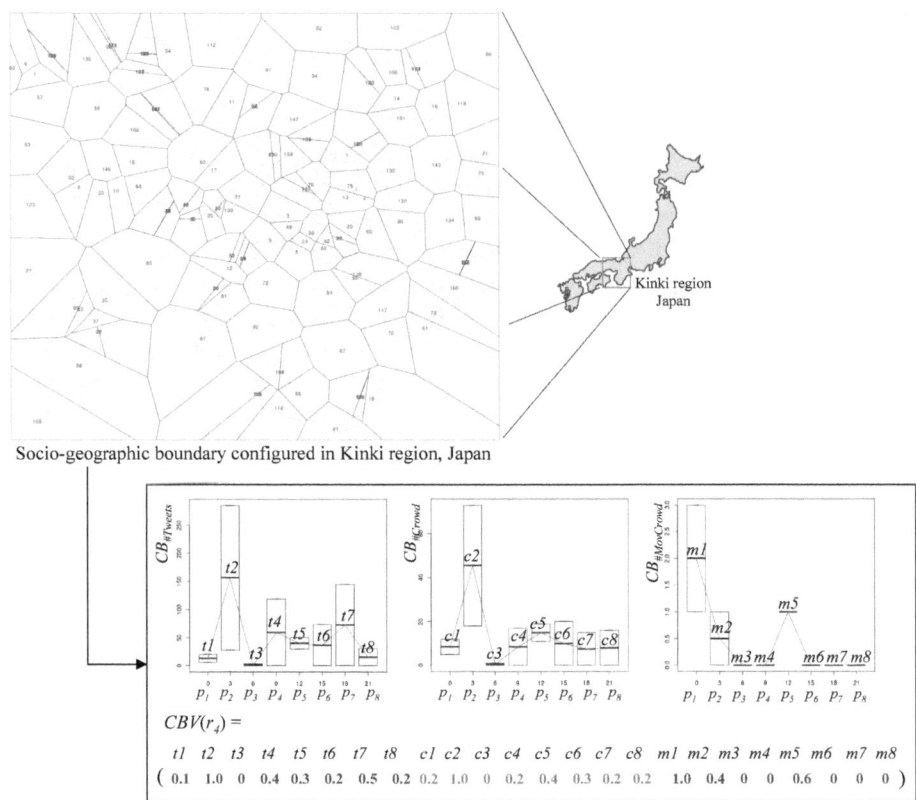

Socio-geographic boundary configured in Kinki region, Japan

Crowd behavioral vector

Fig. 5. An example of crowd behavioral vector in urban areas

4.3 Reasoning Urban Areas Clustered by Crowd Behavioral Patterns

In order to intuitively examine images of city, we investigated the ratios of categories of local facilities in urban regions. For this, we referred to the genre data attached to each local facility in Yahoo! Japan Loco[5]. The most super-ordinate categories categorize most facilities into four; 'life,' 'shop,' 'entertainment' and 'food,' and have subordinate categories. For example, there are some sub-categories of the category 'life' such as 'school,' 'hospital,' or 'post office.' In addition, some local facilities relevant to industry had not been categorized by any genre data. Therefore, we prepared new category 'work' and attached it to the rest facilities manually. In this experiment, for the simplicity, we focused on the most super-ordinate five categories; 'life,' 'shop,' 'entertainment' 'food,' and 'work.'

In this case, we investigated the number of local facilities included in each category in each urban region. However, because urban regions differ, it is not

---

[5] Yahoo! Japan Loco: http://maps.loco.yahoo.co.jp/

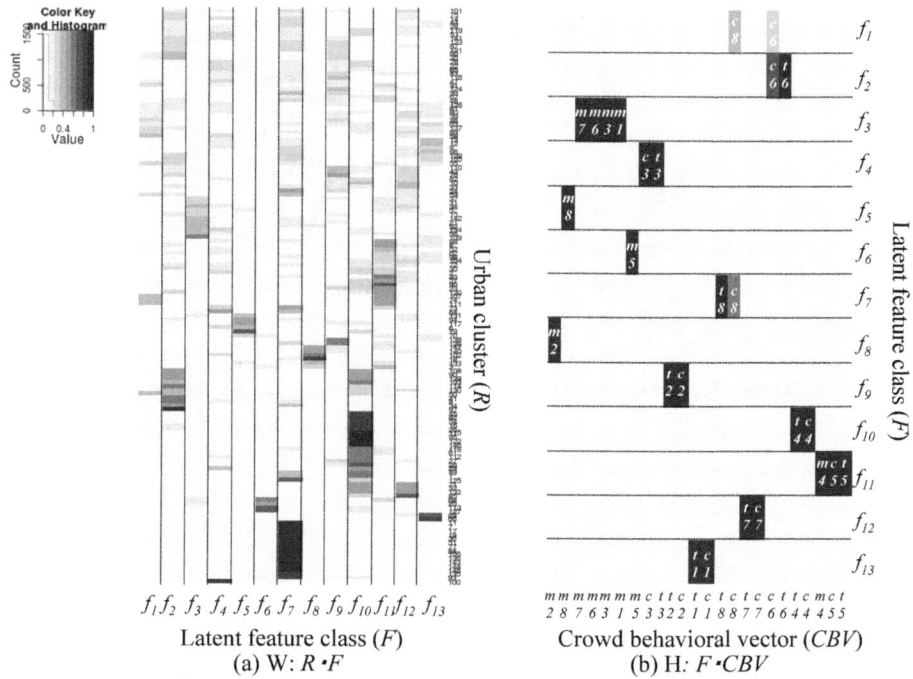

Fig. 6. Extracting crowd behavioral patterns in urban areas using NMF

appropriate to compare with the absolute numbers of the facilities. Hence, by calculating the average of local facilities included in each category for all regions using the total number of facilities in each region, we found the ratios of facilities of categories in urban regions as $(C_{r_i} - avg(C))/(max(C) - min(C)) + 2.0$, where C_{r_i} means a ratio of a category in a region r_i, and C is the total list of ratios in a category over all the regions. Thus, we investigated urban clusters based on cyber-physical crowds in terms of local facilities in the physical-real space. Finally, we show the result using radar chart as shown in Fig. 7 (A)(ii) and (B)(ii) which were illustrated based on the ratio of local facilities in urban areas in Fig. 7 (A)(i) and (B)(i) clustered by latent feature classes.

When investigating the categorical ratio of local facilities in urban clusters based on the latent feature class f_2, the ratio of the category 'work' was relatively higher than others as shown in Fig. 7 (A)(ii). In the case of f_2, we can find that two measures of #Tweets and #Crowd are increasing during the time period from 15:00 to 18:00. In fact, due to this time period including the finish time of working hours or lectures in schools for lots of people and the urban clusters are observed in Kobe and Osaka which are various office buildings and schools. Accordingly, we can reason the urban characteristics in terms of local facilities relevant to 'work.' In urban clusters of f_1, the ratio of the category 'food' was relatively higher than others as shown in Fig. 7 (B)(ii). Because f_1 is that the

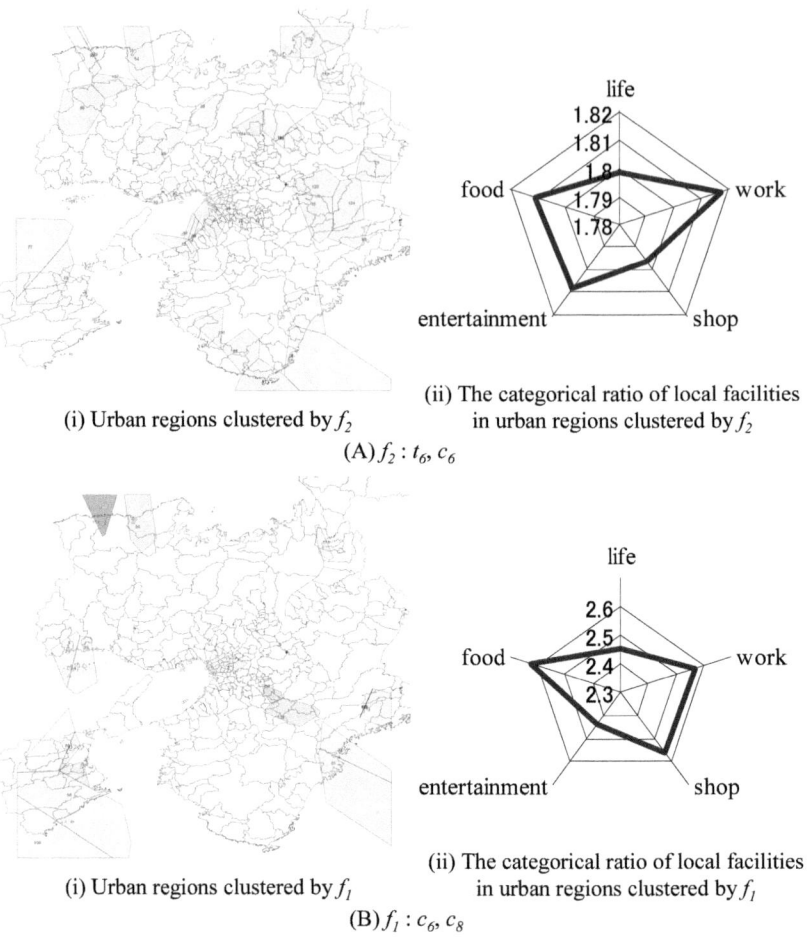

(i) Urban regions clustered by f_2 (ii) The categorical ratio of local facilities
in urban regions clustered by f_2

(A) $f_2 : t_6, c_6$

(i) Urban regions clustered by f_1 (ii) The categorical ratio of local facilities
in urban regions clustered by f_1

(B) $f_1 : c_6, c_8$

Fig. 7. Experimental result : Regions of Latent Urban Characteristics and Categories of Local Facilities

measure #$Crowd$ increases during two time periods from 15:00 to 18:00 and from 21:00 to 24:00, we observed crowds who went to restaurants or cafeterias for dinner after work in the regions.

5   Conclusion and Future Work

In this paper, we proposed a method to extract latent urban characteristics in terms of crowd behavior using cyber-physical crowd lifelogs over LBSN. We also introduced our research model consisting of three important factors in recent society; physical-real space, cyber-social space and cyber-physical crowds, and attempted to explore the relationships and interactions among them by observing crowd behavior. In the experiment with massive dataset, we showed urban characteristics based on crowd behavioral patterns.

In the future work, we will develop various measures representing crowd behavior utilizing crowd-sourced data from LBSN. In fact, it is a critical challenge to invent such kinds of crowd behavior mining functions for the purpose of utilizing the social networks as a fundamental framework to understand the emerging cyber-physical space and furthermore to utilize the collected experiences of crowds for building practical urban life support systems.

Acknowledgements. This research was supported in part by the Microsoft Research IJARC Core Project.

References

1. Dempster, A.P., Laird, N.M., Rubin, D.B.: Maximum likelihood from incomplete data via the EM algorithm. J. Royal Statiscal Soc., Ser. R 39(1), 1–38 (1977)
2. Byers, S., Raftery, A.E.: Nearest-neighbour clutter removal for estimating features in spatial point processes. Journal of the American Statistical Association 93, 577–584 (1998)
3. Cheng, Z., Caverlee, J., Lee, K., Sui, D.Z.: Exploring millions of footprints in location sharing services. In: Proc. of Fifth International AAAI Conference on Weblogs and Social Media, ICWSM 2011, pp. 81–88 (2011)
4. De Longueville, B., Smith, R.S., Luraschi, G.: OMG, from here, I can see the flames!: a use case of mining location based social networks to acquire spatio-temporal data on forest fires. In: Proc. of the 2009 International Workshop on Location Based Social Networks, LBSN 2009, pp. 73–80. ACM (2009)
5. Lee, D.D., Seung, H.S.: Learning the parts of objects by non-negative matrix factorization. Nature 401(6755), 788–791 (1999)
6. Lee, R., Wakamiya, S., Sumiya, K.: Discovery of unusual regional social activities using geo-tagged microblogs. World Wide Web (WWW) Special Issue on Mobile Services on the Web 14(4), 321–349 (2011)
7. Lloyd, S.P.: Least squares quantization in PCM. IEEE Transactions on Information Theory 28(2), 129–137 (1982)
8. Lynch, K.: The Image of the City. The MIT Press (1960)
9. Mcgill, R., Tukey, J.W., Larsen, W.A.: Variations of box plots. The American Statistician 32(1), 12–16 (1978)
10. Sakaki, T., Okazaki, M., Matsuo, Y.: Earthquake shakes Twitter users: real-time event detection by social sensors. In: Proc. of the 19th International Conference on World Wide Web, WWW 2010, pp. 851–860. ACM (2010)
11. Tezuka, T., Lee, R., Takakura, H., Kambayashi, Y.: Cognitive characterization of geographic objects based on spatial descriptions in web resources. In: Proceedings of the Workshop on Spatial Data and Geographic Information Systems (SpaDaGIS) (March 2003)

Mining Social Networks
for Significant Friend Groups

Carson Kai-Sang Leung* and Syed K. Tanbeer

Department of Computer Science, University of Manitoba, Canada
kleung@cs.umanitoba.ca

Abstract. The emergence of Web-based communities and hosted services such as social networking sites has facilitated collaboration and knowledge sharing between users. Hence, it has become important to mine this vast pool of data in social networks, which are generally made of users linked by some specific interdependency such as friendship. For any user, some groups of his friends are more significant than others. In this paper, we propose a tree-based algorithm to mine social networks to help these users to distinguish their significant friend groups from all the friends in their social networks.

Keywords: Advanced database applications, social networks, social media, social computing, knowledge discovery, data mining, social network analysis and mining.

1 Introduction

Rapid growth and exponential use of social digital media has led to an increase in popularity of social networks and the emergence of social network mining, which combines data mining with social computing [6,10,17,20]. As *social networks* [3,15] are generally made of social entities (e.g., individuals, corporations, collective social units, or organizations) that are linked by some specific types of interdependency such as friendship, a social entity can be connected to another entity as his friend. Similarly, a social entity can be linked to another entity as his next-of-kin, friend, collaborator, co-author, classmate, co-worker, team member, and/or business partner via other interdependency such as kinship, friendship, common interest, beliefs, and financial exchange.

Social network mining [2,7,8,12] aims to discover implicit, previously unknown, and potentially useful knowledge from a vast pool of data residing in the social networking sites such as Facebook [4,14], Twitter [13,16,19], and LinkedIn. In LinkedIn, a user can create a professional profile, add connections to other users (as friends, colleagues, and/or classmates), and exchange messages. In addition, he can also join common-interest groups and participate in discussions. Although the number of friends/connections may vary from one user to another, it is not uncommon for a user p to have hundreds of friends/connections. Among

---

* Corresponding author.

H. Yu et al. (Eds.): DASFAA Workshops 2012, LNCS 7240, pp. 180–192, 2012.

them, some groups of friends are more important or significant to p than others. Significance of a friend group G depends on various measures including the connectivity between G and p (e.g., friends in G who frequently view p's profile, send messages to p, and/or make postings to p's discussion are considered to be more significant to p than others) as well as the "rank" of G (e.g., friends in G who have certain experience, skills and expertise, education, and/or role are considered to be more significant to p than others). To elaborate, a group of friends who frequently participate in p's discussion by adding many posts are considered to be significant to p. Moreover, their postings are considered to be more significant if they are "ranked" as experts.

Note that, although we use LinkedIn in the above example, similar observations on the desire of mining significant groups of friends/connections/followers can be made on other social networking sites (e.g., Facebook, Twitter). So, a natural question to ask is how to find these significant friend groups, especially when a user has hundreds of friends/connections in his social network and there are many users in the network? Manually go through the entire friend list seems to be impractical. A more algorithmic approach is needed. In this paper, we propose an algorithm to help users to mine social networks for their significant friend groups. **Key contributions** of this paper are our proposal of a tree structure called **S**ignificant **F**riend-**tree** (SF-tree) to capture important information about friends/connections in social networks and our design of an efficient algorithm to mine significant friend groups from the SF-tree.

This paper is organized as follows. Section 2 states the problem definition. Section 3 describes our algorithm for constructing an SF-tree and mining those significant friend groups from the SF-tree. Experimental results in Section 4 show the effectiveness of our algorithm. Section 5 presents the conclusions.

2   Problem Definition

Before we define the problem of mining social networks for significant friend groups, let us consider Table 1 (which shows an illustrative friend database F_{DB} capturing social information about 10 users in a social network) and Table 2 (which shows the pre-computed confidence values of these 10 users). $F_{DB} = \{L_1, ..., L_7\}$ in Table 1 consists of seven *friend lists*, each lists friends of a user. For example, L_2 lists four friends/connections of Gail—namely, Amy, Don, Ed, and Jeff. Each friend f_i in the friend list $L_{id(p)}$ of a user p is associated with a **weight** $wt(f_i, L_{id(p)})$ that indicates the strength/association/measure value of f_i from p's point of view. For example, "Don(20)" on L_2 indicates that Don added 20 posts to Gail's discussion. Note that the weight is asymmetric (e.g., the weight of Don on Gail's list is different from the weight of Gail on Don's list: $wt(\text{Don},L_2)=20 \neq wt(\text{Gail},L_5)=10$) and can vary from one list to another (e.g., the weight of Don on Gail's and Carl's lists are different: $wt(\text{Don},L_2)=20 \neq wt(\text{Don},L_3)=30$). Then, let $G = \{f_1, f_2, \ldots, f_k\}$ be a group of k common friends (i.e., *friend group*). For example, {Amy, Don} is a (common) friend group of Gail, Carl, and Helen. Let F_{DB}^G denote the set of lists in F_{DB} that contain group G. For example, if $G=\{\text{Amy, Don}\}$, then $F_{DB}^G=\{L_2, L_3, L_4\}$.

Table 1. A sample friend database F_{DB}

Friend list L_{id} with weight $wt(f_i, L_{id})$	$limp(L_{id})$ with $sig=0.2$	
$L_1 \equiv$ Jeff:{Don(30), Gail(40), Ivy(60)}	61	49
$L_2 \equiv$ Gail:{Amy(20), Don(20), Ed(10), Jeff(50)}	56	56
$L_3 \equiv$ Carl:{Amy(10), Bob(10), Don(30), Ed(10), Jeff(20)}	52	44
$L_4 \equiv$ Helen:{Amy(50), Don(30), Ed(30)}	68	68
$L_5 \equiv$ Don:{Amy(20), Fred(10), Gail(10), Jeff(20)}	33	25
$L_6 \equiv$ Ed:{Amy(10), Fred(10), Helen(30)}	30	4
$L_7 \equiv$ Amy:{Carl(20), Don(20), Gail(30)}	45	35

Table 2. Confidence table

Friend f_i	Confidence $conf(f_i)$	Friend f_i	Confidence $conf(f_i)$
Amy	0.40	Fred	0.80
Bob	0.80	Gail	0.70
Carl	0.50	Helen	0.60
Don	0.70	Ivy	0.20
Ed	0.90	Jeff	0.50

While the weight $wt(f_i, L_{id(p)})$ indicates the strength/association/measure value of f_i with respect to a user p, the **confidence value** $conf(f_i)$ shown in Table 2 indicates the "rank" of user f_i (based on his experience, skills and expertise, education, role, importance, reputation, prominence) in social networks. For example, opinions posted by an expert Don (with $conf(\text{Don})=0.70$) are considered to be more important than those posted by a non-expert Ivy (with $conf(\text{Ivy})=0.20$).

Then, the problem of *mining significant friend groups* is to discover from friend database F_{DB} all groups of friends with high significance value (e.g., higher than or equal to a user-specified minimum significance threshold *minSig*). In the following, we define the significance of friend groups in a step-by-step fashion.

Definition 1. *The **importance of a friend** f_i in a friend list L_j—denoted as $\boldsymbol{fimp(f_i, L_j)}$—measures the significance of f_i in L_j and is calculated by* $fimp(f_i, L_j) = wt(f_i, L_j) \times con(f_i)$.

Example 1. The importance of Don in L_2 shown in Table 1 is $fimp(Don, L_2)$ $= 20 \times 0.70 = 14$, which reflects the number of posts Don made on Gail's discussion (20) and Don's individual "rank" (0.70). □

To a further extent, the list importance of a *group* G of friends in a friend list L_j can be computed by summing the $fimp(f_i, L_j)$ values for every $f_i \in G$. For example, the list importance value of $G=\{\text{Amy, Don}\}$ in L_2 is (20×0.40) $+ (20 \times 0.70) = 8+14 = 22$, which indicates the total importance of the group (consisting of both Amy and Don) in Gail's friend list L_2.

Definition 2. *The **importance of a friend group** **G** in friend database $\boldsymbol{F_{DB}}$ is defined as $\boldsymbol{dgimp(G)} = \sum_{L_j \in F_{DB}^G} \sum_{f_i \in G} fimp(f_i, L_j)$.*

Example 2. Recall that, if $G=\{\text{Amy, Don}\}$, then $F_{DB}^G=\{L_2, L_3, L_4\}$ indicating that Amy and Don appear as (common) friends of Gail (L_2), Carl (L_3), and

Helen (L_4). And, $dgimp(G) = 22 + [(10\times0.40) + (30\times0.70)] + [(50\times0.40) + (30\times0.70)] = 22 + 25 + 41 = 88$.   □

Definition 3. *The **importance of a list L_j**—denoted as **$limp(L_j)$**—is the total importance of all friends in a friend list L_j and is defined as $limp(L_j) = \sum_{f_i \in L_j} fimp(f_i, L_j)$.*

Example 3. The importance of list L_2 in Table 1 is $limp(L_2) = 8+14+(10\times0.90) +(50\times0.50) = 56$ as indicated in the second column of Table 1. This value indicates that the overall combined importance of all friends of Gail (L_2).   □

Definition 4. *The **importance of a friend database F_{DB}** is the total importance of all friend lists in F_{DB}, i.e., **$dimp(F_{DB})$** $= \sum_{L_j \in F_{DB}} limp(L_j)$.*

Example 4. Recall from Example 3 that $limp(L_2)=56$. By computing the *limp* values for the remaining six friend lists in F_{DB}, we get $61+56+52+68+33+30 +45 = 345$ (i.e., the sum of the second column in Table 1) as the total importance of all seven friend lists in F_{DB}.   □

Definition 5. *The **significance** of a friend group G—denoted by **$sig(G)$**—is defined as the ratio of the importance of G in F_{DB} to the importance of F_{DB}, i.e., $sig(G) = \frac{dgimp(G)}{dimp(F_{DB})}$.*

Example 5. Recall from Example 2 that $dgimp(\{\text{Amy, Don}\})=88$, and recall from Example 4 that $dimp(F_{DB})=345$. Then, the significance of $\{\text{Amy, Don}\}$ in Table 1 is $sig(\{\text{Amy, Don}\}) = \frac{88}{345} \approx 0.26$.   □

A friend group is *significant* in a social network media if its significance value is no less than the user-specified minimum significance threshold (denoted as *minSig*). If *minSig*=0.20 (which represent 20% of database importance $dimp(F_{DB})$), then $G=\{\text{Amy, Don}\}$ is significant because its significance value $sig(G)\approx0.26 \geq 0.20=minSig$.

Example 6. Recall from Example 5 that $\{\text{Amy, Don}\}$ is a significant friend group because $sig(\{\text{Amy, Don}\})\approx0.26$. Let us consider friend groups $\{\text{Don}\}$ and $\{\text{Amy}\}$. For $\{\text{Don}\}$, its significance value $sig(\{\text{Don}\})= \frac{21+14+21+21+14}{345} = \frac{91}{345} \approx 0.26 \geq 0.20=minSig$. However, $sig(\{\text{Amy}\}) = \frac{8+4+20+8+4}{345} = \frac{44}{345} \approx 0.13 < 0.20=minSig$. In other words, although $\{\text{Amy, Don}\}$ is a significant friend group, its subset $\{\text{Amy}\}$ is *not* a significant friend group.   □

As observed from Example 6, the significance of a friend group does not satisfy the *downward closure property* (cf. support or frequency [1], which is downward closed). This leads to a challenging problem when mining significant friend groups from social networks. To address the issue, we define the following.

Definition 6. *The **containing list importance** of a friend group G in F_{DB}— denoted as **$climp(G)$**—is the sum of importance of all lists in F_{DB} that contain G, i.e., $climp(G) = dimp(F_{DB}^G) = \sum_{G \subseteq L_j \in F_{DB}^G} limp(L_j)$.*

Example 7. Recall that, if $G=\{\text{Amy, Don}\}$, then $F_{DB}^G=\{L_2, L_3, L_4\}$ indicating that Amy and Don appear as (common) friends of Gail (L_2), Carl (L_3), and Helen (L_4). Then, recall from Table 1 that the *limp* values for L_2, L_3 and L_4 are 56, 52 and 68, respectively. Thus, $climp(G) = 56+52+68 = 176$.

Similarly, if $G'=\{\text{Amy}\}$, then $F_{DB}^{G'}=\{L_2, L_3, L_4, L_5, L_6\}$ indicating that Amy appears as a (common) friend of Gail (L_2), Carl (L_3), Helen (L_4), Don (L_5), and Ed (L_6). Thus, $climp(G') > climp(G)$ for $G=\{\text{Amy, Don}\}$. This can be easily verified that $climp(G')=176+33+30=239$ as $limp(L_5)=33$ and $limp(L_6)=30$ in Table 1.   □

Based on above definition, we observed that $climp(G) \le climp(G')$ where $G' \subseteq G$. Hence, if G is a significant group, then G' must be a significant group. In other words, if G' is *not* a significant group, then G cannot be a significant group. Hence, we can maintain the downward closure property when significant friend groups are mined based on $climp(G)$ instead of $dgimp(G)$. On the one hand, an advantage of computing significance based on $climp(G)$ is that we can take advantage of the downward closure property. On the other hand, a drawback is that we may generate some false positives, which can be removed with an additional post-processing scan of F_{DB}.

3   Construction and Mining of SF-Trees

In this section, we first propose a prefix tree based data structure to efficiently capture the database content, and we then design a corresponding pattern-growth based mining algorithm to discover significant friend groups from social network database.

Our proposed tree structure is called **S**ignificant **F**riend **T**ree (SF-tree). It is compact and easy to build. Each node in an SF-tree consists of (i) an item, and (ii) *climp* (i.e., the sum of *limp* values of all lists that pass through or end at the node). In addition, the last node of a list maintains the user information (i.e., p). Note that, for any list, the information we maintain in all nodes of its path is always the same (except the last node, which keeps the user information as well), which makes the SF-tree compact. The key steps for the SF-tree construction algorithm are presented below.

3.1   SF-Tree Construction

The SF-tree construction algorithm starts by scanning both the social network database F_{DB} and the confidence table once to capture the basic information regarding users and their friend lists. With this scan, the algorithm calculates $dimp(F_{DB})$, *climp* and *sig* values of each group with a single friend. Then, it removes friends with low *climp* values and sorts all the remaining friends according to their *climp* values.

To demonstrate the SF-tree construction, let us consider F_{DB} shown in Table 1 and the confidence table shown in Table 2 when mining with $minSig=0.20$. After

Table 3. Significance and *climp* values of friends after the first DB scan

f_i	$sig(f_i)$	$climp(f_i)$	Remove?	f_i	$sig(f_i)$	$climp(f_i)$	Remove?
Amy	0.13	239	No	Fred	0.05	63	Yes
Bob	0.02	52	Yes	Gail	0.16	139	No
Carl	0.03	45	Yes	Helen	0.05	30	Yes
Don	0.26	282	No	Ivy	0.03	61	Yes
Ed	0.13	176	No	Jeff	0.13	141	No

the first scan of F_{DB}, the *sig* and *climp* values of all single friends are calculated (as shown in Table 3). Among the 10 friends, Bob, Carl, Fred, Helen and Ivy are removed from the candidate set due to their low *climp* values. Afterwards, the SF-tree is constructed in descending order of *climp* values. The H-table is then created by arranging the remaining friends in descending order of their *climp* values. Note that even though $sig(\{Amy\})$, $sig(\{Ed\})$, $sig(\{Gail\})$ and $sig(\{Jeff\})$ are all less than *minSig*, we avoid removing them from further consideration. The reason is that their corresponding *climp* values are high, which indicates that they may be significant to other friends.

With the second scan of F_{DB}, an SF-tree is constructed in a similar fashion as the FP-tree [5] by inserting each list of F_{DB}. Before inserting a list into the SF-tree, we remove all *insignificant* friends from the list and adjust its *limp* value to reflect the removal of insignificant friends. Note that friends with low *climp* values would have no influence on the computation of significant friend groups. Hence, removing these friends at this early stage helps reduce the number of false positives in the long run. Let us continue with our example, when *minSig*=0.20, we remove Ivy from L_1 and adjust $limp(L_1)$ from 61 to 49 (i.e., $limp(L_1) - fimp(\text{Ivy}, L_1) = 61 - (60 \times 0.20) = 49$). The adjusted *limp* value for each list is shown in the last column in Table 1.

For each list, the algorithm stores the new *limp* value in the tree. Fig. 1(a) shows contents of the SF-tree after inserting L_1 (for Jeff). Note that the last node in the tree (i.e., "Gail:49") maintains the user information (i.e., p=Jeff) of the list. L_2 in F_{DB} is then inserted with $limp(L_2) = 56$ (ref. Fig. 1(b)). Since L_1 and L_2 share a common prefix (i.e., "Don"), the algorithm increases the *climp* value for nodes in the common prefix (i.e., "Don") from $limp(L_1)$=49 to $limp(L_1)+limp(L_2)$=105 by adding the value of $limp(L_2)$. Nodes in the remaining part of L_2 carry the value of $limp(L_2)$. Since the second list is for Gail,

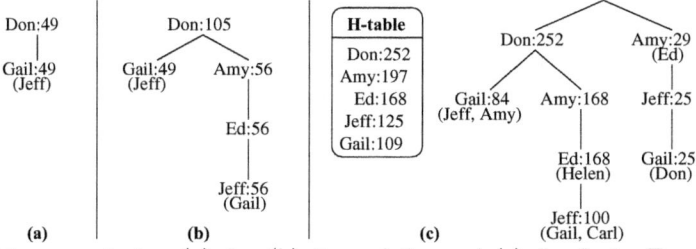

Fig. 1. SF-trees capturing (a) L_1, (b) L_1 and L_2, and (c) L_1–L_7 in F_{DB} in Table 1 when *minSig*=0.20

the last node in the path stores such user information (i.e., "Gail") as shown in Fig. 1(b). Other lists can be inserted in a similar fashion. Fig. 1(c) shows contents of the SF-tree after inserting all seven lists in F_{DB}.

To facilitate a fast tree traversal, in addition to keeping the *climp* value for each friend, the H-table also maintains node pointers to the first occurrence of each friend in the SF-tree. Similar to that of an FP-tree [5], the SF-tree also maintains horizontal node pointers for all nodes having the same friend's name. For simplicity, we do not show these pointers in the figure.

Based on the above description of SF-tree construction, the resulting SF-tree possesses the following important property: The *climp* value in a node x in an SF-tree maintains the sum of *limp* values of all the lists that pass through or end at x for all the nodes in the path from x to the root.

Lemma 1. *Given a friend database F_{DB} and a user-specified minimum significance threshold minSig, the complete set of all significant friend groups can be mined from an SF-tree built when minSig is applied to F_{DB}.*

Proof. An SF-tree keeps a set of significant friends in a list L_j for every list L_j, and the tree stores the accumulated *climp* value for each node. Hence, SF-tree mining based on this *climp* value ensures that no significant friend group will be missed. Moreover, an SF-tree is constructed by considering only the candidate significant friends (based on their *climp* values) in a list. As such, it can be assured that all potentially significant friend groups can be mined from the SF-tree built for a specific *minSig*.   ☐

Based on Lemma 1, we can find all significant friend groups from the constructed SF-tree using a pattern-growth mining algorithm, which will be discussed in the next section.

3.2   SF-Tree Mining

Recall from Section 3.1 that a complete set of significant friend groups can be found with the first scan of F_{DB}. Hence, the SF-tree can be used for finding potentially significant friend groups having number of friends more than one. We follow the usual tree-based [5] pattern mining approach when mining our SF-tree. The basic operations in SF-tree mining are the construction of the projected databases for a potentially significant friend group and the recursive mining of the further potentially significant friend extensions of that group. It does so by examining all the conditional SF-trees consisting of the set of potentially significant friend groups occurring with a suffix group. Hence, the mining proceeds to recursively mine the SF-tree of decreasing size to generate candidate significant friend groups without additional database scan.

To illustrate how to mine the SF-tree, let us revisit our example. Specifically, given the SF-tree in Fig. 1(c) that captures F_{DB} shown in Table 1 with *minSig*=0.20, the SF-tree mining starts with the construction of a projected database for the last friend (i.e., Gail) in the H-table. Such a projected database for Gail is constructed by taking all the branches with suffix Gail as shown in

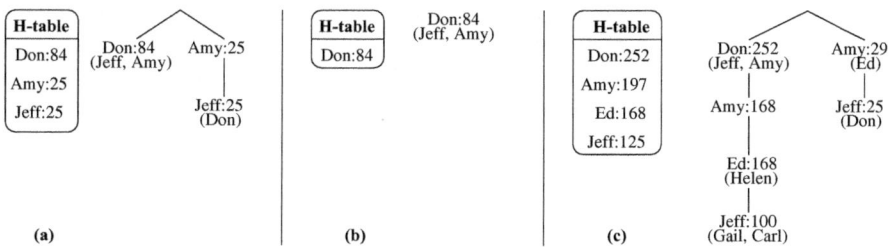

Fig. 2. Applying SF-tree mining to F_{DB} in Table 1 when $minSig$=0.20: (a) Projected DB for {Gail}, (b) Conditional DB for {Gail}, and (c) SF-tree after projecting Gail

Table 4. Significant friend groups

Candidate group:$climp(G)$	Friends of ...	$dgimp(G)$	$sig(G)$	Significant?
{Gail, Don}:84	{Jeff, Amy}	84	0.24	Yes
{Jeff, Ed}:100	{Gail, Carl}	53	0.15	No
{Jeff, Ed, Amy}:100	{Gail, Carl}	65	0.19	No
{Jeff, Ed, Amy, Don}:100	{Gail, Carl}	100	0.29	Yes
{Jeff, Ed, Don}:100	{Gail, Carl}	88	0.26	Yes
{Jeff, Amy}:125	{Gail, Carl, Don}	65	0.19	No
{Jeff, Amy, Don}:100	{Gail, Carl}	82	0.24	Yes
{Jeff, Don}:100	{Gail, Carl}	70	0.20	Yes
{Ed, Amy}: 168	{Gail, Carl, Helen}	77	0.22	Yes
{Ed, Amy, Don}:168	{Gail, Carl, Helen}	133	0.39	Yes
{Ed, Don}:168	{Gail, Carl, Helen}	103	0.30	Yes
{Amy, Don}:168	{Gail, Carl, Helen}	88	0.26	Yes

Fig. 2(a). The table shows the sum of *climp* values of all friends that co-occur with {Gail}. Based on this value for each friend in the SF-tree, we can find the list of friends in the projected database of {Gail} that may generate potentially significant friend group with {Gail}. For example, *climp*(Don) in the projected database of {Gail} is at least *minSig*, while *climp* values for other friends (i.e., Amy and Jeff) are less than *minSig*. Hence, we can safely remove Amy and Jeff from the projected database of {Gail} and construct the conditional database for {Gail}, as shown in Fig. 2(b).

The potentially significant friend groups are generated from the corresponding conditional databases. For example, the set of potentially significant friend groups with {Gail} is generated as {Gail, Don}:84 from the conditional database of {Gail} in Fig. 2(b), where 84 indicates the *climp* value of the group. Along with the potentially significant friend group, we also keep the user information for the group (i.e., {Jeff, Amy}) in the mining result.

After creating the projected database, the original SF-tree is adjusted by pushing the user information at the node up to its parent. For example, the user information of the node "Gail:84" (i.e., Jeff, Amy) and the node "Gail:25" (i.e., Don) are pushed to their respective parent nodes, as shown in Fig. 2(c). Such operation enables us to obtain the correct user information for any node in the SF-tree during the whole mining phase.

Further extension of the potentially significant friend group {Don, Gail} is mined by creating a projected database for {Gail, Don} from the conditional database of {Gail}. In our example, the projected database for {Gail, Don} is empty, which indicates that no further candidate significant friend group will be generated from {Gail}'s conditional database. Hence, mining for {Gail} is terminated. Mining for the remaining friends in the H-table is carried out in a similar fashion. The set of all candidate significant friend groups generated by mining the SF-tree is shown in Table 4. With another database scan, we eliminate all the non-significant friend groups from the set of candidate significant friend groups.

4   Experimental Results

In this section, we evaluated different aspects (e.g., the number of generated candidate groups, runtimes, and scalability) of SF-trees in mining significant friend groups from social media. To the best of our knowledge, SF-tree mining is the first approach to mine significant friend groups from social media databases. However, as the mining of high utility patterns [18] can be considered to be relevant to our mining of significant friend groups, we compared our SF-tree mining with three existing high utility pattern mining algorithms (e.g., Two Phase [11], FUM and DCG+ [9]). All programs were written in Microsoft Visual C++ 6.0 and run with Windows XP operating system on a 2.13 GHz CPU with 2GB main memory. There are some basic differences among our proposed SF-tree mining, Two Phase, FUM, and DCG+. First, the three existing high utility mining algorithms are all Apriori-based [1], i.e., they use the levelwise candidate generation-and-test paradigm. They require N scans of F_{DB} (where N is the maximum size of high utility patterns) and a high computation cost for generating the candidates. In contrast, our SF-tree mining explores the tree-based pattern-growth mining technique, which allows us to mine the complete set of significant friend groups with three scans of F_{DB} without using the levelwise candidate generation-and-test paradigm. Second, unlike our SF-tree, the high utility mining algorithms do not maintain the user information in their respective data structure, which restricts them from providing the knowledge of association between significant friend groups and others.

Since the three high utility pattern mining algorithms were not designed for social network mining, we used the datasets that are mostly used in frequent pattern mining domain for fair comparison. In these datasets that are available at the Frequent Itemset Mining Implementation dataset repository (http://fimi.cs.helsinki.fi/data), each transaction consists of a unique transaction ID and a set of items (which is similar to a list of friends described

Table 5. Characteristics of datasets

Dataset	Type	#transactions	#items	Max trans. len.	Avg trans. len.
Mushroom	Dense	8124	119	23	23
Retail	Sparse	88162	16470	76	10.31
Kosarak	Sparse	990002	41270	2498	8.1

in this paper). Table 5 shows the characteristics (e.g., dense vs. sparse, large vs. small dataset, long vs. short transactions) of these datasets. For example, *Mushroom* is a dense dataset with many long frequent patterns. We use this dataset to demonstrate the type of social media data with fewer number of people but long/large groups in each person's list. Both *Retail* and *Kosarak* are sparse and very large datasets (in terms of number of transactions and domain items) with a combination of long and short transactions. Hence, they may correspond to scenarios when a very large number of individuals use a social network and/or when the connectivity between a user and their friends in a social network vary a lot. For all these datasets, we mapped transaction IDs into user information, sets of items into friend groups in every list, and transactions into friend lists. In addition, to represent $wt(f_i, L_j)$, we associated a random number to each item in a transaction. We also generated random numbers as the confidence values of friends in F_{DB}. As ongoing work, we plan to conduct additional experiments using social network data.

4.1   Candidate Significant Friend Group Generation

We compared the number of candidate significant friend groups (i.e., false positives) generated by our SF-tree mining with the three existing algorithms over different datasets by varying *minSig* values. As shown in Fig. 3, the number of candidates increased when lowering the value of *minSig* for all datasets and algorithms. However, SF-tree outperformed the other three algorithms as SF-tree generated significantly fewer candidates. The reason was that, for dense datasets (e.g., *Mushroom*), there is a very high probability for each friend to occur in every list, which increases the number of potential groups when applying the candidate generation-and-test paradigm. Hence, the three high utility mining algorithms generated comparatively large number of candidates. Conversely, sparse datasets (e.g., *Retail*) contain too many individual friends in many transactions, which further increases the number of candidate groups for the three high utility mining algorithms. In contrast, as SF-tree uses tree-based pattern-growth mining, it avoids generating candidates in dense or sparse datasets.

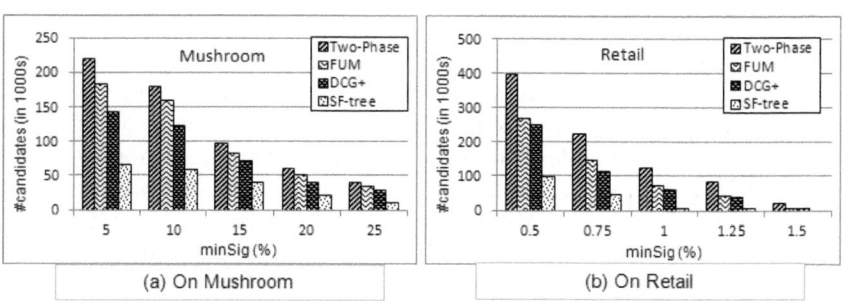

Fig. 3. The number of candidates (i.e, false positives)

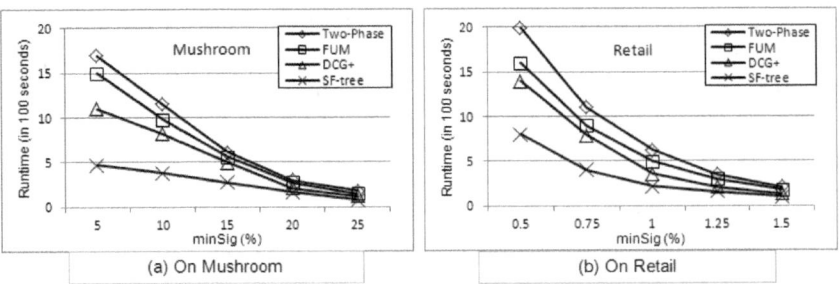

Fig. 4. Runtime

4.2   Runtimes

We measured runtimes of SF-tree and the three high utility mining algorithms. The reported runtimes for SF-tree include times for tree construction (i.e., two scans of F_{DB}), mining, and candidate pruning (i.e., an additional scan of F_{DB}); the reported runtimes for the other three algorithms include times for multiple database scans for candidate generation-and-test.

The number of generated candidates (which increased with the decrease of $minSig$) directly influences the runtimes for all algorithms. The more the generated candidates, the longer were the runtimes. As shown in Fig. 4, SF-tree performed better than the other three algorithms for all datasets. For example, the three algorithms generated very large numbers of candidates for *Mushroom*. To handle a large number of friends and friend lists in *Retail*, the three algorithms required substantially longer runtimes to scan F_{DB} when compared to SF-tree (which required three scans of F_{DB}). Although the performance of all algorithms were similar for higher $minSig$ values (due to a very small number of generated candidate friend groups), the gap was widened for lower $minSig$ values.

4.3   Scalability Test

To test the scalability of our SF-tree mining, we used very large datasets such as *Kosarak*. Fig. 5 shows results on scalability tests of all four algorithms. Due to substantially long runtime required to handle *Kosarak* for some low thresholds, we did not obtain reportable results for DCG+. Moreover, although both Two Phase and FUM completed the mining process for high $minSig$, they suffered from some problem for low $minSig$. Hence, we did not plot their runtimes for low thresholds. In contrast, our SF-tree mining completed the mining process and showed linear scalability with high and low $minSig$ thresholds. SF-tree generated significantly fewer candidate friend groups within very reasonable runtimes due to early pruning of non-significant friends during the tree construction and mining process. This demonstrated that SF-tree mining is scalable.

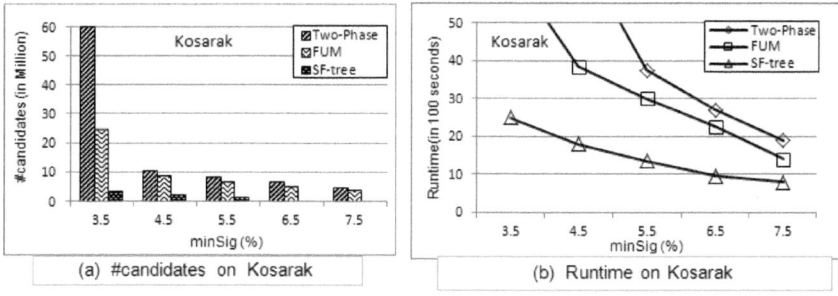

Fig. 5. Scalability test

5 Conclusions

In this paper, we proposed a novel algorithm to mine social networks for significant friend groups. Our social network mining algorithm first constructs a significant friend-tree (SF-tree) to capture important information about linkage between users in the social networks, and it then uses the SF-tree to find significant friend groups among all friends of users in the social networks.

Acknowledgement. This project is partially supported by NSERC (Canada) and University of Manitoba.

References

1. Agrawal, R., Srikant, R.: Fast algorithms for mining association rules in large databases. In: VLDB 1994, pp. 487–499 (1994)
2. Cameron, J.J., Leung, C.K.-S., Tanbeer, S.K.: Finding strong groups of friends among friends in social networks. In: IEEE DASC (SCA) 2011, pp. 824–831 (2011)
3. Carrington, P.J., Scott, J., Wasserman, S. (eds.): Models and Methods in Social Network Analysis. Cambridge University Press (2005)
4. Fan, W., Yeung, K.H.: Virus propagation modeling in Facebook. In: ASONAM 2010, pp. 331–335 (2010)
5. Han, J., Pei, J., Yin, Y.: Mining frequent patterns without candidate generation. In: ACM SIGMOD 2000, pp. 1–12 (2000)
6. Lee, W., Lee, J.J.-H., Song, J.J., Eom, C.S.-H.: Maximum reliable tree for social network search. In: IEEE DASC (CSN) 2011, pp. 1243–1249 (2011)
7. Leung, C.K.-S., Carmichael, C.L.: Exploring social networks: a frequent pattern visualization approach. In: IEEE SocialCom 2010, pp. 419–424 (2010)
8. Leung, C.K.-S., Carmichael, C.L., Teh, E.W.: Visual Analytics of Social Networks: Mining and Visualizing Co-authorship Networks. In: Schmorrow, D.D., Fidopiastis, C.M. (eds.) FAC 2011, HCII 2011. LNCS (LNAI), vol. 6780, pp. 335–345. Springer, Heidelberg (2011)
9. Li, Y.-C., Yeh, J.-S., Chang, C.-C.: Isolated items discarding strategy for discovering high utility itemsets. DKE 64(1), 198–217 (2008)
10. Liu, H., Yu, P.S., Agarwal, N., Suel, T.: Guest editors' introduction: social computing in the blogosphere. IEEE Internet Computing 14(2), 12–14 (2010)

11. Liu, Y., Liao, W.-k., Choudhary, A.K.: A Two-Phase Algorithm for Fast Discovery of High Utility Itemsets. In: Ho, T.-B., Cheung, D., Liu, H. (eds.) PAKDD 2005. LNCS (LNAI), vol. 3518, pp. 689–695. Springer, Heidelberg (2005)
12. Obradovic, D., Pimenta, F., Dengel, A.: Mining shared social media links to support clustering of blog articles. In: CASoN 2011, pp. 181–184 (2011)
13. Pennacchiotti, M., Popescu, A.-M.: Democrats, republicans and starbucks afficionados: user classification in Twitter. In: ACM KDD 2011, pp. 430–438 (2011)
14. Tang, C., Ross, K., Saxena, N., Chen, R.: What's in a Name: A Study of Names, Gender Inference, and Gender Behavior in Facebook. In: Xu, J., Yu, G., Zhou, S., Unland, R. (eds.) DASFAA Workshops 2011. LNCS, vol. 6637, pp. 344–356. Springer, Heidelberg (2011)
15. Wasserman, S., Faust, K.: Social Network Analysis: Methods and Applications. Cambridge University Press (1994)
16. Weng, J., Lim, E.-P., Jiang, J., He, Q.: TwitterRank: finding topic-sensitive influential twitterers. In: ACM WSDM 2010, pp. 261–270 (2010)
17. Xu, G., Zong, Y., Pan, R., Dolog, P., Jin, P.: On Kernel Information Propagation for Tag Clustering in Social Annotation Systems. In: König, A., Dengel, A., Hinkelmann, K., Kise, K., Howlett, R.J., Jain, L.C. (eds.) KES 2011, Part II. LNCS (LNAI), vol. 6882, pp. 505–514. Springer, Heidelberg (2011)
18. Yao, H., Hamilton, H.J., Butz, C.J.: A foundational approach to mining itemset utilities from databases. In: SDM 2004, pp. 482–486 (2004)
19. Ye, S., Wu, S.F.: Measuring Message Propagation and Social Influence on Twitter.com. In: Bolc, L., Makowski, M., Wierzbicki, A. (eds.) SocInfo 2010. LNCS, vol. 6430, pp. 216–231. Springer, Heidelberg (2010)
20. Yumoto, T., Sumiya, K.: Measuring Attention Intensity to Web Pages Based on Specificity of Social Tags. In: Yoshikawa, M., Meng, X., Yumoto, T., Ma, Q., Sun, L., Watanabe, C. (eds.) DASFAA 2010. LNCS, vol. 6193, pp. 264–273. Springer, Heidelberg (2010)

Ranking Structural Parameters
for Social Networks

Nidhi R. Arora, Wookey Lee, and Simon Soon-Hyoung Park

INHA University, Incheon, South Korea
{trinity,dhini27,fgm0626}@inha.ac.kr

Abstract. The emergence of various social networks on the web have transformed the way people share, and search for, information. A key distinguishing feature of social networks is that the link between the users are more important than the users themselves. There are numerous studies on the link analysis of Social Network for efficient and effective retrieval of user desired information. In this paper,we study the link analysis of social networks to rank individual links so that we have established a bridge between keyword search based from the information retrieval and the structurs from the social networks successfully.

Keywords: ranking, social networks, graph.

1   Introduction

The term social networks implies networks that represent inter-connection between users based on their common interests, friendships, and sharing relationship. Thus,social networks provide a platform for users to establish relations with other users on the web the mobile devices, and SNS , based on their mutual interaction like online messaging or other forms of communication, sharing common interests , events or activities. For example, Youtube [27] is a well known social networking site for on-line video sharing; Flickr [7] is popular for photo sharing and Facebook/Orkut[20] for finding friends and establishing social relationship. A recent study revealed that social networks produce up to one-third of the current web content and that is increasing tremendously.

The study of various characteristics of social networks, called social networking analysis, has received considerable attention in research and development recently. Social networking analysis is achieved by using network theory or graph theory, similar to Web as graph approach. In a social network, nodes represent individual users or events and links represent relationship like friendship, kinship or common interests, like-dislike, comments, feedback etc. In contrast to Web as a graph, there are various kinds of nodes and numerous types of relation exist between the nodes.

Some of the extensively studied characteristics of social networks are discovery of potential friends, centrality and cliques, recommendation systems, detecting influential bloggers, and derive trust relationship. Along with these features, another important feature of social networks is their rich content, i.e., large amount

H. Yu et al. (Eds.): DASFAA Workshops 2012, LNCS 7240, pp. 193–203, 2012.

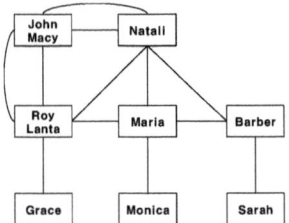

Fig. 1. Example social network described

of structured content present in various nodes connected by links. However, the social network search to pinpoint networked information has not been explored fully. In this paper, we aim to apply the social network analysis in the context of information retrieval systems.

A particularly important aspect of merging IR to social networks is to decide for the statistical relevance of the, content and, trust or relationship. In contrast to using a search engine, such as Google, people prefer to pursue their friends for suggestions or recommendations for relevant information to their queries. But certain similarities with Web search like huge document collection (large number of users, in case of social networks) and link connectivity enables us to adopt web based search algorithms to social information retrieval. In this paper, we would like to extend the Cohesive Arc Measure [19] for enumerating content and link relevance in response to user search. As mentioned before, social network is modeled as directed graph, $G=(V,E)$; V represents a set of users and E represents association between the users in form of e-mails or on-line messages. Keyword query, information need in the form of set of keywords, is used to model the search requirement. The answer to user query is the node (user), whose message contains the desired information. We create a feature vector, corresponding to the query keywords, for each of the Network nodes. Hence, the size of the feature vector is same dimension as the user query and the values contained in it describe the relevance of the keywords. In case, a Network node contains all the queried keywords, system returns Network nodes with maximum relevance, else system finds a pair of nodes that contains all the queried keywords.

In traditional information retrieval models, documents are taken as the re-trieval units and the content of a document is considered reliable, based on the relevance of the node. However, this reliability assumption is no longer valid in the context of social networks. Here, the relevance of the content, is not only determined by the relevance of the node but also by the recommendation from other nodes (users). This is due to the underlying assumption of social networks: "links between the nodes is more important than the actual physi-cal nodes themselves". Hence, the ranking models developed over the years, for information retrieval for the networked information are not directly applicable for social networks. In this paper, we propose a ranking measure based on the structural layout of the network, similar to, but distinct from, the well known Page Rank or HITS algorithm from the conventional search systems.

2   Social Network Ranking Functions

In this section, we present the scoring function to evaluate the relevance of a node and/or link for returning relevant information to the user. Keyword search mechanism over Network structured data is extensively studied over the past few decades. The general practice is to assign, each node, a tf-idf based relevance score and then combine the scores using an aggregation function, such as SUM [3]. The problem with this approach is that single node score does not acknowledge the presence of multiple query keywords.

Also, Social Networks refers to a network of collaborative authors; users create content by typing text, or inserting images, or URLS in their messages. Hence, accommodating a single node relevance score is not appropriate for deciding the relevance of a given node. We therefore, assign a feature vector to accommodate individual relevance of each of the query keywords for the respective nodes. There are many content relevance measures available in the IR literature, with tf-idf as the baseline measure. Recently, language model based content relevance measure has also been studied to account for the content relevance [4][13]. But language based models require the probabilistic models for both query and document relevance. Since tf-idf measure forms the baseline measure for majority of IR systems, we prefer the use of tf-idf for creating the feature vector of each node as follows:

$$f_u = (c_u^1, c_u^2, ..., c_u^t) \tag{1}$$

Here, f_u refers to the feature vector for node u and c_u^t denotes the content relevance for query keyword t, $\forall t \in Q$. c_u^t represents the content relevance, for each of the keywords, enumerated using the following equations:

$$c_u^t = tf_{u,t} \times idf_t = \frac{tf_{u,t}}{dl_u} \times \frac{\ln(d_t)}{\ln(N)} \tag{2}$$

The relevance of a node, in the form of feature vector, is also in line with the current trend of information discovery in both IR and database communities. The tf-idf scores are pre-computed and stored off line during the creation of inverted index. Therefore, feature vectors can be generated efficiently at run time. The information required by users, is mostly present in multiple levels, like group of nodes linked to each other. Hence, relevant results, in response to user query, consist of a mix of single nodes and pair of nodes (in the form of sub-graphs or sub-trees). This structured retrieval requires change in the ranking fundamentals adopted by IR literature.

We propose a novel measure to rank result structures, with query keywords contained by multiple linked nodes, in Cohesive Arc Measure[19]. The relevance of a link incident on nodes u and v, $w_{l_{u,v}}$, is enumerated by the following equation:

$$w_{l_{u,v}} = f_u \otimes f_v = \sum_{x=1}^{t} c_u^x \times \left[\sum_{y=1}^{t} c_v^y - c_v^x \right] \tag{3}$$

The edge computations, using the above equation, have the following advantage: "links incident on nodes, that share different query keywords", is assigned higher relevance as compared to links incident on nodes with same set of query keywords. Table 1 presents the edge weight calculations for feature vectors, of size 2, for a random set of feature vectors.

Table 1. Edge weight calculations for feature vectors

fu	fv	arc weight
(0.9, 0.9)	(0.9, 0.9)	0.810
(0.9, 0)	(0, 0.9)	0.486
(0.9, 0.9)	(0, 0.9)	0.486
(0.7, 0)	(0, 0.9)	0.378
(0, 0.9)	(0.9, 0)	0.324
(0.7, 0)	(0, 0.7)	0.294
(0.5, 0)	(0, 0.9)	0.270
(0.9, 0.9)	(0.7, 0)	0.252
(0.5, 0)	(0, 0.7)	0.210

3   Query Processing

An information retrieval process begins with the user asking a query. The query is generally expressed in terms of *set of keywords* which indicate the users search requirements. Though, other forms of query like, structural queries, for example SQL, also exists but they are not as user friendly as keyword queries. Numerous literary works exist for efficiently processing keyword queries over graph structured data [5][10][16][21]. The baseline representation for social networks is also graph structure, we can easily adopt the available query processing algorithms for finding vertices, or set of vertices that match users requirement.

Given the graph G and keyword query Q, we wish to generate answer vertices or graph sub-structures using Network exploration based heuristics. The main focus of our work is to devise a novel ranking function and not a new search algorithm. We therefore use the current best Steiner tree heuristic, i.e., bidirectional expansion strategy using indexing mechanism adopted from BLINKS [15]. We briefly state the search algorithm used to generate the answer trees for our experimental evaluation.

3.1   Social Network Indexing

The process of creating indexes for social networks is included for an essential pre-processing step for keyword query processing. These kinds of indexes help to identify the set of vertices, that match the query keywords, called *content nodes*. For notational simplicity, we represent the set of vertices that match query keywords as V_t, where t denotes one of the query keyword. Since content based relevance is an inherent part of Web search, we also compute the content based relevance scores for finding relevant answers. Similar to the trend of information distribution on Web, we believe that the user desired information might be dispersed among a set of vertices, in case of social networks also.

Hence, for efficient Network exploration at query processing time, we use two kinds of indexes: (1) Keyword Index (KI) and (2) Node-Keyword Index (NKI). Keyword Index is similar to the conventional inverted-index used by search engines. For every index able keyword t, KI(t) returns the set of content nodes with their tf-idf scores, for a keyword t, ordered by their content relevance. Node-Keyword Index (NKI), adapted from BLINKS [15], provides additional connectivity information from a node to each of the index able keyword. For a given node u and keyword t, NKI(u,t) returns a list of vertices, that contains the keyword t, ordered by their distance (d) from u. For efficient query processing, we further arrange the nodes, within same distance, in decreasing order of their content relevance. In contrast to BLINKS, where graph is partitioned into multiple blocks for materializing the index, we only store the list of nodes up to maximum distance 5. The distance based threshold is in line with current approaches in the literature, where answer trees up to a fixed size are generated by the system.

3.2 Social Network Search Algorithm: $SNSA$

Given a graph G and keyword query Q, the social network search algorithm: $SNSA$ first retrieves the set of content nodes using KI index (described above) [line 3]. For each $u \in V$, assign a content based relevance score, u_c using the content relevance scores from the inverted index [line 4]. Additionally, to each u, we also associate a bit vector of length l to accumulate distance based information from a node to all the query keywords [line 5]. Initially this vector contains the entry 0 for the keywords contained by u and infinity otherwise. The cardinality of bit vector helps to determine whether all the query keywords are satisfied by a node. Also, the AND operation among the bit vectors of the incident nodes of a link helps to determine whether the two nodes share same or different query keywords.

Each node, $u \in V_t$, is prioritized as per the content relevance of vertices, in the priority queue (P_Q). An essential requirement of a good search system is to generate all the possible answers. Therefore, to generate all the answer *subgraph*, the algorithm recursively processes all the nodes until the priority queue is empty [line 6]. However, finding all the answers and enumerating their relevance score is an expensive operation not suitable for on-line query processing. We therefore, compute a lower bound for the future answer *subgraph* and output top-k answer *subgraphs* if new answer is not better than previously found answers.

Line 7-8 extracts the highest priority node u from P_Q and examines its status. If it contains all the query keywords, it is added to answer heap ordered by the final rank. Otherwise, it is excuted for identifying structural connectivity with other *keyword* nodes. NKI() returns the best node to be explored for each keyword i not contained by u. Note that in case v is already a member of P_Q, its priority is increased by adding $1/\epsilon$ of p_u to p_v, where $\epsilon = l - m_u$. Otherwise, v is

Algorithm 1. *SNSA*

$Input : Q(t_1, \ldots, t_n),\ top - k$
$Output : top - k\ trees$

 1: $V_t \leftarrow \mathrm{KI}(t),\ \forall t \in Q$
 2: **for** $u \in V_t$ **do**
 3:     Store in P_Q, ordered by their relevance scores
 4: **end for**
 5: **while** $(P_Q \neq \phi)$ **do**
 6:     $u \leftarrow$ highest priority node from P_Q
 7:     **if** u contains all the query keywords **then**
 8:         Compute $score(u)$ and store in Ans
 9:     **else**
10:         **for all** $i \in [1 \ldots l]$ such that $dist_u[i] = \infty$ **do**
11:             $v \leftarrow \mathrm{NKI}(u, t_i)$
12:             Add v to P_Q
13:         **end for**
14:         **if** u contains all the query keywords **then**
15:             Create answer tree T, add to Ans and remove from P_Q
16:         **end if**
17:     **end if**
18:     **if** $|Ans| \geq$ top-k and stopping condition is met **then**
19:         Output top-k answer trees from Ans
20:     **end if**
21: **end while**

---

added to P_Q with assigned priority $1/\epsilon$ of p_u. Though prioritizing vertex selection helps to determine content relevant vertices, the order in which k answers are generated is fairly different from the final $top - k$ result. For a new answer to be present in top-k, the unexplored nodes in P_Q must have the same *query amplitude* as some of the nodes in the answer generated so far. We compute the lowest query amplitude of nodes present in Ans, represented as m_{Ans}, and highest query amplitude of P_Q, represented as m_{P_Q}. If $m_{Ans} \geq m_{P_Q}$, answer *subgraphs* generated so far are presented to the user; otherwise, system generates more *subgraph* till the condition is satisfied.

4   Case Study on Twitter Data

In this section, we present the empirical analysis done on a subset of the real-world social networking site Twitter. We downloaded the tweets of 1,701 users in all. Hence, the total number of vertices, $|V|$=1,710 and total number of edges, $|E|$ is approximately 2 million. However, out of these 2,405,566 edges, there are only 264 unique edges; majority of edges are regular interactions between a group of friends.

The experiments are performed on an Intel Core 2 Duo personal computer with 2.13 GHz processing power and 1 GB of RAM. The query processing and indexing algorithms are implemented in Java programming language. We selected a set of 9 queries as listed in Table 2. The queries are selected based on the popularity and distribution of query keywords. Also, due to popular nature of the twitter website, many users publish some content every few minutes. Even though we associate the text publish by a user to the corresponding node, we consider each individual message as one document while enumerating tf-idf using Eq.(2).

We first present the query processing time for generating top-k answers for the queries listed in Table2. The query processing time includes the time required for generating feature vectors and then producing the top-k result. Fig.2 presents the query processing time elapsed in generating top-k result for queries listed in Table2.

It can easily be observed that for almost all the queries, the query processing time for generating top-k result is within the time frame of upto 100ms. Intuitively, the query processing time increases with increase in the value of k. An interesting observation is that for queries related to celebrities, like Q1=Lindsay Lohan or Q4=American Idol, there is a high degree of overlap in the query processing time. This is because most of the users use these keywords together. Hence, query keywords are not distributed across multiple nodes. Another interesting observation is the peak in the curves for queries, Q3 and Q7. The reason for higher query processing time, for top-20 result, is that while freely discussing with friends, people publish their views/recommendations on related topics, which contain query keywords distributed across multiple nodes.

Table 2. Query set used for evaluating Twitter Data

Query ID	Query Keywords
Q1	Lindsay Lohan
Q2	Power Track
Q3	Winter Festival
Q4	American Idol
Q5	Black Friday Shopping
Q6	Tiger Woods Twitter
Q7	Organic Wheat Bread
Q8	Hurricane Atlantic Ocean
Q9	Youtube Updates iphone

We first present the query processing time for generating top-k answers for the queries listed in Table2. Fig.2 presents the query processing time elapsed in generating top-k result for queries listed in Table2. It can easily be observed that for almost all the queries, the query processing time for generating top-k result is within the time frame of up to 100ms. Unlike currently prevalent IR query processing in Databases, relevant information is not distributed across multiple nodes or nodes with link distance greater than about 3 hops.

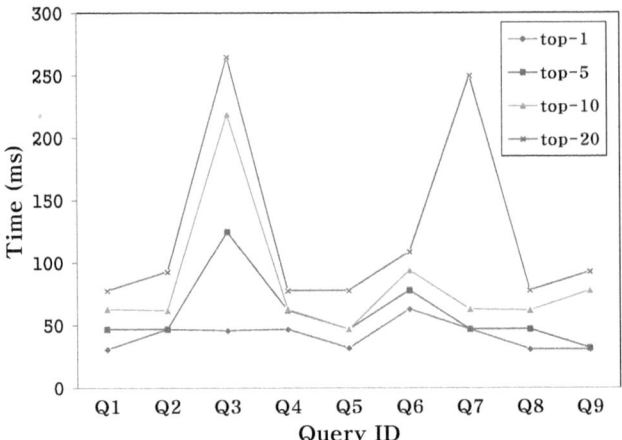

Fig. 2. Query Processing time for generating top-k result

5   Related Works

There have been two kinds of social network ranking measurements: the node based measurement and the arc based measurement. The one is called global measure since they are engaged in the number of arcs, including in-links and out-links, such as PageRank[2], SimRank[18], and HITS[12], etc. PageRank is one of the most popular alternatives in the information retrieval community which assumes that the graph G is a strongly coupled component that is, all other nodes can be reachable from any one node of the directed graph and the method uses a damping factor in order to guarantee the non-negativity, irreducibility and aperiodic properties of nodes weights when computing the weights using the Markov transition technique. The HITS algorithm can be an alternative for a global weight, but none has been implemented as a web search engine. There have been many node measures, called the local measures for objects, such as cosine measure[4][16], probabilistic measure[9], and anchor text[6], hierachical measure[14][15], entity[26], etc. These kinds of measures have similar limitations exposed on content manipulations, called content spamming, so we merged the global and the local measure to devise a bidirectional edge measure, extended from cohesive measure[19].

The structured information of social network can enhance the search effectiveness concerning social network entities highly relevant. For the possible alternatives for the social network search, computation and structuring of every arc usually focus on expressing the whole social network graph [1][14][22] or emphasizing a part of it, named 'small world' or 'topic level social network' [8][23][24]. Nevertheless, overly complicated process of considering the circuit path problem which causes inefficiency of the social network search[16][23]. Hence, unlike typical network analysis, it is enormously significant in social network search to generate an efficient graph and for its structure search[11][14]. The studies related to social network are primarily comprised of building social network site for mobile devices, including subjects such as system of user interface design of social network navigation[14][25]. However, not many social network searches which reflect the characteristics of mobile devices currently exist.

However, recently they have become almost insignificant as full browsing emerged as the prevalent trend. Coupled with the introduction of Apple's iPad or Samsung's GalaxyTab, researches regarding the applications adoptable to social network[27][20][26] are beginning to be conducted. Moreover, studies centered on the virtuous cycle of market's food chain related to these applications, i.e., the Digital Ecosystem[14] are projected to be largely feasible in the future. In order to overcome the limit of conventional searches, studies based on graphs are extensively initiated these days. To be specific, exemplary researches (e.g., Information Unit[17], EASE[16], BLINKS[10], STAR[11]) delve into these Network oriented topics. In Information Unit[17], while the group Steiner tree method was applied, regarding the candidate sets of obtained solutions, MST(Minimum Spanning Tree) was adopted. Extending from a similar research subject, in EASE[16], the complexity of problem was simplified by replacing the method with r-radius Steiner graph, and searching speed was hugely accelerated than in Information Unit applying a certain "graph index". However, since this method is based on matrix operation, its disadvantage is that excessive time consumption results from in case the adjacency matrix generally grows. In addition, in case of BLINKS[10], the method was applied after reducing the search storage space using clusters of keyword set. On the other hand, in STAR[11], the technique of finding the Steiner graph from Wikipedia, and shortening the tree's height connecting leaf nodes with keywords was used. In these studies, assuming the graph is undirected, certainly social network environment was not put into consideration.

6   Conclusion

Social networks provide a platform for users to share and upload various kinds of information content on-line. Through the various kinds of relationship that exists between users, people connect with one another much more easily and with trust. Hence, in order to search for a new piece of information, people prefer to take recommendations from their friends or socially connected other users. In this work, we have tried to bridge the gap between keyword search based Information Retrieval and Social networks due to our novel arc measurement. An interesting observation in the experiment is the peaks in the curves for queries that while freely discussing with friends, people used to publish their views and opinions on the related topics, which contain query keywords distributed across multiple nodes.

Acknowledgments. This research is partially supported by Basic Science Research Program through the NRF of Korea funded by the MEST (2011-0026441).

References

1. Adamic, L., Adar, E.: How to search a social network. Social Networks 27(3), 187–203 (2005)
2. Bressan, M., Pretto, L.: Local computation of PageRank: the ranking side. In: CIKM, pp. 631–640 (2011)
3. Cohen, S., Mamou, J., Kanza, Y., Sagiv, Y.: XSEarch: a semantic search engine for XML. In: VLDB (2003)
4. Croft, W.B.: Language models for information retrieval. In: International Conference on Data Engineering (2003)
5. Dalvi, B.B., Kshirsagar, M., Sudarshan, S.: Keyword search on external memory data graphs. In: VLDB, pp. 1189–1204 (2008)
6. Dou, Z., et al.: Using anchor texts with their hyperlink structure for web search. In: Proc. SIGIR, pp. 227–234 (2009)
7. Flickr, http://www.flickr.com
8. Granovetter, M.: SOCIAL SCIENCE Ignorance, Knowledge, and Outcomes in a Small World. Science 301(5634), 773–774 (2003)
9. Hammami, M., Chahir, Y., Chen, L.: WebGuard: A Web Filtering Engine Combining Textual, Structural, and Visual Content-Based Analysis. IEEE TKDE 18(2), 272–284 (2006)
10. He, H., Wang, H., Yang, J., Yu, P.S.: BLINKS: ranked keyword searches on graphs. In: SIGMOD, pp. 305–316 (2007)
11. Kasneci, G., Ramanath, M., Sozio, M., Suchanek, F.M., Weikum, G.: STAR Steiner-Tree Approximation in Relationship Graphs. In: ICDE, pp. 868–879 (2009)
12. Kleinberg, J.M., Tardos, E.: Balanced outcomes in social exchange networks. In: Proc. STOC, pp. 295–304 (2008)
13. Lavrenko, V., Croft, W.B.: Relevance based language models. In: Proceedings of the 24th Annual International ACM SIGIR Conference on Research and Development in Information Retrieval (2001)
14. Lee, W., Leung, C.S., Lee, J.J.: Mobile Web Navigation in Digital Ecosystems Using Rooted Directed Trees. IEEE Transactions on Industrial Electronics 58(6), 2154–2162 (2011)
15. Lee, W., Song, J.J., Leung, C.K.-S.: Categorical Data Skyline Using Classification Tree. In: Du, X., Fan, W., Wang, J., Peng, Z., Sharaf, M.A. (eds.) APWeb 2011. LNCS, vol. 6612, pp. 181–187. Springer, Heidelberg (2011)
16. Li, G., Ooi, B.C., Feng, J., Wang, J., Zhou, L.: EASE an effective 3-in-1 keyword search method for unstructured, semi-structured and structured data. In: SIGMOD, pp. 903–914 (2008)
17. Li, W., Candan, K.S., Vu, Q., Agrawal, D.: Retrieving and organizing web pages by "information unit". In: Proc. WWW, pp. 230–244 (2001)
18. Lizorkin, D., Velikhov, P., Grinev, M., Turdakov, D.: Accuracy estimate and optimization techniques for SimRank computation. VLDB Journal 19(I.1), 45–66 (2010)
19. Nidhi, A., Lee, W., Kim, Y.: Cohesive Arc Measures for Web navigation. In: International Conference on Contemporary Computing (2008)
20. Orkut, http://www.orkut.com
21. Qin, L., Yu, J.X., Chang, L., Tao, Y.: Querying Communities in Relational Databases. In: ICDE, pp. 724–735 (2009)
22. Raz, O., Gloor, P.A.: Size Really Matters–New Insights for Start-ups' Survival. Management Science, 169–177 (February 2007)

23. Sun, B., Mitra, P., Giles, C.L.: Independent informative subgraph mining for graph information retrieval. In: Proc. CIKM, pp. 563–572 (2009)
24. Tang, J., Wu, S., Gao, B., Wan, Y.: Topic-level social network search. In: Proc. KDD, pp. 769–772 (2011)
25. Wang, P., González, M.C., Hidalgo, R.C., Barabási, A.: Understanding the spreading patterns of mobile phoneviruses. Science, 1071–1076 (2009)
26. You, G., et al.: SocialSearch: enhancing entity search with social network matching. In: Proc. EDBT, pp. 515–519 (2011)
27. Youtube, http://www.youtube.com

Collaborative Similarity Measure
for Intra Graph Clustering

Waqas Nawaz, Young-Koo Lee, and Sungyoung Lee

Department of Computer Engineering, Kyung Hee University, Korea
{wicky786,yklee}@khu.ac.kr, sylee@oslab.khu.ac.kr

Abstract. Assorted networks have transpired for analysis and visualization, including social community network, biological network, sensor network and many other information networks. Prior approaches either focus on the topological structure or attribute likeness for graph clustering. A few recent methods constituting both aspects however cannot be scalable with elevated time complexity. In this paper, we have developed an intra-graph clustering strategy using collaborative similarity measure (IGC-CSM) which is comparatively scalable to medium scale graphs. In this approach, first the relationship intensity among vertices is calculated and then forms the clusters using k-Medoid framework. Empirical analysis is based on density and entropy, which depicts the efficiency of IGC-CSM algorithm without compromising on the quality of the clusters.

Keywords: Graph clustering, collaborative similarity, k-Medoid clustering, entropy, density, Jaccard similarity coefficient.

1   Introduction

Graph as an expressive data structure is most widely used to model real life objects and their relationships in many application domains like social network, web, sensor network, and telecommunication. Graph clustering is very challenging and interesting research areas. Numerous researchers have focused different aspects of it, discussed in [1], [2], [3], and [4].

The key objective of most of the graph clustering algorithms is to identify strongly connected vertices within a graph with similar neighborhood. The vertices inside the sub-graph are highly cohesive while sparsely connected with other vertices of the graph. All the un-supervised clustering algorithms [5], [6], have the ability to find out natural number of clusters within the whole graph, whereas semi-supervised techniques [7] require extra information regarding the clusters such as number of clusters as an input parameter. We have used term node and vertex interchangeably throughout this paper.

Many researchers utilized the relational context of a network to discern interesting groups of entities, generally known as clusters [1]. In relational aspect, the connectivity among vertices or with similar neighborhood, harmonized topological structure, and characteristic resemblance are playing key role in graph clustering.

H. Yu et al. (Eds.): DASFAA Workshops 2012, LNCS 7240, pp. 204–215, 2012.

Classical applications of graph clustering include community detection in social networks, identification of functional related protein modules in large biological network (protein-protein interaction networks), etc. Numerous prior approaches either focus on the topological structure; attribute likeness or both for graph clustering [10], [11], and [14]. However they compromise either on quality or time.

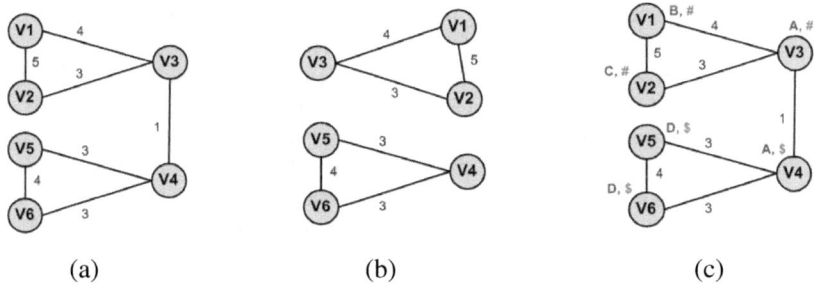

(a)   (b)   (c)

Fig. 1. Graph Vertex Structures (a) connected, weighted and contextual graph, weighted graph (b) disconnected, weighted graph (c) connected, weighted, and multi-labeled

In this paper, we introduce the algorithm which belongs to a graph vertex clustering, an un-supervised method, whereas it requires the number of clusters as an input parameter prior to cluster the graph. It also provides an opportunity to control the connectivity or similarity strength among the clustered nodes. The deviation from existing state-of-the-art approaches in terms of the following facets: (a) a new two way node to node similarity measure is utilized which is more powerful because it considers both structural and contextual aspects (b) iterative clustering is applied which has low time complexity without compromising on cluster's quality, (c) also considers three basic scenarios for each vertex in graph as shown in Fig. 1. Key contributions of this work are as follows:

— A new graph vertex clustering strategy is proposed which is simple in nature. It can capture both structural and contextual similarities among nodes in the graph simultaneously.
— Similarity among the vertices is calculated once, which is symmetric and utilized successively.
— A customized k-Medoid framework is adopted for clustering.
— Disconnected graph nodes can be clustered together based on collaborative similarity measure.
— It is easily scalable for small and medium scale graphs because of its simplicity.

The rest of the paper is organized as follows. Section 2 briefly explains the related work. The proposed idea is elaborated in section 3. Subsequently its empirical analysis is carried out and then conclusion and future direction are drawn in section 5 followed by references.

2   Related Works

In traditional relational data clustering, the distance measure is primarily based on attribute similarity, e.g., Euclidian distance among two attribute vectors. On the other hand, in graph clustering the vertex closeness is measured on the basis of (i) connectivity, i.e., possible number of paths between two vertices, (ii) neighborhood similarity, i.e., number of common neighbors between them, or (iii) attribute or contextual similarity, i.e., number of common vertex's features or attributes.

In various real applications, both the vertex properties and the graph topological structure are important. For instance, in a social network, topological structure represents the relationship among different people in a group, while vertex properties describe the role of person. Most of the graph clustering and summarization approaches, mentioned in this and former section, only deal with one aspect and ignore the other.

Numerous existing graph clustering algorithms consider the topological structure of a graph so that every sub-graph attains the cohesive internal structure. It includes clustering based on normalized cut, max flow min-cut problem, structural density, modularity respectively [8], [9], [10]. Recently a new approach is proposed which deals with the attributes of an arbitrary pair of vertices [11], as a result vertices with the similar attribute values are grouped into respective partition or cluster.

A new probabilistic algorithm, i.e. Top Graph Clusters (TopGC) is devised in [12], which finds the best clique like clusters inside the large graphs. It works with both directed and undirected graph and support overlapping based on the given percentage. The space complexity is relatively high, because it is required to maintain the random permutations, hash table, and signatures for clustering. Time and search space complexity has been reduced by introducing the pruning phase. This approach cannot find the definite clusters in a graph which contains the topological as well as contextual information due to the fact that it only considers the structural facet.

Similarly, a three phase Transitive Node Similarity [13] approach has been developed for graph node clustering which consider only single similarity aspect. It is obvious that the search space get reduced incrementally because of the fact that already clustered nodes are not considered in the later iterations. This helps to reduce the space complexity and search space, which also helps to minimize the time complexity. In case of disconnected graph, each isolated component forms a separate cluster.

Conversely in [21], an efficient graph summarization technique based on two database-style operations has been proposed. The first operation, called SNAP, produces a summary graph by grouping nodes based on user-selected node attributes and relationships. Secondly in k-SNAP operation, it further allows users to control the resolutions of summaries.

Recently, a Unified distance measure is introduced in [14], [18] for graph node clustering. It captures both structural and attributes similarities simultaneously which is based on attribute augmented graph, by adding attribute nodes to the original graph and link all the respective nodes. The attribute augmented graph contains more edges in the graph, at least greater than the number of vertices ($> |V|$), which leads to more space complexity of the algorithm. Unified distance measure calculation (involves matrix multiplication) and clustering process is repeated, with iterative weight updates for convergence, which requires more computation time.

All of the methods in this section have the deficiency either with respect to cluster quality or algorithm time complexity. As you have seen none of them is able to fulfill both requirements simultaneously. In subsequent section, we have elaborated new technique which take care both aspects concurrently without any additional dependency.

3   An Intra Graph Clustering (IGC)

We need to construct a graph nodes clustering (i.e. intra graph clustering) method which can handle both structural and contextual similarity (collaborative in nature) in an efficient manner.

3.1   Problem Statement Formulation

An undirected, weighted, multi-labeled graph $G = \{V, E, W, A\}$, not necessarily connected, where V and A is the set of all the vertices and attributes respectively, $E = \{(v_i, v_j) \mid v_i, v_j \in V\}$ are the set of undirected links, and two vertices, v_i and v_j, may be connected with single link having cost $w_{ij} > 0$. Each vertex may contains one or more attributes, $A = \{a_1, a_2, \ldots a_m\}$. The set of possible values for any arbitrary attribute a_i is $Dom(a_i) = \{a_{i1}, a_{i2}, \ldots a_{inp}\}$ where $n_p = |Dom(a_i)|$. The total number of links of a vertex v_i is called the degree of v_i and is represented as $deg(v_i)$. If there is a direct link between any two vertices, e.g. v_i and v_j, in the graph, we call them as direct neighbor otherwise indirect neighbors in case of indirect link. Mostly used symbols along their descriptions are given in Table 1.

Intra graph clustering is the process of the relative partitioning the vertices of the graph into **k** disjoint sub-graphs where $G_i = \{V_i, E_i, W_i, A_i\}$ and $V = \bigcup_i^k (V_i)$ and for any $i \neq j$, $V_i \cap V_j = \emptyset$. The ultimate goal is twofold; attain high quality clusters, and time efficiency.

Table 1. Frequently used symbols and their brief description

Symbol	Description
G	A positive weighted, multi-label, and undirected graph
V	Set of the graph vertices
E	Set of edges/links in between nodes
N	No. of nodes inside the graph G
C	Set of the centroids in the graph
M	No. of possible attributes which can be associated with the node in the graph
K	No. of clusters to be found within the graph space
$SIM(X,Y)$	Similarity between two arbitrary sets X and Y, each set contains variable number of elements
$\mid ? \mid$	Magnitude of the inside item
N_a	A neighborhood vector of a vertex a, all the nodes which are directly connected are considered to be in the neighborhood of the node
$deg(v_a)$	Number of links which are incident on the vertex a
w_{ab}	Weight associate on the link among node a and b
$v_a \leftrightarrow v_b$	If there exist a direct link between two vertices v_a and v_b
$v_a \cdots v_b$	If two vertices are indirectly connected with each other through an arbitrary path, when no direct link is present
$CSim(v_a, v_b)$	Collaborative similarity measure between two vertices in the graph
$\mathcal{F}_{obj}$	Objective function
$c_i, cList[i]$	i^{th} cluster centroid, i^{th} centroid contained in cList

3.2   System Architecture

The proposed system architecture in Fig. 2 consists of two major components. Firstly, it requires estimating the relationship among nodes in the un-clustered graph. It will depict that how much relevance among nodes exists or correlated with each other topologically and contextually or semantically. During the clustering process the topological or contextual information remains unchanged so similarity among nodes is calculated once, in contrast to [14] and [13] which incrementally explore the search space and calculate the similarity as the new node encountered, and utilized multiple times in successive steps.

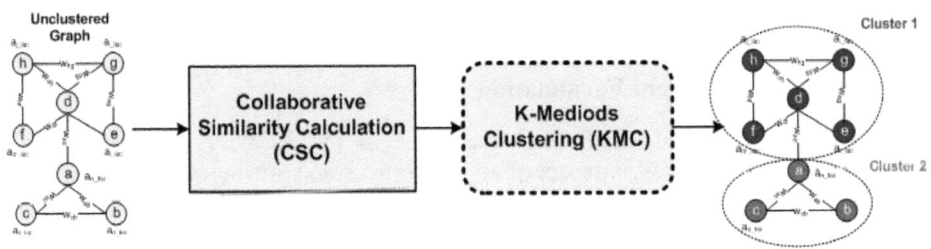

Fig. 2. Proposed Architecture

The similarity measures among nodes along the original graph information are the pre-requisites for the clustering component which is iterative in nature. Additionally, number of clusters k to be found is provided in advance. At the prior stage, we need influence factor α which can control the association among nodes based on topological structure and contextual relevance.

3.3   Collaborative Similarity Measure (CSM)

The similarity measure between graph vertices reflects the strength of their relationship or connectivity. In this approach, we utilize this strength for clustering similar vertices in one cluster and dissimilar to another. We have considered weighted, undirected, and multi-labeled vertex graphs, so relationship among two vertices can be found in the any of the following forms: (1) Directly Connected, (2) In-directly Connect, (3) Disconnected.

In case of direct connection, Fig. 3b, two vertices must share a single link in between them. When there is no direct link between two vertices, e.g. vi and vj, however from vertex vi we can reach on the other vertex vj by following an arbitrary links then both vertices are indirectly connected with each other, as shown in Fig. 3c. Thirdly, in absence of direct and indirect link, vertices are stated as disconnected. In this paper we assume that two vertices are connected only with single link.

$$\text{SIM}(X, Y) = \frac{|X \cap Y|}{|X \cup Y|} = \frac{|X \cap Y|}{|X| + |Y| - |X \cap Y|} \tag{1}$$

In literature, Jaccard similarity coefficient measure [16], given in Eq. (1), is generally acceptable and most widely used especially in data mining applications [17]. Due to its simplicity, it has been utilized in many application areas to find out the relevance among the objects. Even though it is application independent, but in order to incorporate this similarity measure inside the proposed approach, we need to re-define the notions of the objects due to the presence of structural behavior. The important structural factor of the graph is the links between the vertices to form an association.

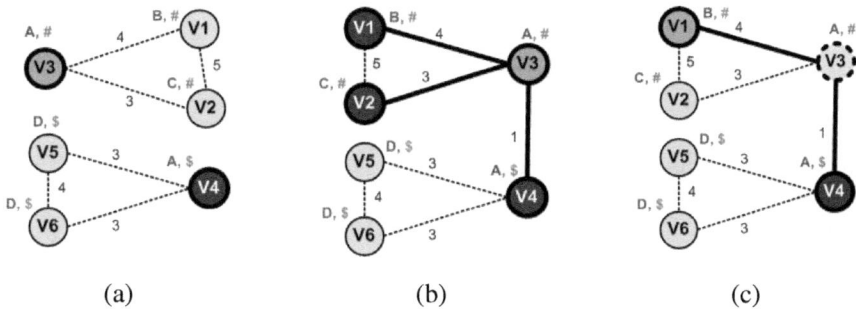

        (a)   (b)   (c)

Fig. 3. Scenarios for similarity between source (green) and destination(red) nodes following some intermediate nodes (yellow) (a) No direct path exist (b) Directly connected (c) In-directly connected, shortest path

    In Eq. (2), $SIM(v_a, v_b)^{weighted}_{struct}$ represents the similarity between two vertices by exploiting the topological structure of the graph. It can capture all three possible scenarios, e.g. connected, indirectly connected, and disconnected. We need to define another similarity measure which can cover contextual information contained by vertices of the node. Here contextual information or semantics means that node can be participating in multiple contexts. Consider the example of social network, where each user is represented with a vertex and there are some associated contexts, like a person's role can be Teacher, Friend, Researcher, author/co-author of a paper.

$$SIM(v_a, v_b)^{weighted}_{struct} = \begin{cases} \dfrac{|N_a \cap N_b|}{|N_a \cup N_b|} * (w_{ab}), & v_a \leftrightarrow v_b \\ 0, & otherwise \end{cases} \tag{2}$$

Our strategy is equally acceptable for weighted and un-weighted graphs. The weights on edges along the path are considered explicitly which is not the case is in [13]. It will enhance the intensity of the relation among nodes based on the weights associated with edges. If two nodes are sharing edges with high weight will obviously have higher similarity as compared to low weight edge sharing nodes. In the absence of edge weights, constant value is expected to be associated with each link in the graph.

    One of the key aspects of this approach is to consider contextual similarity which is defined in Eq. (3) along the structural cohesiveness of the nodes. Its importance is evident from the applications where the nodes emerge in different contexts. For example, in social network, the persons can be represented by nodes and edges among nodes reflect their relationships. Each person can have different roles or contexts like occupation as student, doctor, engineer, and designer etc. Similarly in co-author network, each author is reflected with a node and if there are two authors of the same paper then a link is attach among those nodes in [14], [18]. Accordingly authors may have contributions in different areas of research at different time intervals; usually they used the labels or attribute associated with the corresponding nodes in the graph to reflect this information.

$$SIM(v_a, v_b)_{context}^{weighted} = \frac{\prod_{i=1, v_a \&\& v_b \leftarrow a_i}^{M}(w_{a_i})}{\prod_{j=1, v_a|| v_b \leftarrow a_j}^{M}(w_{a_j})}, \quad v_a \leftrightarrow v_b \; || \; v_a \cdots v_b \quad (3)$$

Collaborative Similarity $= CSim(v_a, v_b) =$

$$= \begin{cases} \alpha * SIM(v_a, v_b)_{struct} + (1 - \alpha) * SIM(v_a, v_b)_{context}, & v_a \leftrightarrow v_b \\ \prod_{i=1}^{q} CSim(v_{pi}, v_{pi+1}), \; v_a \cdots v_b, & v_p \text{ is on path } v_a \text{ and } v_b \\ (1 - \alpha) * SIM(v_a, v_b)_{context}, & otherwise \end{cases} \quad (4)$$

The semantics or attributes associated with the vertices which define the context of each vertex can also be prioritized by an associated weight w_{a_i} in Eq. (3). At the beginning of the algorithm, these values get initialized and remain fixed throughout all computations.

In order to find the similarity based on the set of shared features, Jaccard similarity coefficient measure gives us required functionality, elaborated in Eq. (1). Conclusively the similarity measure through which we consider three scenarios is given in Eq. (4). Both first and third scenarios are quite simple compared to the case when we have set of possible paths from source vertex to destination instead a direct or no link respectively. In order to extend the similarity for indirect path, we must utilize the desirable property, i.e. longer the path among two arbitrary vertices smaller the value of the similarity measure. Additionally, we should consider another property that the similarity measure value between two vertices along the entire path must be less than the intermediate vertices similarity.

Lemma 1 (Transitivity): Let $p = \{v_{p1}, v_{p2}, \cdots v_{pq+1}\}$ be a path from source vertex v_{p1} to target vertex v_{pq+1}. Then for all the intermediate vertices i =1, 2..., q

$$sim(v_a, v_b) = \prod_{i=1}^{q} sim(v_{pi}, v_{pi+1}) \leq sim(v_{pi}, v_{pi+1})$$

Proof: It is based on the fact that the similarity value $sim(v_{pi}, v_{pi+1})$ lies in the interval [0, 1].

For any pair of vertices, many different paths may exist from initial vertex v_i to final vertex v_f. The similarity value may vary based on the selected path. In order to avoid this variation and to be consistent, we have adopted shortest path between that pair of vertices. Conceptually inside the dense community the similarity value is high because of transitivity property. However there is possibility to have smaller value even for shortest path as compared to other path options. This option is expected to have high computational cost. Mostly the shortest path produces larger value [13] unless an intermediate transition may cause significant similarity decrease, as given in Eq. (4).

3.4   Algorithm Details

The proposed method, Collaborative Similarity based Graph Clustering, can operate on undirected, weighted or un-weighted, and multi-attributed graph which requires two parameters as prior knowledge; number of clusters to be found even though it's unsupervised and factor which controls the importance of the topological structure and contextual associations. It does not require altering the structure of the graph. Systematically, it consists of two main components similarity calculation and clustering which is iterative in nature. The pseudo code is omitted due to space limit. For the proof of the concept, we have illustrated the similarity and clustering results in Table (2).

Initialization module considers the issues related to memory allocation, variable management for temporary and permanent storage. Un-clustered graph data is retrieved and stored in the main memory for later processing.

Similarity value estimation is core part of this algorithm because the subsequent processing is entirely dependent on the results of this module. Basically, the similarity value is anticipated among all possible pair of vertices in terms of their topological and behavioral or contextual resemblance, using Eq. (4). At this point, the most important factor which needs to be considered is the symmetry of the values due to undirected graph. For example if there are two vertices v_a and v_b then $\text{CSim}(v_a, v_b) = \text{CSim}(v_b, v_a)$.

At the last stage, the partitioning of the graph vertices is done by utilizing the pre-calculated similarity values among vertices by employing the inherent features of the k-Medoid clustering infrastructure. In start vertices are randomly selected as the centroids (expected center points) or initial seeds to represent the hidden clusters. Then we associate neighboring vertices to the nearest centroids to make a partition based on their similarity distance. Finally, the quality of each cluster is analyzed based on the density and entropy. The ultimate goal of this iterative process is to approximate the objective function, given in Eq. (5).

The crucial part in this algorithm is to accurately measure the similarity value which can reflect the true relationship among vertices in the graph. Normally in a graph structure from one vertex to another there might be various paths available. In the heart of similarity calculation, the shortest path strategy has been incorporated to get rid of diverse paths dilemma.

As you can observe, the relationship among all possible pairs of vertices is estimated, so in accordance with this if we have V the total number of vertices in the graph then the time complexity for estimating the collaborative similarity (which consider both aspects simultaneously) among each pair of vertices will be $O(|V|2)$. On the other hand, similar task, i.e. similarity estimation, in SA approach [14] costs us O $(|V|3)$ due to matrix multiplication. However, shortest path calculation is done efficiently in order of O $(|E| + |V| \log |V|)$ using the Fibonacci heap [22].

Table 2. (a) Collaborative Similarity among vertices given in Fig. 3 using Eq. (4), (b) Clustering results on the graph (in Fig. 3) by varying number of clusters (K), quality of each measure is calculated using Density and Entropy

$CSim(v_a, v_a)$	vertex	V1	V2	V3	V4	V5	V6
	V1	1	2.67	1.17	0.20	0.18	0.18
	V2	2.67	1	0.92	0.15	0.14	0.14
	V3	1.17	0.92	1	0.17	0.15	0.15
	V4	0.2	0.15	0.17	1	0.92	0.92
	V5	0.18	0.14	0.15	0.92	1	2.5
	V6	0.18	0.14	0.15	0.92	2.5	1

K	Clustered Vertices	Density	Entropy
2	{V1,V2,V3},{V4,V5,V6}	0.42	0.133
3	{V1,V3},{V2},{V4,V5,V6}	0.28	0.084
4	{V5},{V6},{V4},{V1,V2,V3}	0.21	0.084

(a)   (b)

Where $\mathcal{F}_{obj}$ an objective function which depends on two sub functions D(...), E(...) as density and entropy respectively. The detail description for each function can be obtained from [14]. A well know k-Medoid clustering strategy is adopted which have the tendency to explore the best centroid candidates iteratively. At the beginning of the cluster phase, the centroids picked randomly, and in subsequent iterations the summated distance is computed using similarity measures and compared to select vertex with maximum value as new centroid. The termination criteria for clustering process is the decline of the objective function described in Eq. (5) or small improvement in the function value over subsequent iterations.

$$\mathcal{F}_{obj} = \max_k \left(\left(\alpha * D\left(\{V_q\}_{q=1}^k\right) \right) - \left((1-\alpha) * E\left(\{V_q\}_{q=1}^k\right) \right) \right) \qquad (5)$$

In our clustering scheme, the objective function $\mathcal{F}_{obj}$ needs to be maximized for clusters quality enhancement. High density corresponds to tight connection among vertices and low entropy ensures that most of the vertices in the cluster have similar contextual aspects or labels. In this respect higher objective function value always leads us towards better performance. If there is no improvement with respect to objective function in consecutive iterations it is supposed to be converged or terminated.

Influence factor α (alpha) is the controlling parameter which is exploited in Eq. (4) and Eq. (5) to balance the impact of both connectivity and semantics. Its value range is from 0 to 1. When alpha is 0 then vertices having similar attributes get clustered in one region irrespective of their interconnection and associated weights. However, value 1 has opposite impact to group densely connected regions of vertices instead their context. We have given an equal importance to both factors by taking its value 0.5. In this paper, its value remains fixed throughout the clustering process.

4   Empirical Analyses

The experiments were carried out on single 32-bit machine having 2.40GHz Intel dual core processor with 4GB main memory, and windows 7 as an operating system. The proposed method and SA-Cluster methods have been implemented using open source matrix manipulation library JAMA[1] in Java. The comparison has been carried out among the following methods:

- **IGC-CSM.** The proposed efficient algorithm which considers both contextual and structural similarity without compromising its effectiveness.
- **SA-Cluster.** [14] It also considers both aspects but lack in scalability to huge graph due to iterative time consuming random walk strategy.
- **S-Cluster.** It is focused on topological structure of the graph.
- **W-Cluster.** It has combined both aspects, i.e. structural and attribute, however is based on random walk strategy which requires more computation [18].
- **K-SNAP.** It's a top down methodology which is concerned about attribute similarity [21].

4.1   Datasets

We have analyzed the proposed strategy on real and synthetic datasets. The political blogs network[2] dataset contains 1490 web blogs on United States politics with 19,090 hyperlinks between these blogs. This dataset contains one attribute associated with the web-blogs which has two possible values (in other words the domain of this attribute)

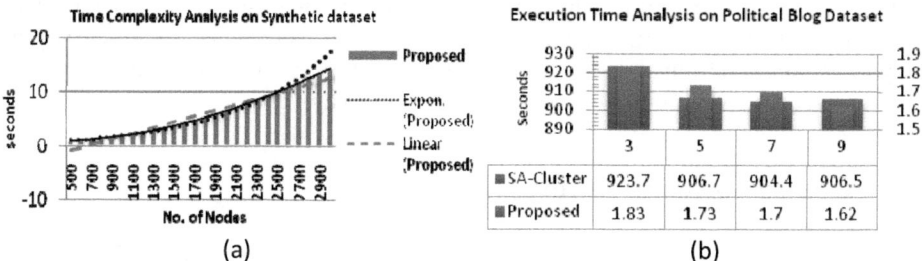

Fig. 4. (a) Graph size vs. time, (b) No. of Clusters vs. Time (seconds)

as either liberal or conservative. The real-life datasets are most of the times limited to small variation in attributes, links, and associated weights.

In all experiments α value set to 0.5. Our proposed method has the time complexity quadratic in nature which is acceptable for medium scale graphs as analyzed with respect to the graph size and number of clusters in Fig. 4. In order to develop a time efficient algorithm, it should not compromise on quality or effectiveness.

4.2   Cluster Evaluation Criteria

The structural and contextual quality of the clusters has been analyzed in terms of Density and Entropy respectively [14]. Combined effect can also be analyzed using F-measure [20], however due to space limit it has been omitted.

---

[1] http://math.nist.gov/javanumerics/jama
[2] http://www-personal.umich.edu/mejn/netdata

Strong connection among vertices can be easily analyzed by utilizing the density function. It is the ratio between number of edges present in a cluster and total number of edges in the whole graph. The ratios get accumulated for all clusters to evaluate the overall impact. Its values lie in the interval of [0, 1].One of the key aspects to measure the quality of the cluster is to determine the relevancy among vertices based upon their attributed nature.

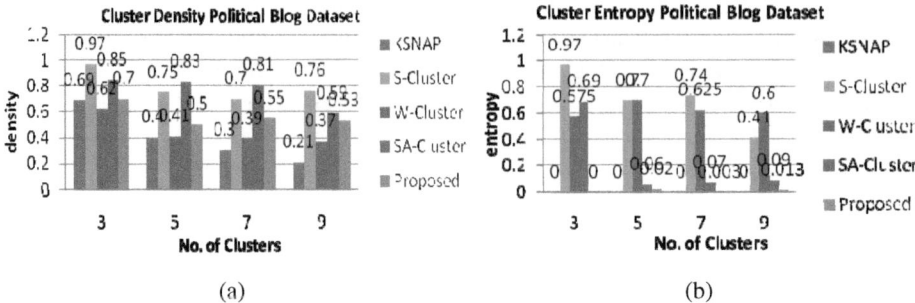

Fig. 5. (a) Number of Clusters vs. Density (b) Number of Clusters vs. Entropy

Fig. 5 (a) and (b) shows the competitive results against density and entropy respectively. Our algorithm maintains the quality in terms of density and entropy by varying number of clusters. It is obvious from Fig. 5 (b) that overall entropy difference is large which shows good clustering tendency.

5   Conclusions and Future Directions

In this paper, we study the problem of graph clustering which enables us to efficiently partition the vertices based on homogeneous characteristics in terms of context and topology. We have carefully designed the collaborative similarity measure to reflect the relational model among pair of vertices. Subsequently, well known k-Medoid clustering framework is adopted for iterative evolution of all the clusters. The quality estimation of each cluster is concurrently done based on density, and entropy measures. Experiments on synthetic and real datasets depict the time efficiency of the proposed methodology while sustaining the cluster quality.

Our idea works in polynomial time (i.e. quadratic in nature) however it's still hard to scale up for very huge datasets. The automatic adaptation of the tuning parameter, alpha, should also require extensive analysis which is beyond the scope of this paper.

References

1. Cook, D.J., Holder, L.B.: Mining Graph Data. Wiley and Sons (2006)
2. Drineas, P., Frieze, A., Kannan, R., Vempala, S., Vinay, V.: Clustering Large Graphs via the Singular Value Decomposition. Machine Learning 56(1-3) (2004)

3. Dongen, S.: Graph Clustering by Flow Simulation, Ph.D. thesis, University of Utrecht (2000)
4. Flake, G.W., Tarjan, R.E., Tsioutsiouliklis, K.: Graph Clustering and Minimum Cut Trees. Journal of Internet Mathematics 1(4), 385–408 (2003)
5. Newman, M.: Detecting Community Structure in Networks. The European Physics Journal B 38, 321–330 (2004)
6. Huang, X., Lai, W.: Clustering Graphs for Visualization via Node Similarities. Journal of Visual Languages and Computing 17, 225–253 (2006)
7. Anand, R., Reddy, C.K.: Graph-Based Clustering with Constraints. In: Huang, J.Z., Cao, L., Srivastava, J. (eds.) PAKDD 2011, Part II. LNCS, vol. 6635, pp. 51–62. Springer, Heidelberg (2011)
8. Shi, J., Malik, J.: Normalized cuts and image segmentation. IEEE Trans. Pattern Analysis and Machine Intelligence 22(8), 888–905 (2000)
9. Xu, X., Yuruk, N., Feng, Z., Schweiger, T.A.J.: Scan: a structural clustering algorithm for networks. In: Proc. 2007 Int. Conf. Knowledge Discovery and Data Mining (KDD 2007), San Jose, CA, pp. 824–833 (August 2007)
10. Newman, M.E.J., Girvan, M.: Finding and evaluating community structure in networks. Phys. Rev. E 69, 026113 (2004)
11. Tian, Y., Hankins, R.A., Patel, J.M.: Efficient aggregation for graph summarization. In: Proc. 2008 ACM-SIGMOD Int. Conf. Management of Data (SIGMOD 2008), Vancouver, Canada, pp. 567–580 (June 2008)
12. Marcopol, K., et al.: Scalable Discovery of Best Clusters on Large Graphs. Proceedings of VLDB Endowment 3(1) (2010)
13. Tiakas, E., et al.: Graph Node Clustering via Transitive Node Similarity. In: 14th Panhellenic Conference on Informatics (2010)
14. Zhou, Y., et al.: Graph Clustering Based on Structural/Attribute Similarities. In: Proceedings of VLDB Endowment, France (2009)
15. Larsen, B., Aone, C.: Fast and effective text mining using linear-time document clustering. In: Proceedings of the Fifth ACM SIGKDD International Conference on Knowledge Discovery and Data Mining, pp. 16–22 (1999)
16. Jaccard, P.: Etude Comparative de la Distribution Floraledansune Portion des Alpes et des Jura. Sociètè Vaudoise des Sciences Naturelles 37, 547–579 (1901)
17. Everitt, B.: Cluster Analysis, 3rd edn. Edward Arnold, London (1993)
18. Cheng, H., et al.: Clustering Large Attributed Graphs: A Balance between Structural and Attribute Similarities. ACM Transaction on Knowledge Discovery from Data 5(2) (February 2011)
19. Fredman, M., Tarjan, R.: Fibonacci Heaps and their Uses in Improved Network Optimization Algorithms. Journal of ACM 34, 596–615 (1987)
20. Witsenburg, T., et al.: Improving the Accuracy of Similarity Measures by Using Link Information. In: International Symposium on Methodologies for Intelligent Systems, 9th edn., Poland (2011)
21. Tian, Y., Hankins, R.A., Patel, J.M.: Efficient aggregationfor graph summarization. In: Proc. 2008 ACM-SIGMOD Int.Conf. Management of Data (SIGMOD 2008), Vancouver, Canada, pp. 567–580 (June 2008)
22. Fredman, M., Tarjan, R.: Fibonacci Heaps and their Uses in ImprovedNetwork Optimization Algorithms. Journal of the ACM 34, 596–615 (1987)

A Grid-Based Index and Queries
for Large-Scale Geo-tagged Video Collections

He Ma[1], Sakire Arslan Ay[1], Roger Zimmermann[1], and Seon Ho Kim[2]

[1] School of Computing, National University of Singapore, Singapore 117417
[2] Integrated Media Systems Center, University of Southern California, LA, CA 90089
{mahe,dcssakir,rogerz}@comp.nus.edu.sg, seonkim@usc.edu

Abstract. Currently a large number of user-generated videos are produced on a daily basis. It is further increasingly common to combine videos with a variety of meta-data that increase their usefulness. In our prior work we have created a framework for integrated, sensor-rich video acquisition (with one instantiation implemented in the form of smartphone applications) which associates a continuous stream of location and direction information with the acquired videos, hence allowing them to be expressed and manipulated as spatio-temporal objects. In this study we propose a novel multi-level grid-index and a number of related query types that facilitate application access to such augmented, large-scale video repositories. Specifically our grid-index is designed to allow fast access based on a bounded radius and viewing direction – two criteria that are important in many applications that use videos. We present performance results with a comparison to a multi-dimensional R-tree implementation and show that our approach can provide significant speed improvements of at least 30%, considering a mix of queries.

1 Introduction

Due to recent developments in video capture technology, a large number of user-generated videos are produced everyday. For example, smartphones, which are carried by users all the time, have become extremely popular for capturing and sharing online videos due to their convenience, good image quality and wireless connectivity. Moreover, a number of sensors (*e.g.*, GPS and compass units) are commonly integrated in smartphones. Consequently, some useful meta-data, especially geographical properties, can easily be captured while video is being recorded. This association of video scenes and their geographic meta-data has opened interesting research areas, for example, the meta-data can aid in the indexing and searching of geo-tagged videos.

In our earlier work [3] we proposed to model the viewable scene area (*i.e.*, the field-of-view, or FOV) of video frames as pie-shaped geometric figures using geospatial sensor data, such as camera location and direction. This approach describes a set of video frames as a series of spatial objects. Consequently, video search can be rephrased as a known spatial data selection problem. The objective then is to index the spatial objects and to search videos based on their geographic

H. Yu et al. (Eds.): DASFAA Workshops 2012, LNCS 7240, pp. 216–228, 2012.

properties. Our previous study demonstrated the feasibility of geographic sensor meta-data for searching large video repositories. To make a geospatial search engine practical, there remain critical issues to be solved. Specifically the search efficiency is a key goal which requires a high performance index structure to manage large sets of meta-data. To the best of our knowledge, there has been no such study for geo-tagged video search.

YouTube has recently archived 13 million hours of video in a single year (2010) and its growth is still increasing [1]. Assuming all videos in a large repository are represented using sensor meta-data, *i.e.*, streams of geospatial objects, efficiently searching among the meta-data becomes interesting. An important observation in this context is that the geo-space is bounded while the number of videos is open-ended. Based on this observation, we propose a new multi-level grid-based index structure for geo-tagged videos. Specifically, the introduced structure allows efficient access of FOVs based on their distance to the query location and the cameras' viewing directions. We introduce a number of related query types which aim to support applications that focus on video-specific searches.

One of the unique query types proposed in this work is the *k-Nearest Video Segments query* (k-NVS). A k-NVS query can significantly enhance human perception and decision making in identifying requested video images, especially when search results return a large number of videos in a high-density data area. The query can additionally be specified with a bounded radius range to only return results within a maximum distance. Alternatively, the query may consider a certain viewing direction that show the query point from the given direction. An example application which can utilize k-NVS is to automatically build a panoramic (360 degree) view of a point-of-interest. The application needs to search for the nearest videos that observe the query point from different viewing directions and that are within a certain distance from it.

In the remaining sections of the manuscript we describe our geo-tagged video indexing and search approach, and report on an extensive experimental study with synthetic datasets. The results illustrate that the grid index efficiently scales to large datasets and significantly speeds up the query processing (compared to an R-tree) especially for directional queries. The rest of this paper is organized as follows. Section 2 provides background information and summarizes the related work. Section 3 details the proposed data structure. Section 4 introduces the new query types and details the query processing algorithms. Section 5 reports the results of the performance evaluation of the proposed algorithms. Finally, Section 6 concludes the paper.

2   Background and Related Work

2.1   Modeling of Camera Viewable Scene

The *camera viewable scene* is what a camera in geo-space captures. This area is referred to as *camera field-of-view* (FOV for short) with a shape of a pie-slice [3]. The FOV coverage in 2D space can be defined with four parameters:

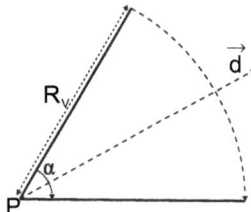

Fig. 1. Illustration of the FOV model in 2D space

camera location P, camera's orientation vector $\vec{d}$, viewable angle α, and maximum visible distance R_V (see Fig. 1). The location P of the camera is the $\langle latitude, longitude \rangle$ coordinate read from a positioning device (*e.g.*, GPS). The camera's orientation $\vec{d}$ is obtained based on the orientation angle (θ), which can be acquired from a digital compass. The camera viewable angle (α) is calculated based on the camera and lens properties for the current zoom level [8]. The visible distance R_V is the one at which a large object within the camera's FOV can be recognized. Then, the camera viewable scene at time t is denoted by the tuple $FOV(P\langle lat, lng \rangle, \theta, \alpha, R_V)$. The tuple values, *i.e.*, the geospatial meta-data, are acquired from embedded sensors during video capture.

2.2   Related Work

Associating geo-location and camera's orientation information for video retrieval has become an active topic. Hwang et al. [10] and Kim et al. [11] proposed a mapping between the 3D world and the videos by linking the objects to the video frames in which they appear, using GPS location and camera's orientation. Liu et al. [13] presented a sensor enhanced video annotation system (referred to as SEVA). Our prior work [3] proposed a viewable scene model to link the video content and sensor information. However, none of the above methods addresses indexing and searching issues, on the large scale.

Our approach represents each video frame as a spatial object. There exist two categories of spatial data indexing methods: *data-driven structures* and *space-driven structures* [17]. The R-tree family (including R-tree [9], R$^+$-tree [18], R*-tree [5]) belongs to the category of data-driven structures. However, these methods are designed mainly for supporting efficient query processing while the construction and the maintenance of the data structure is computationally expensive. The space-driven structures include methods such as quadtree [7] and Voronoi diagram [16]. Recent researches use either the skip quadtree [6] or Voronoi diagram [15] to process multiple types of queries. However, these data structures consider spatial objects as points or small rectangles, and none of them are appropriate to index our FOV model.

Our prior work [12] proposed a vector-based approximation model to efficiently index and search videos based on the FOV model. It mapped an FOV to two individual points in two 2D subspaces using a space transformation. This model works well on supporting the geospatial video query features, such as point query with direction and bounded distance between the query point and camera position. However, it does not investigate query optimization issues.

Vector model works effectively for basic query types, such as point and range query, but does not support the k-NVS query. Moreover, there was no consideration in scalability. Next we will introduce the proposed three-level index structure.

3   Grid-Based Indexing of Camera Viewable Scenes

We present our design of the memory-based grid structure for indexing the coverage area of the camera viewable scenes. The proposed structure constructs a three-level index (see Fig. 2). Note that, each level of the index structure stores only the ID numbers of the FOVs for the efficient search of the video scenes. The actual FOV meta-data (*i.e.*, P, θ, α, and R_V values) are stored in a MySQL database where the meta-data for a particular FOV can be efficiently retrieved through its ID number. Fig. 3 illustrates the index construction with an example of a short video file. In Fig. 3 (c), only the index entries for the example video file are listed.

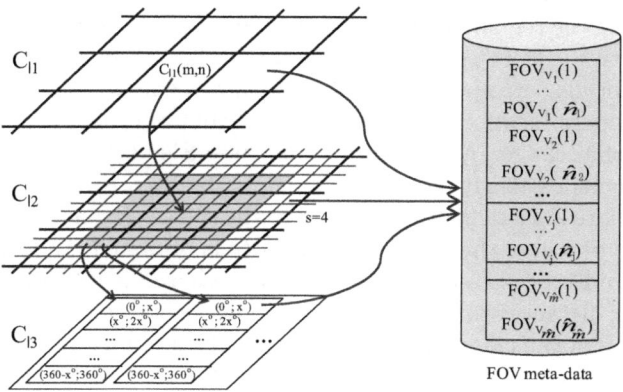

Fig. 2. Three-level grid data structure

The first level organizes the embedding geo-space as a uniform coarse grid. The space is partitioned into a regular grid of $M \times N$ cells, where each grid cell is an $\delta \times \delta$ square area, and δ is a system parameter that defines the cell size of the grid. A specific cell in the first-level grid index is denoted by $C_{\ell 1}(row, column)$ (assuming the cells are ordered from the bottom left corner of the space). Specifically, FOVs are mapped to the grid cells that overlap with their coverage areas and each grid cell maintains the IDs of the overlapping FOVs.

The second-level grid index organizes the overlapping FOVs at each first-level cell based on the distance between the FOV camera locations and the center of the cell. To construct the second-level grid, each $C_{\ell 1}$ cell is further divided into $s \times s$ subcells of size $\left(\frac{\delta}{s} \times \frac{\delta}{s}\right)$, where each subcell is denoted by $C_{\ell 2}(f, g)$ (see Fig. 2). s is a system parameter and defines how fine the second-level grid index is.

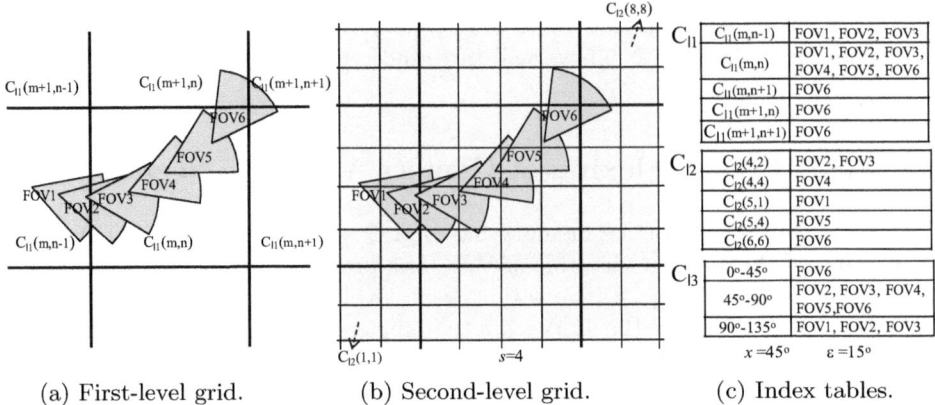

(a) First-level grid.   (b) Second-level grid.   (c) Index tables.

Fig. 3. Index construction example

For each first level grid cell $C_{\ell1}(m,n)$, we maintain the range of the second-level subcells, covering the region in and around $C_{\ell1}(m,n)$ and containing all the FOVs that overlap with the cell $C_{\ell1}(m,n)$. In Fig. 2, the shaded region at $C_{\ell2}$ shows the range of $C_{\ell2}$ subcells corresponding to the first-level cell $C_{\ell1}(m,n)$. Note that the FOVs whose camera locations are at most R_V away from cell $C_{\ell1}(m,n)$, will also be included in that range. In the example shown in Fig. 3, the second-level range for $C_{\ell1}(m,n)$ includes all subcells $C_{\ell2}(1,1)$ through $C_{\ell2}(8,8)$. In order to retrieve the FOVs closest to a particular query point in the cell $C_{\ell1}(m,n)$, first, the second-level cell $C_{\ell2}(f,g)$ where the query point resides is obtained, and then the FOV IDs in and around subcell $C_{\ell2}(f,g)$ are retrieved, where the FOVs are ordered according to their distances to the query point. The second-level index enables the efficient retrieval of the closest FOVs in the execution of k-NVS queries.

The first- and second-level grid cells hold the location and distance information only, therefore cannot fully utilize the collected sensor meta-data, such as direction [12]. To support the directional queries we construct a third-level in the index structure that organizes the FOVs based on the viewing direction. The 360° angle is divided into $x°$ intervals in clockwise direction, starting from the North (0°). We assume an error margin of $\pm\varepsilon°$ around the FOV orientation angle $\theta°$. Each FOV is assigned to one or two of the view angle intervals that its orientation angle margin ($\theta°\pm\varepsilon°$) overlaps with. In Fig. 3, the third table lists the third-level index entries for the example video for $x=45°$ and $\varepsilon=15°$.

For a video collection with about 2.95 million FOVs, the index size for the three-level index structure is measured as 1.9GB. As the dataset size gets larger the index size grows linearly. For example, for datasets with 3.9 million and 5.4 million FOVs, the index size is measured as 2.5GB and 3.3 GB, respectively. In our experiments in Section 5, we report the results for a dataset of 5.4 million FOVs. Next we will describe the query processing for various query types.

4   Query Processing

We represent the coverage of a video clip as a series of FOVs where each FOV corresponds to a spatial object. Therefore the problem of video search is transformed into finding the spatial objects in the database that satisfy the query conditions. In searching video meta-data, unlike a general spatial query, the query may enforce additional parameters. For example, it may search with a range restriction for the distance of the camera location from the query point (*query with bounded radius*) or it may ask only for the videos that show the query location from a certain viewpoint (*query with direction*). In this section we introduce several new spatial query types for searching camera viewable scenes. All the queries work at the FOV level. We formulate them in Section 4.1 and explain the query processing in Section 4.2.

4.1   Query Definitions

Let $FOV_{v_j} = \{FOV_{v_j}(i), i = 1, 2, ..., \widehat{n_j}\}$ be the set of FOV objects for video v_j and let $\mathbb{FOV}=\{FOV_{v_j}, j = 1, 2, ..., \widehat{m}\}$ be the set of all FOVs for a collection of $\widehat{m}$ videos. Given $\mathbb{FOV}$, a query q returns a set of video segments $\{VS_{v_j}(s, e)\}$, where $VS_{v_j}(s, e)=\{FOV_{v_j}(i), \ s \leq i \leq e\}$ is a segment of video v_j which includes all the FOVs between $FOV_{v_j}(s)$ and $FOV_{v_j}(e)$, where i stands for the ith frame.

Definition 1. *Point Query with Bounded Radius* (PQ-R): Given a query point q in geo-space and a radius range from MIN_R to MAX_R, the PQ-R query retrieves all video segments that overlap with q and whose camera locations are at least MIN_R and at most MAX_R away from q, i.e.,

PQ-R(q,MIN_R,MAX_R) :
$q \times \mathbb{FOV} \rightarrow \big\{VS_{v_j}(s, e),$ where $\forall j \ \forall i \ s \leq i \leq e, \ FOV_{v_j}(i) \cap q \neq \varnothing$
$\qquad\qquad$ and $MIN_R \leq dist(P(FOV_{v_j}(i)), q) \leq MAX_R\big\},$
where P returns the camera location of an FOV and function $dist$ calculates the distance between two points.

Definition 2. *Point Query with Direction* (PQ-D): Given a query point q in geo-space and viewing direction β, the PQ-D query retrieves all FOVs that overlap with q and that were taken when the camera was pointing towards β with respect to the North. Using only a precise direction value β may not be practical in video search, therefore a small angle margin ε is introduced. The query searches for the video segments whose directions are between $\beta - \varepsilon$ and $\beta + \varepsilon$.
PQ-D(q,β): $q \times \mathbb{FOV} \rightarrow \big\{VS_{v_j}(s, e),$ where $\forall j \ \forall i \ s \leq i \leq e, \ FOV_{v_j}(i) \cap q \neq \varnothing$
$\qquad\qquad$ and $\beta - \varepsilon \leq D(FOV_{v_j}(i)) \leq \beta + \varepsilon\big\},$
where D returns an FOV's camera direction angle θ with respect to the North.

Definition 3. *Range Query with Bounded Radius* (RQ-R): Given a rectangular region q_r in geo-space and a radius range from MIN_R to MAX_R, the RQ-R query retrieves all video segments that overlap with q_r and whose camera locations are at least MIN_R and at most MAX_R away from the border of q_r.

Definition 4. *Range Query with Direction* (RQ-D): Given a rectangular region q_r in geo-space and a viewing direction β, the RQ-D query retrieves all video segments that overlap with region q_r and that show it with direction β.

Definition 5. *k-Nearest Video Segments Query* (k-NVS): Given a query point q in geo-space, the k-NVS retrieves the closest k video segments that show q. The returned video segments are ordered based on their distance to q.

$$\text{k-NVS}(q,k): q \times \mathbb{FOV} \rightarrow \{(VS_{v_j}(s_1, e_1), .., VS_{v_j}(s_k, e_k)), \text{ where } \forall s_t, e_t(t = 1, .., k),$$
$$\text{and } \forall j \forall i \ s_t \leq i \leq e_t, \ FOV_{v_j}(i) \cap q \neq \varnothing$$
$$\text{and } dist(VS_{v_j}(s_t, e_t), q) \leq dist(VS_{v_j}(s_{t+1}, e_{t+1}), q)\},$$

The function *dist* calculates the minimum distance between the camera locations of a video segment and the query point.

Definition 6. *k-Nearest Video Segments Query with bounded Radius* (k-NVS-R): Given a query point q and a radius range from MIN_R to MAX_R, the k-NVS-R query retrieves the closest k video segments that show q from a distance between MIN_R to MAX_R. The returned video segments are ordered by distance.

Definition 7. *k-Nearest Video Segments Query with Direction* (k-NVS-D): Given a query point q and a viewing direction β, the k-NVS-D query retrieves the closest k video segments that show q with the direction β.

4.2   Algorithm Design

The query processing is performed in two major steps. First, the FOVs (i.e., the video frames) that satisfy the query conditions in the set $\mathbb{FOV}$ are retrieved. The returned FOVs are grouped according to the video files that they belong to. Next, the groups of adjacent FOVs from the same videos are post processed to retrieve as the video segments in the query results. We argue that, the length of the resulting video segments should be larger than a certain threshold length for visual usefulness. In our implementation, for the point and range queries, the returned FOVs are processed to find out the consecutive FOVs that form the segments. For the k-NVS query, the video segments are formed simultaneously as the closest FOVs are retrieved. If two separate segments of the same video file are only a few seconds apart, they are merged and returned as a single segment. In the following paragraphs, we will describe these queries under three groups: Point query (PQ-R and PQ-D), Range query (RQ-R and RQ-D) and k-NVS query (k-NVS and k-NVS-D). Within each query group, we will further elaborate on the direction and bounded radius queries. Please refer to our technical report [14] for the detailed description of the query processing algorithms.

We retrieve the FOVs that match the query requirements in two steps: a *filter step* followed by a *refinement step*. First, in the filter step, we search the three-level index structure starting from the first level and then moving down to the second and third levels, if needed. We refer to the resulting set of FOVs from the filter step as the *candidate set*. In the refinement step, an exhaustive method is carried out to check whether an FOV actually satisfies the query conditions.

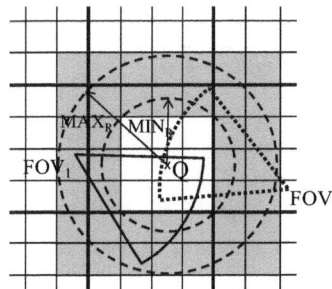

Fig. 4. Applying the distance condition for the PQ-R query

Point Query. The PQ-R query searches the video frames with a certain radius range restriction for the distance of the camera locations from the query point, according to the required level of details in the video. And the PQ-D query exploits the collected camera directions to retrieve the video segments that show the the query point from the requested viewing direction. When processing the point query, we first calculate the first-level cell ID $C_{\ell 1}$ where the query point is located. For the typical point query (PQ), the set of candidate FOVs would include all FOVs indexed at the cell $C_{\ell 1}$. For PQ-R, we additionally apply the distance condition given by the radius range (MIN_R, MAX_R). We reduce the search area for the candidate FOVs in the second-level index by eliminating the subcells outside of the radius range. For example, in Fig. 4, according to the minimum (MIN_R) and maximum (MAX_R) distance conditions, only the FOVs located between the two dotted circles will be returned. Both video frames FOV_1 and FOV_2 overlap with q. However, since the camera location of FOV_2 is outside of the shaded region, indicating that the distance is farther than MAX_R, FOV_2 will not be included in the candidate set. For PQ-D, we check the third–level index cell to find cells that cover the query angle range given by (β-ε, β+ε). After the candidate FOVs are retrieved, we run the refinement step to get the actually matching FOVs.

Range Query. When the search application asks for the videos that show a region q_r, rather than a point location q, it may issue a range query. The RQ-R query queries the closeness to the query region, and the RQ-D query searches for the scenes of the query region from different view points. In our range query processing algorithm, we use a hybrid approach where we try to cover the query region with a mixture of $C_{\ell 1}$ and $C_{\ell 2}$ cells. We try to minimize the uncovered regions in cells (i.e., minimizing the false positives) and at the same time, we also try to minimize the duplicate FOV IDs in the candidate set, by using as many $C_{\ell 1}$ cells as possible. The goal is to reduce the size of the candidate set, so that the time required to process the FOVs in the refinement step is minimized. Hence, among the $C_{\ell 1}$ cells that overlap with q_r, we choose the cells whose overlap areas are larger than a certain threshold value ϕ. If the overlap area is less than ϕ, we cover the overlap region with the $C_{\ell 2}$ subcells.

k-NVS Query. In our geo-tagged video search system, we propose the *k-Nearest Video Segments* query as, "For a given point q in geo-space, find the nearest k video segments that overlap with q." Taking Fig. 5 as an example, the camera locations of the video segment V_1 at time t are closer to the query point q than that of V_2. Due to the camera location and viewing direction, the FOVs of V_1 cannot cover q while the FOVs of V_2 can. In the k-NVS query, V_2 will be

---

Algorithm 1. k-Nearest Video Segments (k-NVS) Query

Input: query point: $q\langle lat, lng\rangle$, number of output video segments: k,

Output: vector $segments \langle VS_{v_j}(s_t, e_t)\rangle$

1   $C_{\ell 1} = $ getCellID(q), $C_{\ell 2} = getSubCellID(q)$;

2   priority_queue $sortedFOVs \langle FOVID$, distance to $q \rangle = \emptyset$;

3   $subCellsInR = $ applyRadius(q, $C_{\ell 2}$, 0, R_V); i=0; distClose=δ/s;

4   **while** *not enough FOVs AND nextSubCells=getNeighbors(q,subCellsInR,i++)* *is not empty* **do**

5      $candidateFOVs = $ fetchData($nextSubCells$);

6   **for** *all the FOVs in the candidateFOVs* **do**

7   **if** *pointInFOV(q, FOV)* **then**

8          $sortedFOVs$.push($\langle FOVID, dist(q, FOV)\rangle$);

9   **end**

10   **end**

11   **while** *sortedFOVs.top()* $\leq$ *distClose AND numsegments* $<$ *k* **do**

12       $topFOV = sortedFOVs$.pop();

13   **if** *isNewSegment(topFOV,res)* **then**

14         $numsegments$++;

15   **end**

16   res.push($topFOV$);

17   **end**

18   i++; $distClose+ = \delta/s$;

19 **end**

20 **return** $segments = getVideoSeg(res)$

---

selected as the nearest video considering the visibility. Additional radius range and viewing direction requirements can be added through the k-NVS-R and k-NVS-D queries. Algorithm 1 formulates the k-NVS query processing. We first

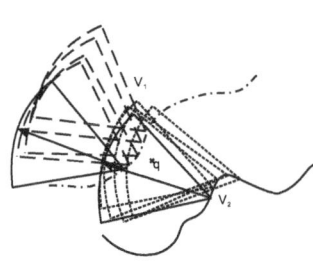

retrieve the $C_{\ell 1}$ and $C_{\ell 2}$ cells where q is located and then search the neighboring subcells around $C_{\ell 2}$ from which the FOVs can see q. In Algorithm 1, we first retrieve the candidate FOVs in the subcells closest to $C_{\ell 2}$. And at each round we gradually increase the search distance by δ/s and retrieve the FOVs in the next group of cells within the enlarged distance. We apply the refinement step on these candidate FOVs and store them in a priority queue, in which the FOVs are sorted based on their distance to q. After each round, the resulting FOVs are organized as videos segments. The search for resulting FOVs ends either when the number of video segments reaches k or when there are no more subcells

Fig. 5. Illustration of k-NVS query

that need to be checked.

5   Experimental Evaluation

5.1   Data Generation and Experimental Settings

Due to the difficulty of collecting large set of real videos, synthetic dataset for moving cameras with positions inside a $75km \times 75km$ region in the geo-space is used to test the performance of the grid-based index structure. We generate camera trajectories using the Georeferenced Synthetic Meta-data Generator [4]. The generated synthetic meta-data exhibit equivalent characteristics to the real world data. The camera's viewable angle is $60°$ and the maximum visible distance (R_V) is $250m$ [3]. In the experiments, we generated a dataset with about 5.4 million video frames. To simulate real-world case, we set the maximum speed of moving cameras as $60km/h$, with the average speed as $20km/h$. Besides the speed limit, we also set the camera's maximum rotation limit as $30°$ per second, which guarantees that the camera rotates smoothly. The detailed parameters for generating the dataset can be found in our technical report [14].

For all the experiments we constructed a local MySQL database and stored all the FOV meta-data, as well as the index structure tables. All the experiments were conducted on a server with two quad core Intel(R) Xeon(R) $X5450$ 3.0GHz CPUs and 16GB memory running under Linux RedHat 2.6.18. All the comparisons are based on the geo-coordinates (latitude and longitude). The experiment results show the cumulative number of FOVs returned for $10,000$ randomly generated queries within the experiment region. In our experiments, we mainly measure the Processing Time (PT for short) and the number of Page Access (PA for short). PT includes the total amount of time for searching for the candidate set through the index structure in the filter step and the time for using the exhaustive method to process the refinement step. We assume that even if the index structure is in memory, when we access to it, we count PA as it was on disk. This helps to analyze the performance if the index structure is disk-based instead of memory-based. In the following experiments, if not specifying, the default value of k is 20. The important parameters are shown in Table 1 and more details are summarized in our technical report [14].

5.2   Comparisons

R-tree is one of the basic index structures for spatial data which is widely used. In our experiments, we insert the Minimum Bounding Rectangle (MBR for short) of all FOVs into an R-tree and process the query types based on this R-tree for comparison. We use the R-tree implementation by Melinda Green [2] which is highly optimized and achieves good performance. On the R-tree, we first search for all the FOVs whose MBRs overlap with the query input (filter step) and then use the exhaustive method to calculate the actual overlap (refinement step). Consequently, some of the parameters (e.g., value of k for k-NVS query, distance condition, etc.) have no effect on PA for the R-tree.

Table 1. Experiment parameters and their values

Parameter	Value
Page Size	4096
Cache Size	4096
Non-Leaf Node Size	64
Leaf Node Size	36
$C_{\ell 1}$ Node Size	68
$C_{\ell 2}$ Node Size	36
$C_{\ell 3}$ Node Size	4
FOV Meta-data	32
Grid Size δ	$250m$
Cell Division Factor s	4
Angle Error Margin ε	$15°$
Overlap Threshold ϕ	30%

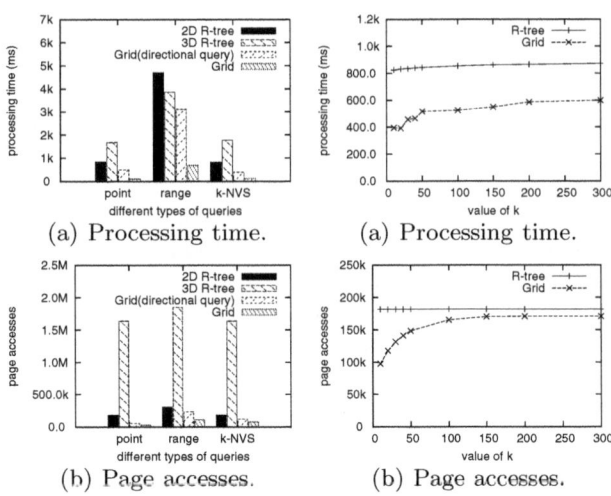

(a) Processing time.

(a) Processing time.

(b) Page accesses.

(b) Page accesses.

Fig. 6. Effect of direction   **Fig. 7.** Effect of value of k

Effect of the Direction Condition. We first proceed to study the efficiency of the proposed grid-based index structure with directional queries. The $2D$ and $3D$ R-trees used in Fig. 6 denote the R-tree for processing queries without and with direction condition, respectively. In this experiment, the query datasets used for directional queries are same as those used in queries without direction, except the additional viewing direction constraint. Fig. 6 (a) shows that, in the processing of PQ-D and k-NVS-D queries, PT of the $3D$ R-tree is almost two times of that of the $2D$ R-tree. The reason is that searching one more dimension slows down the performance of the R-tree. When processing RQ-D query, although searching for candidate sets costs more time, PT of the $3D$ R-tree is smaller because of less number of candidates obtained from the filter step, which accelerates the refinement step. On the contrary, the grid-based approach accesses the third-level grid ($C_{\ell 3}$) to narrow down the search for a small amount of meta-data. Fig. 6 (b) shows that PA for the $3D$ R-tree is over eight times larger than that of the $2D$ one. In contrast, PA for the grid-based approach processing directional queries shrinks to about half of the ones without direction restriction.

Effect of the k Value. Next we study the effect of the k value for k-NVS query. As k increases, PT increases for the grid-based index at the beginning and keeps nearly unchanged when k is larger than 200, which is close to the maximum number of FOVs found in PQ. PT for R-tree is almost the same with different k values because all the results are found and sorted once. When k is larger than 150, PA for the proposed approach is almost the same since the searching radius is enlarged to the maximum according to the design of the structure. In Figs. 7, the gap between the two structures in PT and PA shows that even if the datasets are large and k is big, the proposed structure performs better than R-tree.

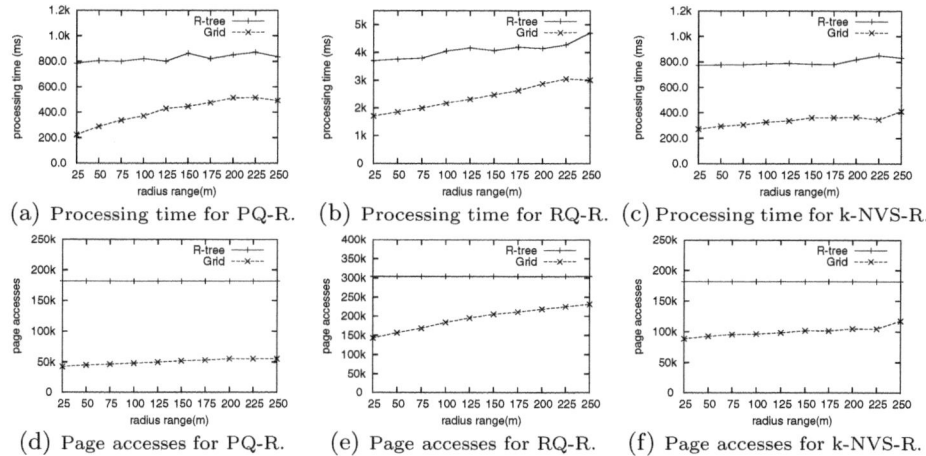

(a) Processing time for PQ-R. (b) Processing time for RQ-R. (c) Processing time for k-NVS-R.

(d) Page accesses for PQ-R.   (e) Page accesses for RQ-R.   (f) Page accesses for k-NVS-R.

Fig. 8. Effect of distance condition

Effect of the Distance Condition. We finally evaluate the effect of the distance condition by varying the radius range from $25m$ to $250m$. The results shown in Fig. 8 are the averages of the different radius ranges. In general, our grid-based index structure greatly outperforms the R-tree on both PT and PA. As shown in Figs. 8 (d), (e) and (f), PA of R-tree remains the same because R-tree finds out all the FOVs whose MBRs overlap with the query in the filter step, regardless of the radius range. In contrast, PA of the proposed method grows as the radius range becomes larger.

6   Conclusions

In this study we proposed a novel multi-level grid-based index structure and a number of related query types that facilitate application access to augmented, large-scale video repositories. Our experimental results show that this structure can significantly speed up query processing, especially for directional queries, compared to the typical R-tree spatial data index structure. The grid-based approach successfully supports new geospatial video query types such as queries with bounded radius or queries with direction restriction. We also demonstrate how to form the resulting video segments from FOVs retrieved.

Acknowledgments. This research is supported by the Singapore National Research Foundation under its International Research Centre @ Singapore Funding Initiative and administered by the IDM Programme Office.

References

1. http://www.youtube.com/t/press_statistics
2. http://superliminal.com/sources/RTreeTemplate.zip

3. Arslan Ay, S., Zimmermann, R., Kim, S.: Viewable Scene Modeling for Geospatial Video Search. In: ACM Int'l Conference on Multimedia, pp. 309–318 (2008)
4. Arslan Ay, S., Zimmermann, R., Kim, S.H.: Generating Synthetic Meta-data for Georeferenced Video Management. In: ACM SIGSPATIAL Int'l Conference on Advances in Geographic Information Systems, GIS (2010)
5. Beckmann, N., Kriegel, H., Schneider, R., Seeger, B.: The R*-tree: An Efficient and Robust Access Method for Points and Rectangles. In: ACM SIGMOD Int'l Conference on Management of Data, pp. 322–331 (1990)
6. Eppstein, D., Goodrich, M., Sun, J.: The Skip Quadtree: A Simple Dynamic Data Structure for Multidimensional Data. In: Annual Symposium on Computational Geometry (2005)
7. Finkel, R., Bentley, J.: Quad Trees: A Data Structure for Retrieval on Composite Keys. Acta Informatica 4(1), 1–9 (1974)
8. Graham, C.: Vision and Visual Perception (1965)
9. Guttman, A.: R-Trees: A Dynamic Index Structure for Spatial Searching. In: ACM SIGMOD Int'l Conference on Management of Data (1984)
10. Hwang, T., Choi, K., Joo, I., Lee, J.: MPEG-7 Metadata for Video-based GIS Applications. In: IEEE Int'l Geoscience and Remote Sensing Symposium (2004)
11. Kim, K., Kim, S., Lee, S., Park, J., Lee, J.: The Interactive Geographic Video. In: IEEE Int'l Geoscience and Remote Sensing Symposium (IGARSS), vol. 1, pp. 59–61 (2003)
12. Kim, S., Arslan Ay, S., Yu, B., Zimmermann, R.: Vector Model in Support of Versatile Georeferenced Video Search. In: ACM SIGMM Conference on Multimedia Systems (2010)
13. Liu, X., Corner, M., Shenoy, P.: SEVA: Sensor-Enhanced Video Annotation. In: ACM Int'l Conference on Multimedia, pp. 618–627 (2005)
14. Ma, H., Arslan Ay, S., Zimmermann, R., Kim, S.H.: Metadata Organization and Query Optimization for Large-scale Geo-tagged Video Collections. NUS/SoC Technical Report TR10/11, National University of Singapore (October 2011)
15. Nutanong, S., Zhang, R., Tanin, E., Kulik, L.: The V*-Diagram: A Query-dependent Approach to Moving KNN Queries. Proceedings of the VLDB Endowment 1(1), 1095–1106 (2008)
16. Okabe, A.: Spatial Tessellations: Concepts and Applications of Voronoi Diagrams. John Wiley & Sons Inc. (2000)
17. Rigaux, P., Scholl, M., Voisard, A.: Spatial Databases with Application to GIS. SIGMOD Record 32(4), 111 (2003)
18. Roussopoulos, N., Faloutsos, C., Timos, S.: The R+-tree: A Dynamic Index for Multi-dimensional Objects. In: Int'l Conference on Very Large Databases (VLDB), pp. 507–518 (1987)

Indexing Partial History Trajectory and Future Position of Moving Objects Using HTPR*-Tree

Ying Fang[1], Jiaheng Cao[1], Yuwei Peng[1], and Nengcheng Chen[2]

[1]School of Computer, Wuhan University, China
{fangying,jhcao}@whu.edu.cn
[2]State Key Laboratory of Information Engineering in Surveying, Mapping and Remote Sensing,
Wuhan University, China
cnc_dhy@hotmail.com

Abstract. Currently, most indexing methods of moving objects are focused on the past position, or the present and future one. In this paper, we propose a novel indexing method, called History TPR*-tree(HTPR*-tree), which not only supports predictive queries but also partial history ones involved from the most recent update instant of each object to the last update time of all objects. Based on the TPR*-tree, our Basic HTPR*-tree adds creation or update time of moving objects to leaf node entries. In order to improve the update performance, we present a bottom-up update strategy for the HTPR*-tree by supplementing a hash table, a bit vector and a direct access table. Experimental results show that the update performance of the HTPR*-tree is better than that of the Basic HTPR*-and TPR*-tree. In addition to support partial history queries, the update and predictive query performance of the HTPR*-tree are greatly improved compared with those of the R^{PPF}-tree.

Keyword: Basic HTPR*-tree, History TPR*-tree, bottom-up update strategy.

1 Introduction

Traditional spatial index structures are not appropriate for indexing moving objects because the constantly changing locations of objects requires constant updates to the index structures and thus greatly degrades their performance. Numerous researchers have studies index structures for moving objects. They can be classified into two major categories depending on whether they deal with past position query or future prediction.

In general, indexing about past positions or trajectories of moving objects only stores history information from some past time until the time of the most recent position sample of each object $o(t^o_{mru})$. However, indexing of the current and anticipated future positions can only supports the query from the last update time ($t_{lu}=max(t^o_{mru}|o \in O)$) to the future. Simply combining the two kinds of indices does not solve the indexing problem: for any object, its position for times in-between the time of the most recent

H. Yu et al. (Eds.): DASFAA Workshops 2012, LNCS 7240, pp. 229–242, 2012.

sample and the last update time of all objects cannot be indexed readily with most of existing techniques.

Figure 1 shows the trajectories of three one-dimensional moving objects. The first type of index supports the solid parts, and the second type supports the dashed parts. How to store the dash-dotted parts is the most important issue in the index structures of moving object. For example, because the query time of timeslice query $Q2$ is earlier than the last update time (t_{lu}) of all objects but later than the most recent update instant of all objects $(t_{mru}=min(\ t^o_{mru}\ |o\in O)$, querying the indexing about past trajectories of moving objects only achieves partial objects such as O_3. However, querying index of the current and anticipated future positions cannot get any object although objects O_1 and O_2 should be obtained in query $Q2$. In order to realize query $Q2$ completely, indexing of the dash-dotted parts is a very important issue and we will focus on it in this work.

In order to completely query near-past positions and even from the past to the future positions of moving objects, several indices are proposed in the literature. For example, applying partial persistence to the TPR-tree, Pelanis et al. [1] propose R^{PPF}-tree which can index the past, present and anticipated future position of moving objects. However, because of *time split* of node, a history trajectory segment may be stored in several entries and even in several nodes, which greatly increases query and update cost.

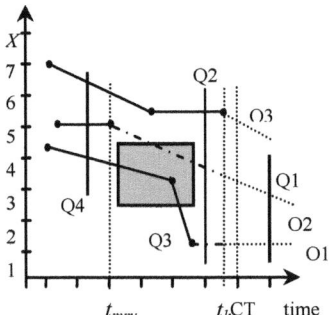 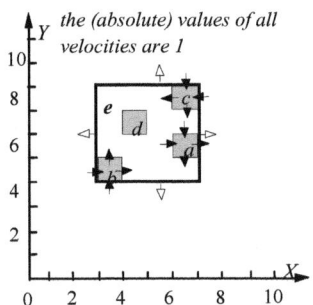

Fig. 1. Querying the Positions of Moving Objects   **Fig. 2.** MBRs & VBRs at reference time 0

As we know, the TPR-tree [2] and the TPR*-tree [3] are the most typical indexing methods in all indices of the current and anticipated future positions of moving object. However, in both the TPR- and TPR*-tree, partial history trajectories of moving objects which do not update at the last update time (t_{lu}) are implicit (such as $O1$ and $O2$ in Figure 1) and thus they cannot be queried. In order to query history trajectories in the TPR*-tree, the index structure should be modified. Traditionally an update operation in an R-Tree based index involves a top-down deletion followed by a top-down insertion. In an environment where updates are frequent, top-down update may result in a huge performance bottleneck. So, update of the TPR- and TPR*-tree which adopts top-down approach is inefficient. Lee et al. [4] propose bottom-up update approach for R-Tree. Liao et al. [5] apply bottom-up update strategy to the TPR-tree and propose HTPR-tree.

This paper proposes a novel index structure based on the TPR*-tree, named History TPR*-tree (HTPR*-tree), which takes into account moving object creation time or update time in the leaf node entry. HTPR*-tree not only supports predictive queries, but also supports partial history queries involved from the most recent update instant of each object (t^o_{mru}) to the last update time (t_{lu}) of all objects. At the same time, bottom-up update strategy is applied to the new index in order to support frequent update. This new index structure is the foundation of indexing the past, present and future positions of moving objects.

The rest of the paper organized as follows. Section 2 presents related works. Section 3 describes the TPR-tree and the TPR*-tree. The structure of the Basic HTPR*-tree, the dynamic maintenance, and query algorithms are described in section 4. Section 5 discusses the bottom-up update of the HTPR*-tree. Section 6 contains an extensive experimental evaluation, and we present our conclusions and future work in section 7.

2   Related Works

A number of index structures have been proposed for moving object database. Most of these index structures are classified into two categories; one of them is to handle past positions or trajectories, and the other is to handle current and future positions.

History trajectories of moving object are generally given by polylines. Because the R-tree is easily capable of indexing line segments, some variations of the R-tree are adopted for polylines indexing. The STR-tree [6] attempts to group segments according to their trajectory memberships, also taking spatial locations into account. The TB-tree [6] aims only for trajectory preservation, leaving other spatial properties aside. Kumar et al. [7] apply partial persistence to the R-tree (PPR-tree), the objective being to support spatio-temporal applications. Based on the multi-version B-tree (MVB-tree), Tao and Papadias [8] propose the MV3R-tree which consists of an MVR-tree and a 3D R-tree to index past trajectory data.

All the above indices capture only the positions of objects from some past time up until the time of the most recent update. However, they could not describe past positions for times in-between the most recent update instant of each object (t^o_{mru}) and the last update time (t_{lu}) of all objects (described as the dash-dotted parts in Figure 1).

The representations of the current and near-future positions of moving objects are quite different, as are the indexing challenges and solutions. Tayeb et al. [9] use the PMR-quadtree as their underlying spatial access methods for indexing the future trajectories. Papadopoulos et al. [10] use the duality transformation to transform a line segment (e.g., trajectory) from the time-space domain into a point in the two-dimensional space. Patel et al. [11] propose an indexing method, called STRIPES, which indexes predicted trajectories in a dual transformed space. By introducing parametric bounding rectangles in R-tree, the TPR-tree [2] provides the capability to answer the queries about current positions and future positions. The TPR*-tree [3] improved upon the TPR-tree by introducing a new set of penalty functions based on a revised query cost model. This leads to a different grouping of objects into index tree nodes.

Based on the B$^+$-tree, indices for moving objects not only supporting queries efficiently but also supporting frequent updates are proposed. Jensen et al. propose the B^x-tree [12], which employs space partitioning and data/query transformations to index object positions and has good update performance. Chen et al. propose the ST^{2}B-tree [13], a Self-Tunable Spatio-Temporal B+-Tree index for moving object database, which is amenable to tuning.

Several indices support both the past and the future movement of the objects. Sun et al. [14] propose a method for approximate query based on multidimensional histograms. The BB^x-tree proposed by Lin et al. [15] retains the old phases so that past, present, and future positions of moving objects are indexed, but it only indexes broken "polylines".

Among many indexing techniques for moving objects, the R^{PPF}-tree [1] is based on the TPR-tree and support query from the past to the future positions of moving objects. It can not only accurately index position for times in-between the most recent instant of each object and the last update time of all objects, but also describe connected trajectories of objects. However, since a history trajectory segment stored in several entries and even in several nodes leads to very complicated query and update procedure, a more practical indexing method should be developed.

3   TPR- and TPR*-Tree

3.1   TPR-Tree

The TPR-tree [2] is essentially a time parameterized R*-tree. The index stores velocities of the elements along with their positions in nodes. The index structure as well as the algorithms for search, insert and delete used are very similar to that of R*-tree. A two-dimensional moving object is represented with *MBR* $o_R=\{o_{R1-}, o_{R1+},o_{R2-},o_{R2+}\}$ where o_{Ri-} (o_{Ri+}) describes the lower (upper) boundary of o_R along the i-th dimension ($1\leq i\leq 2$), and *VBR* $o_V=\{o_{V1-},o_{V1+},o_{V2-},o_{V2+}\}$ where o_{Vi-} (o_{Vi+}) describes the velocity of the lower (upper) boundary of o_R along the i-th dimension ($1\leq i\leq 2$). Figure 2 shows the *MBRs* and *VBRs* of 4 objects a, b, c and d at reference time 0. The arrows (numbers) denote the directions (values) of their velocities. The *MBR* and *VBR* of b are $b_R=\{3,4,4,5\}$ and $b_V=\{1,1,1,1\}$ respectively. In Figure 2, the objects are clustered into one leaf node e whose *MBR* and *VBR* are $e_R=\{3,7,4,8\}$ and $e_V=\{-1,1,-1,1\}$ respectively. The *MBR* of a non-leaf entry always encloses those of the objects in its subtree, but it is not always tight. For example, e at timestamp 2 is much larger than the tightest bounding rectangle for 4 objects a, b, c and d.

3.2   TPR*-Tree

Tao et al. [3] propose a cost model, and a hypothetical optimal tree for predictive indices using a TPR-tree style of indexing, and replace integral penalty metrics with the area of sweeping region. Given a moving object o and a time interval T, the sweeping region $SR(o,T)$ is the region swept by o during T. The area of sweeping region is described as $A_{SR}(o,q_T)$. The TPR*-tree improves the TPR-tree by employing a new set of insertion and deletion algorithms that aim at minimizing equation: $Cost(q)=\sum_{\text{every node }o}A_{SR}(o',q_T)$. At the same time, it proposes a novel *ChoosePath* algorithm (finding the best insertion path globally) to determine the node at any level that has the least cost of deterioration.

Insertion to a full node generates an overflow, in which case both the TPR- and TPR*-trees re-insert a fraction of the entries from the node. In TPR- tree, the removed entries have a high chance to be re-inserted back into node which contains entries need to be removed again. The TPR*-tree uses *Pick Worst* to return a set of entries whose removal reduces the *MBR* or *VBR* of the parent node, which greatly decreases the chance of node split, and only those nodes that still overflow after being re-inserted entries should be split. In addition, the TPR*-tree proposes an improved *active tightening* technique that allows adjusting multiple *MBRs* in a single deletion when the object to be removed falls in the overlapping region of these *MBRs*.

Because of the above improvement, the TPR*-tree is nearly optimal and significantly outperforms the TPR-tree under all conditions. Neither the TPR-tree nor the TPR*-tree, however, can support partial history queries involved from the most recent update instant of each object to the last update time of all objects. In order to query history trajectories hided in the TPR*-tree, the index structure should be modified.

4   Basic HTPR*-Tree

In this section, the index structure of the Basic HTPR*-tree is shown and the insertion, deletion and search algorithms are discussed.

4.1   Index Structure

Different from the TPR*-tree, leaf node entry of the Basic HTPR*-tree includes creation or update time of object, and non-leaf node entry includes the minimal (and the maximal) creation or update time of all objects in leaf nodes pointed by itself. The structure of each leaf node entry is of the form (oid, tpp, st). Here oid is the identifier of the moving object, st is creation or update time of object, and $tpp=(x; v)=(x_1,\ldots, x_m; v_1,\ldots, v_m)$, with the x_i and v_i being the position and velocity in i dimension, respectively, of the object at time st.

The structure of each non-leaf node entry is in the form of $(ptr, tpbr, st1, st2)$. Here ptr is a pointer that points to the child node. $St1$ is the minimal creation or update time of moving objects included in the child node pointed by ptr, and $st2$ is the maximum value compared with $st1$. The $tpbrs$ of the HTPR*-tree are bound time-parameterized points which bound objects since time $st1$. The resulting $tpbr$ has $4m$ coordinates:

$$([x_1^{\vdash},x_1^{\dashv}],\ldots,[x_m^{\vdash},x_m^{\dashv}];[v_1^{\vdash},v_1^{\dashv}],\ldots,[v_m^{\vdash},v_m^{\dashv}])$$

Here, $v_i^{\vdash} = \min_{o\in Node}\{o.v_i\}$ $x_i^{\vdash} = \min_{o\in Node}\{o.x_i - (o.st - \min_{o\in Node}(o.st))\times \min_{o\in Node}(o.v_i)\}$

$v_i^{\dashv} = \max_{o\in Node}\{o.v_i\}$ $x_i^{\dashv} = \max_{o\in Node}\{o.x_i - (o.st - \min_{o\in Node}(o.st))\times \max_{o\in Node}(o.v_i)\}$

Figure 3 shows a leaf node e including four point objects $\{o_1, o_2, o_3, o_4\}$ in one-dimensional Basic HTPR*-tree. Here $o_1=\{o_1, 4, 0.1, 2\}$, $o_2=\{o_2, 3.5, 0.2, 3\}$, $o_3=\{o_3, 2.8, 0.05, 3\}$, $o_4=\{o_4, 3.4, -0.2, 4\}$. So the corresponding *TPBR* in entry describing leaf node e is $([3, 4], [-0.2, 0.2])$, $st1=2$ and $st2=4$.

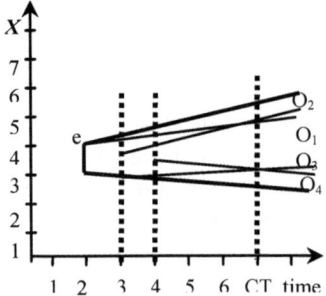   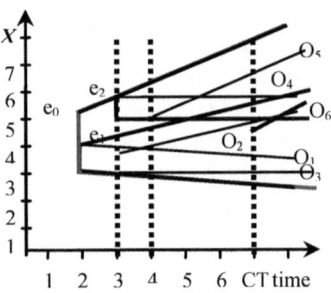

Fig. 3. A leaf Node *e* Including Four Point   **Fig. 4.** Insert a Moving Point Object *o6*
Objects

4.2   Insertion

Because the creation or update time of moving objects is included in leaf node entries of the Basic HTPR*-tree, and non-leaf node entry is different from leaf node entry, the insertion algorithm is a bit more complicated than that of TPR*-Tree.

Algorithm 1 shows the insertion of a moving object in the Basic HTPR*-tree. First, the insertion algorithm initializes two empty re-insertion lists $L_{reinsert1}$ and $L_{reinsert2}$ to accommodate re-insertion moving objects and non-leaf node entries, respectively. Then, the algorithm calls different functions according to whether the root of HTPR*-tree is a leaf node or not. Finally, the algorithm inserts each object in $L_{reinsert1}$ and each entry in $L_{reinsert2}$ to the Basic HTPR*-tree.

Algorithm 2 describes inserting a moving object to a leaf node, and algorithm 3 describes inserting a moving object to the Basic HTPR*-tree rooted by a non-leaf node. **Non-Leaf Node Insert_e** is used to insert non-leaf node entry in the Basic HTPR*-tree. Because space limited, we omit it in this paper.

Algorithm 1. Insert (*r, o*)
/***Input**: o is a moving object with *oid, MBR, VBR, st*; r is the Basic HTPR*-tree*/
1. root=Root(r)   /*achive the root of the Basic HTPR*-tree
2. *re-inserted$_i$*=false for all levels 1≤*i*≤*h*−1 (*h* is the tree height)
3. initialize two empty re-insertion list $L_{reinsert1}$ and $L_{reinsert2}$
4. if root is leaf node   invoke **Leaf Node Insert** (*root, o*)
5. else   invoke **Non-Leaf Node Insert** (*root, o*)
6. for each data *o'* in the $L_{reinsert1}$
7.   if root is leaf node   invoke **Leaf Node Insert** (*root, o'*)
8.   else   invoke **Non-Leaf Node Insert** (*root, o'*)
9. for each entry *e* in the $L_{reinsert2}$
10.   invoke **Non-Leaf Node Insert_e** (*root, e*)
End Insert

Algorithm 2. Leaf Node Insert (N, o)
/* **Input**: N is the leaf node where object o is inserted */
1.   enter the information of o
2.   if N overflows
3.   if $re\text{-}inserted_0$=false   //no re-insertion at leaf level yet
4.   invoke **Pick Data Worst** to select a set S_{worst} of objects
5.   remove objects in S_{worst} from N; add them to $L_{reinsert1}$
6. $re\text{-}inserted_0$=true
7.   else
8.   invoke **Leaf Node Split** to split N into itself and N'
9.   obtain entry e describe node N'
10.   invoke **Non-Leaf Node Insert_e** (P,e) /*P be the parent of N*/
11.   adjust the MBR/VBR/st1/st2 of the node N
End Leaf Node Insert (N, o)

Algorithm 3. Non-Leaf Node Insert (N, o)
/* **Input**: N is the root node of tree rooted by N */
1.   obtain the son node N' of N to insert o through the path achieve by **Choose Data Path**
2.   if N' is the leaf node   invoke **Leaf Node Insert** (N', o)
3.   else   invoke **Non-Leaf Node Insert** (N', o)
4.   adjust the MBR/VBR/st1/st2 of the node N
End Non-Leaf Node Insert(N, o)

Similiar to that in the TPR*-tree, algorithm *Choose Path* in the Basic HTPR*-tree aims at finding the best insertion path globally with a minimum cost increment (minimal increase in equation 1). If a moving object is inserted, *Choose Path* is instantiated by *Choose Data Path*, and non-leaf node entry inserting calls *Choose Entry Path*. Because the creation or update time of moving objects is included in leaf node entries, the enlarge algorithm of entry (caused by inserting a moving object in the Basic HTPR*-tree node) involves history information, and the enlarged entry can support history query. This is the major difference between the insertion of the Basic HTPR*-tree and that of the TPR*-tree.

The query cost model of the Basic HTPR*-tree is the average number of node accesses for answering query q:

$$Cost(q)=\sum_{every\ node\ N} A_{SR}(N', q_T) \qquad (1)$$

where N is the moving rectangle (interval for one-dimensional object) representing a node, N' is the transformed rectangle (interval for one-dimensional object) of N with respect to q, and $A_{SR}(N',q_T)$ is the extent of region swept by N' during q_T.

Figure 4 shows two leaf nodes $e1$ and $e2$ those are sons of the root $e0$. Here the entry corresponding to $e1$ is $\{pt1, \{3.2,4\},\{-0.2,0.3\},2,3\}$, and the entry to $e2$ is $\{pt2, \{5,6\},\{0,0.4\},3,4\}$. Consider the insertion of point object $O_6= \{O_6, 4.6, 0.5, 7\}$ at current time 7. *Choose Data Path* returns the insertion path with the minimum increment in equation 1. The cost increment is 1.2 and 0.9 when $o6$ is inserted to $e1$ and *e2*, respectively. Figure 5(a) describes insertion *o6* to *e1*, and figure 5(b) describes insertion *o6* to *e2*, which is the best insertion node.

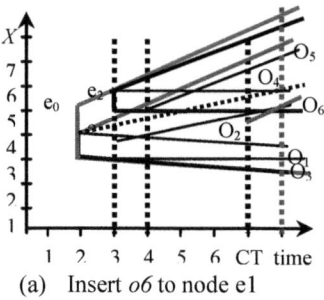 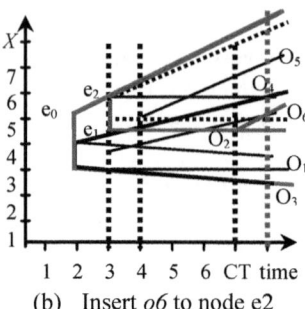

(a)   Insert *o6* to node e1   (b)   Insert *o6* to node e2

Fig. 5. Two Different Insertion Methods

Insertion to a full node N generates an overflow, in which the Basic HTPR*-tree uses *Pick Worst* algorithm that selects a fraction of the entries from the node N and re-inserts them. *Node Split* algorithm splits a full node N into $N1$ and $N2$. The split algorithm selects split axis and split position minimizing equation 2:

$$\Delta A_{SR}=A_{SR}(N_1',q_T)+A_{SR}(N_2',q_T)-A_{SR}(N',q_T) \tag{2}$$

4.3   Deletion

Algorithm 4 describes deleting a moving object in Basic HTPR*-tree. To remove an object o, the deletion algorithm first identifies the leaf node that contains o. In algorithm 4, two empty re-insertion lists $L_{reinsert3}$ and $L_{reinsert4}$ are initialized to accommodate re-insertion leaf node entries (moving objects) and non-leaf node entries, respectively.

Algorithm 4. Delete (r, o)

/***Input**: o is a moving object with *oid, MBR, VBR, st*; r is the Basic HTPR*-tree */
1.   root=Root(r)   /*achive the root of the HTPR*-tree
2.   initialize an empty re-insertion list $L_{reinsert3}$ and $L_{reinsert4}$
3.   if root is leaf node   invoke **Leaf Node Delete** (*root, o*)
4.   else   invoke **Non-Leaf Node Delete** (*root, o*)
5.   for each data o' in the $L_{reinsert3}$ invoke **Non-Leaf Node Insert** (*root, o'*)
6.   for each entry e in the $L_{reinsert4}$ invoke **Non-Leaf Node Insert_e** (*root, e*)
     End Delete (r, o)

Deletion of moving object in leaf node N may generate an underflow, in which case the Basic HTPR*-tree removes all objects in node N to $L_{reinsert3}$, and deletes entry e

describes N in parent node N'. If moving object o is deleted in the Basic HTPR*-tree root by non-leaf node N, the deletion algorithm calls *Leaf Node Delete* or *Non-Leaf Node Delete* in all son node N' of N until o is deleted. In algorithm *Leaf Node Delete* or *Non-Leaf Node Delete*, if deletion of object o in the Basic HTPR*-tree root by node N changes node N, adjustment is needed from parent node N' of N to root. Because of limited space, we omit detailed algorithm *Leaf Node Delete* and *Non-Leaf Node Delete*.

4.4   Search Procedure

The Basic HTPR*-tree supports three kinds of predictive queries: timeslice query $Q=(R, t)$, window query $Q=(R, t_1, t_2)$, and moving query $Q=(R_1, R_2, t_1, t_2)$. At the same time, the Basic HTPR*-tree supports partial history query. Figure 1 describes timeslice query and spatio-temporal range query of moving objects. The dashed parts and the dash-dotted parts of objects trajectories are stored in the Basic HTPR*-tree, and support predictive queries and partial history queries.

For example, query $Q1$ is predictive timeslice query, and gets object $O1$ and $O2$. Query $Q2$, $Q3$, and $Q4$ are history queries. Query $Q2$ intersects with partial history trajectories after the most recent update instant (t^o_{mru}) of objects $O1$ and $O2$ stored in the Basic HTPR*-tree. However, the Basic HTPR*-tree couldn't completely realize query $Q2$ because the query time is earlier than the last update time of all objects (t_{lu}). In order to realize the queries involved from the past to the future, the Base HTPR*-tree should be combined with some indices for describing history trajectories such as TB-tree.

Algorithm 5 is an illustration of spatio-temporal range query in the Basic HTPR*-Tree.

Algorithm 5. RQuery(r, w, T_1, T_2)
1.   root=Root(r)   /*achive the root of the Basic HTPR*-tree*/
2.   get *st1* and *st2* of root
3.   if   T2<*st1* return null
4.   else if root is leaf node   invoke **LeafNodeRQuery** (root, w, T_1, T_2)
5.   else   invoke **nonLeafNodeRQuery** (root, w, T_1, T_2)
ENDRQuery

Algorithm 6. Leaf NodeRQuery (N, w, T1, T2)
1.   for each o of N
2.   if **IN**(o, w, T1, T2)   add o to the result_set
3.   return result_set
ENDLeaf Node RQuery

Algorithm 7. nonLeafNodeRQuery (N, w, T1, T2)
1.   for each son node N' of N
2.   if N' is leafnode   **Leaf NodeRQuery** (N', w, T1, T2)
3.   else   **nonLeafNodeRQuery** (N', w, T1, T2)
ENDnonLeaf Node RQuery

Here, $IN(o, w, T_1, T_2)$ determines whether object o is located in range w from time T_1 to T_2.

5   Bottom-Up Update in HTPR*-Tree

Because the Basic HTPR*-tree update procedures work in a top-down manner, the update is inherently inefficient. In order to support frequent update, bottom-up update strategy is adopted by the HTPR*-tree.

5.1   Whole Structure of the HTPR*-Tree

In addition to the Basic HTPR*-tree, the HTPR*-tree also includes a hash table, a bit vector and a direct access table. Figure 6 shows the whole structure of the HTPR*-tree.

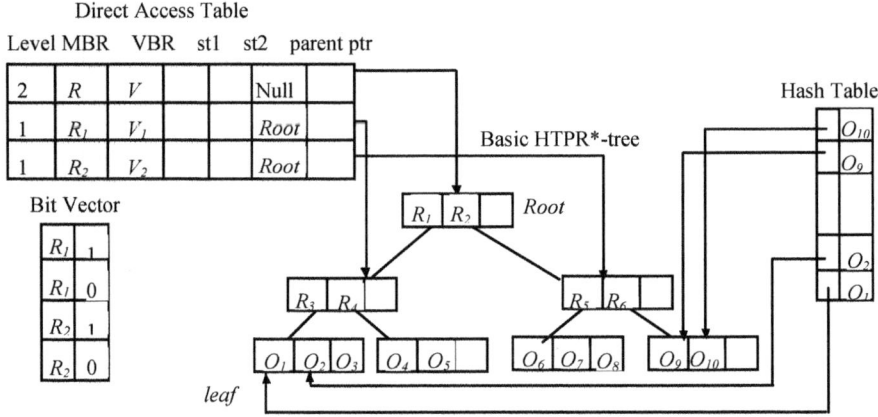

Fig. 6. Whole Structure of the HTPR*-Tree

The item in hash table is defined as vector $<oid, ptr>$, where oid denotes the identifier of moving object, and ptr denotes physical offset of the leaf node in which object entry locates. Hash table allows us to locate the leaf node where the updated object o resides in one disk I/O instead of doing the expensive query on the tree.

Direct access table facilitates quick access to a node's parent in the Basic HTPR*-tree. An entry of the direct access table corresponds to a non-leaf node of the Basic HTPR*-tree, and all the entries are organized according to the levels of the internal nodes they correspond to. An entry in the direct access table is a 7 tuple of the form $< Level, MBR, VBR, st1, st2, parentptr, ptr>$, where $Level$ is the level of the node, MBR and VBR is the bounding box and the velocity bounding rectangle of the node at time $s1t$, respectively, $parentptr$ is a pointer to the node' parent, ptr is a pointer to the node itself, $st1$ is the minimal creation or update time of moving objects included in node, and $st2$ is the maximum value compared with $st1$. Bit vector on the leaf nodes indicates whether they are full or not. The bit vector allows us to determine whether a sibling S of leaf node R is full or not without reading the actual node from disk.

The maintenance cost for the main-memory summary structure is relatively inexpensive. Since most of the node splits occur in the leaf level due to the high node fan-out, inserting a new entry into the direct access table will be very infrequent.

Meanwhile, only when the leaf node is split or deleted, a new entry need be inserted into bit vector or be deleted.

5.2   Bottom-Up Update

The bottom-up update strategy aims to offer a comprehensive solution to support frequent updates in the HTPR*-tree. The main idea of bottom-up update algorithm is described as follows: when an object issues an update request, the algorithm adopts different update method according to object position and velocity after updating, and update time. The detailed update procedure is as follows:

1. If new position lies outside *MBR* (compute in update time) of root or new velocity lies outside *VBR* of root, the algorithm issues a top-down update.

2. If new position and velocity of moving object lie in the *MBR* (compute in update time) and *VBR* of current leaf node, the algorithm modifies the object entry in leaf node directly. At the same time, the algorithm constructs update path from leaf to root using the direct access table and tightens all nodes on that path.

3. If new position and velocity of moving object lie outside the *MBR* (compute in update time)and *VBR* of current leaf node, and the removal of moving object causes leaf node to underflow, the algorithm issues a top-down update.

4. If new position and velocity of moving object lie in the *MBR* (compute in update time) and *VBR* of non-null sibling node, and the removal of moving object couldn't cause leaf node to underflow, the algorithm deletes old entry and inserts new entry in right sibling node. At the same time, the algorithm constructs update path from leaf to root using the direct access table and tightens all nodes on that path.

5. If the new position and velocity of moving object lie in the *MBR* (compute in update time) and *VBR* of a subtree (intermediate node), the algorithm ascends the Basic HTPR*-tree branches to find a local subtree and performs a standard top-down update.

6   Performance Study

6.1   Experimental Setting and Details

In this section, we evaluate the query and update performance of the HTPR*-tree with the TPR*-tree, the Basic HTPR*-tree (TD_ HTPR*-tree), and the R^{PPF}-tree. Due to the lack of real datasets, we use synthetic data simulating moving aircrafts like [3]. First 5000 rectangles are sampled from a real spatial dataset (LA/LB) [16] and their centroids serve as the "airports". At timestamp 0, 100k aircrafts (point objects) are generated such that for each aircraft o, (i) its location is at one (random) airport, (ii) it (randomly) chooses a destination airport, and (iii) its velocity value $o.Vel$ uniformly distributes in [20,50], and (iv) the velocity direction is decided by the source and destination airports.

For each dataset, we construct a HTPR*-tree, a TPR*-tree, a Basic HTPR*-tree and a R^{PPF}-tree whose horizons are fixed to 50, by first inserting 100k aircrafts at timestamp 0. Since the HTPR*-tree only stores history trajectories after the most recent update of each object, and the R^{PPF}-tree can capture the positions of moving objects at all points in time, we only compare predictive query performance of HTPR*-tree with that of the R^{PPF}-tree in our experiments. The cost is measured again as the average number of node accesses in executing 200 predicted window queries with the same parameters q_Rlen, q_Vlen, q_Tlen.

6.2   Performance Analysis

• *Update cost comparison*

Figure 7 compares the average update cost as a function of the number of updates. The node accesses needed in the HTPR*-tree update operation are much less than in the TPR*- and the Basic HTPR*-tree. This is due to the fact that the HTPR*-tree adopts bottom-up update to avoid the excessive node accesses for top-down deletion and insertion search. Since node overlap in the Basic HTPR*-tree is larger than that in the TPR*-tree, the query cost increasing with the number of updates improves the update cost of Basic HTPR*-tree.

Figure 7 shows that the HTPR*-tree has nearly constant update cost, while the update cost of the R^{PPF}-tree increases significantly. Performing update operations on the HTPR*-tree is much fast than doing the same updates on the R^{PPF}-tree. Since a history trajectory after the most recent update instant is stored in several entries and even in several nodes, which should be modified when an update occurs. This greatly enhances the cost of update operation. At the same time, a new trajectory inserted into the R^{PPF}-tree also causes *time split* of node, which also reduce the update performance.

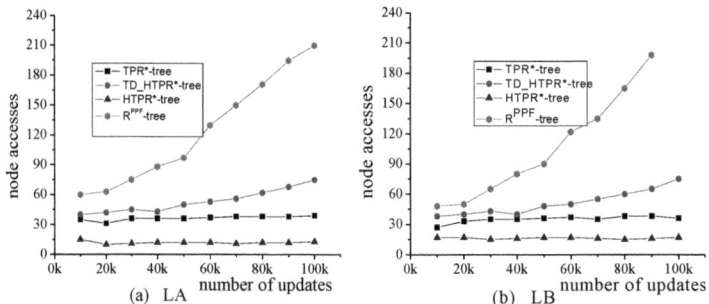

Fig. 7. Update Cost Comparison

• *Query cost comparison*

In Figure 8, we plot the query cost as a function of the number of updates. Figure 8(a) and 8(b) show that the query cost increases slowly with the number of updates in the HTPR*- and Basic HTPR*-tree. The query cost of the HTPR*-tree is a little less than that of the Basic HTPR*-tree. Since the node overlap in the HTPR*-tree is larger than that in the TPR*-tree, the query cost of the HTPR*-tree is a bit higher than that of the TPR*-tree.

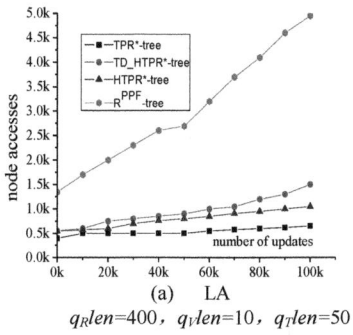
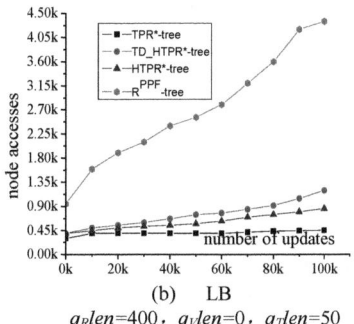

(a)   LA
q_Rlen=400, q_Vlen=10, q_Tlen=50

(b)   LB
q_Rlen=400, q_Vlen=0, q_Tlen=50

Fig. 8. Query Cost Comparison

We know from Figure 8 that the predicted query cost of the R^{PPF}-tree increases significantly with the number of updates and is much more than that of the HTPR*-tree. The reason is that the leaf nodes of the R^{PPF}-tree not only store *alive* entries but also *dead* entries. However, every entry of the HTPR*-tree leaf nodes can describe an effective predicted trajectory, which illustrates the fanout of the live part of the R^{PPF}-tree is much lower than that of the HTPR*-tree. Compared with the R^{PPF}-tree, the predictive query performance of the HTPR*-tree are thus improved greatly.

7   Conclusion

In this paper, we develop a novel index structure named the HTPR*-tree which not only supports predictive queries but also partial history ones. At the same time, we propose a bottom-up update approach to support frequent update operation of the HTPR*-tree. Extensive experiments prove that the update performance of the HTPR*-tree is better than that of the Basic HTPR*-tree (TD_ HTPR*-tree), and TPR*-tree. Compared with the R^{PPF}-tree, the update and the predictive query performance of the HTPR*-tree are greatly improved. Moreover, the HTPR*-tree can supports partial history queries. For the future work, we will combine the HTPR*-tree with history trajectory indices to implement past, current and future information retrieval.

Acknowledgments. This work is supported by the National Natural Science Foundation of China (Grant No.90718027 and Grant No.41171315).

References

[1] M. Pelanis, S. Saltenis, C. S. Jensen. Indexing the Past, Present and Anticipated Future Positions of Moving Objects. In: ACM TODS. (2006) 255-298
[2] S. Saltenis, C. S. Jensen, S. T. Leutenegger, and M. A. Lopez.: Indexing the Positions of Continuously Moving Objects. In: ACM SIGMOD. (2000) 331-342

[3] Y. Tao, D. Papadias, and J. Sun. The TPR*-Tree: An Optimized Spatio-Temporal Access Method for Predictive Queries. In: VLDB. (2003) 790-801

[4] M. Lee, W. Hsu, C. S. Jensen, B. Cui, and K. Teo. Supporting Frequent Updates in R-Trees: A Bottom-Up Approach. In: VLDB. (2003) 608-619

[5] W. Liao, G. F. Tang, N. Jing, Z. N. Zhong. Hybrid Indexing of Moving Objects Based on Velocity. Distribution, Chinese Journal of Computers, Vol. 30, No. 4, 2007

[6] D. Pfoser, C. S. Jensen, and Y. Theodoridis. Novel Approaches to the Indexing of Moving Object Trajectories. In: VLDB. (2000) 395-406

[7] A. Kumar, V. J. Tsotras, and C. Faloutsos. Designing Access Methods for Bitemporal Databases. In: TKDE. 10, 1, (1998) 1–20.

[8] Y. Tao and D. Papadias. MV3R-Tree: A Spatio-Temporal Access Method for Timestamp and Interval Queries. In: VLDB. (2001) 431–440.

[9] J. Tayeb, O. Ulusoy, and O. Wolfson. A Quadtree-Based Dynamic Attribute Indexing Method. The Computer Journal, 41(3): (1998) 185–200

[10] D. Papadopoulos, G. Kollios, D. Gunopulos, and V. J. Tsotras. Indexing Mobile Objects on the Plane. In :MDDS. (2002) 693–697.

[11] Jignesh M. Patel, Yun, Chen V., and Prasad Chakka. STRIPES: An Efficient Index for Predicted Trajectories. In: ACM SIGMOD. (2004) 637--646

[12] C. S. Jensen, D. Lin, B. C. Ooi. Query and Update Efficient B+-Tree Based Indexing of Moving Objects. In: VLDB. (2004) 768–779

[13] S. Chen, B. C. Ooi, K. L. Tan, M. Nacismento. ST^2B-tree: A Self-Tunable Spatio-Temporal B+-tree Index for Moving Objects. In: ACM SIGMOD. (2008) 29–42

[14] J. Sun, D. Papadias, Y. Tao, and B. Liu. Querying about the Past, the Present and the Future in Spatio-Temporal Databases. In: ICDE. (2004) 202–213.

[15] D. Lin, C. S. Jensen, B. C. Ooi, and S. Saltenis. Efficient Indexing of the Historical, Present, and Future Positions of Moving Objects. In: MDM. (2005). 59–66.

[16] Http://www.census.gov/geo/www/tiger

A Linear Broadcast Indexing Scheme
in Road Environments with Sensor Networks

Soo Kang[1], Dongkyo Hwang[1], Junho Park[1],
Dongook Seong[2], and Jaesoo Yoo[1,*]

[1] Department of Information and Communication Engineering,
Chungbuk National University,
410 Seongbong-ro, Heungdeok-gu, Cheongju, Chungbuk, Korea
[2] Korea Advanced Institute of Science and Technology,
291 Daehak-ro, Yuseong-gu, Daejeon, Korea
{biue31u,corea1985,seong.do}@gmail.com,
{junhopark,yjs}@chungbuk.ac.kr

Abstract. Various broadcast schemes have been proposed to provide moving objects on the road network with efficient location-based services. However, they were mainly concerned with the implementation of a broadcast index and did not consider road network environments that moving objects change frequently. In this paper, we propose a novel bi-directional linear broadcast scheme which takes the consideration of road network characteristics. The proposed scheme splits a service area into the sensor cluster-based segments in order to process a query efficiently. The proposed scheme also performs the optimized data update based on road information collected in sensor networks by considering the frequently changed moving object environments. As a result, it can minimize the broadcast of unnecessary data. To show the superiority of our proposed scheme, we compare it with the existing broadcast scheme. Our experimental results show that our proposed scheme reduces about 13% tuning time and about 22% access latency over the existing schemes on average.

Keywords: Road Network, Moving Object, Broadcast, Index, Sensor Networks, Sensor Clustering, Database.

1 Introduction

In recent, with the developments of the wired or wireless internet and communication technologies, the demand for location based services(LBS) has been increased on a large scale. LBSs mean providing a variety of services using location and geographical information such as illegal parking area information services, warning services that set off alarms when vehicles enter children protection areas, and information services on shops selling promotional items[1,2]. To provide this location-based service, a technique to control the effective location information and perform the spatial query is necessary.

---

* Corresponding author.

H. Yu et al. (Eds.): DASFAA Workshops 2012, LNCS 7240, pp. 243–249, 2012.

One of such LBS request methods is the broadcast method. The broadcast technique is the one that the query of clients are not received by the server but all the usable data for clients are broadcasted and each client receive the necessary data to treat the query individually based on the relevant data. The location-based broadcast service based on broadcast technique is appropriate for the environment that there are many clients for the service and the amount of data to provide is fixed. It is not influenced by the number of clients and can provide a regular quality of performance.

The performance of broadcast service can be presented with the tuning time that the client takes to listen to the channel to get the data and the access latency that is taken from request of client for data till reception of relevant data. By optimizing these two elements, the effective broadcast service can be obtained.

In recent, with the development of the high-performance and low-cost sensors, many applications with the sensor network have been studied. Furthermore, some real applications such as smart city for the urban environment information tend to increase. With the sensor network, it is possible to comprehend velocity, density, directions of the vehicles and other information in the road environment with frequent changes. Therefore, it is necessary to find an integrated application system that the information is collected with real-time monitoring through the sensor network and the optimum broadcasting data is generated based on it.

In this paper, we propose a novel bi-directional linear broadcast scheme which takes the consideration of road network characteristics to reduce the access latency and tuning time. The proposed scheme splits a service area into the sensor cluster-based segments in order to process a query efficiently. The proposed scheme also performs the optimized data updates based on road information collected in sensor networks by considering the frequently changed moving object environments.

The remainder of this paper is organized as follows. In Section 2, we present our efficient bidirectional linear broadcast indexing with sensor networks in road environments. Section 3 shows the simulated experiments and compares the existing broadcast scheme with the proposed scheme. Finally, we present concluding remarks in Section 4.

2   The Proposed Method

In this paper, we propose a bi-directional linear broadcast scheme that can recognize various road conditions. Fig.1, shows a broadcast scheme that recognizes road conditions using sensor networks and splits the entire road into segments using the distributed sensor nodes. Each sensor node collects real-time information regarding the objects speed and density located in the area it belongs to. The collected data are sent to the header nodes and they are broadcasted according to a pre-established broadcast strategy optimized for the road conditions based on them.

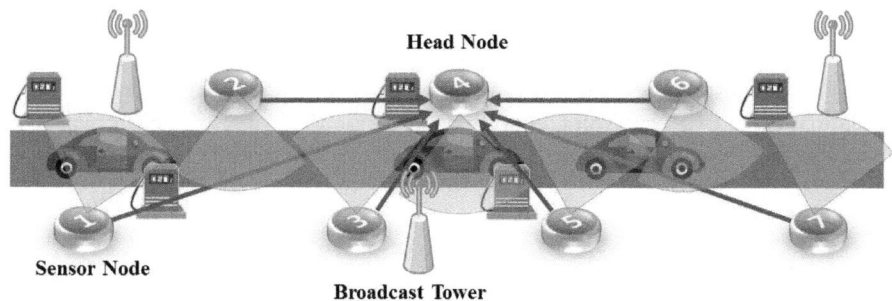

Fig. 1. System Architecture

The broadcast tower is to broadcast data by optimal broadcast strategy. The quality of broadcast schemes can be improved through the exclusion of geographically unrelated geographical data by dispersively broadcasting road information. The sensor clusters are configured by using the linear hop-count in the proposed scheme.

First, select the initial sensor nodes and them as a progressive step. It broadcast only the data that is related in the segment. Fig. 2 shows the road segment separation using the sensor networks. The split road segments can be divided into outbound and inbound lanes according to the moving directions of the objects and the pieces of information on the objects moving in the two directions are not identical. In this regard, when constructing a broadcast index, we should take into account two way indices.

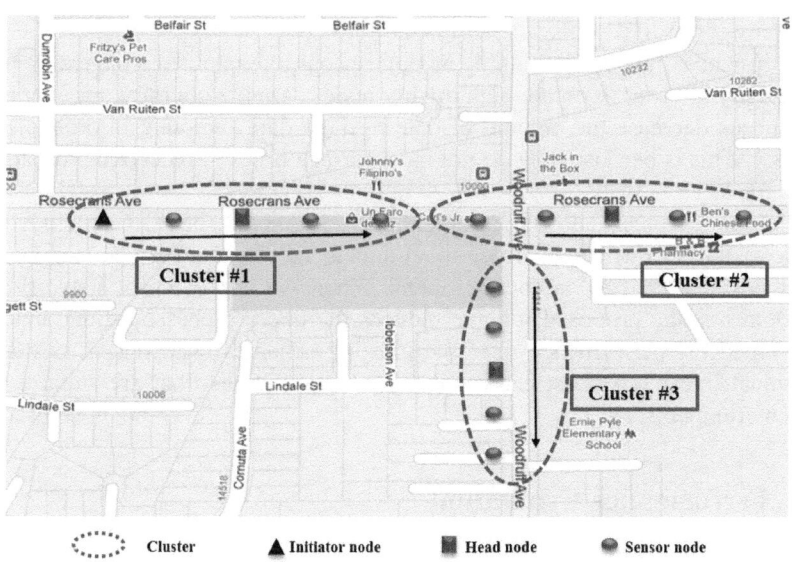

Fig. 2. Road Segment Separation Using Sensor Networks

The proposed bi-directional broadcasting data allocation should linearly be arranged by considering the moving directions of the objects and the access order.

Once the bi-directional broadcasting data is arranged, the proposed scheme establishes a final broadcast strategy by merging the data in a zigzag. Bi-directional road environment cannot interfere if a similar situation. However, the actual application of a two-way road conditions are not the same situation will occur frequently. As shown in Fig. 3, if road situations in both directions are different, sample data are constructed by sampling from the broadcast data in both directions.

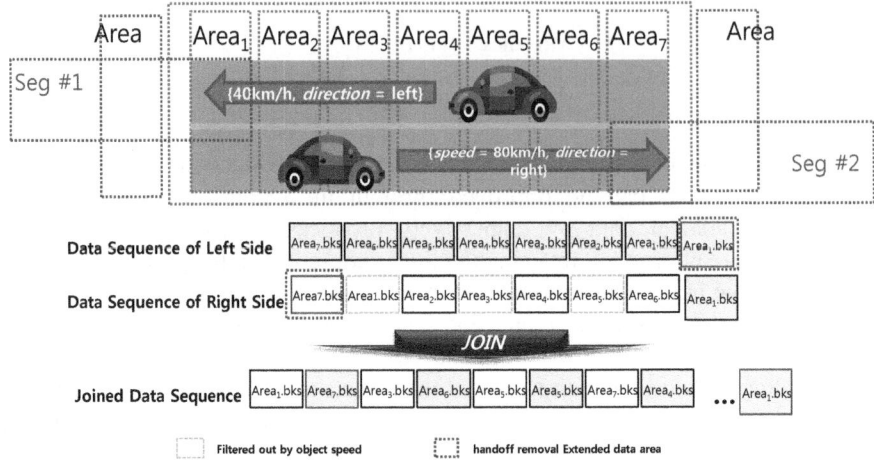

Fig. 3. Bi-directional Broadcast Index Configuration

For instance, the objects running at slow speeds due to traffic congestion have high data density in order to obtain high quality results. On the contrary, moving objects at high speeds decrease the amount of data through data sampling in order to improve accuracy. This is because they do not have enough times to receive the broadcast data on the roads.

Some of the local data of the neighboring segments should be included in each broadcast data sequence in order to eliminate the hand-off phenomenon when an object is transferred to another segment. When we apply sensor networks to road environments, the proposed scheme reduces the query processing time and achieves good broadcast performance over the existing schemes. The reason is that the bi-directional linear broadcast can recognize road situations, that are suitable for fluid road environments.

3   Performance Evaluation

We have developed a simulator based on JAVA to evaluate our proposed scheme and the existing broadcast scheme based on P2P[5]. The performance evaluation was carried out through the evaluation environment in Table 1.

Table 1. Evaluation Environment

Parameters	Value
Total Length of Road Network (Km)	21
Length of Segment (Km)	7
Length of Area (Km)	7
Size of Bucket (Bytes)	1000
Size of Data (Bytes)	1
Number of Data (EA)	128
Bandwidth of Broadcast Channel (Mbps)	2

Fig. 4 and Fig. 5 show the tuning time access latency according to the amount of data. The tuning time increases as the number of objects increases in existing broadcast scheme. But, in this proposed scheme, the whole service area is divided into segments of sensor cluster and only the data of segment is broadcasted, excluding the locally non-related data to minimize the tuning time. The amount of broadcasting data increases as the number of objects increases. With this, the unnecessary data with the query is increased in the broadcast data. Therefore, the existing broadcast scheme increases the access latency. However, the distributed broadcast is carried out to receive the data directly related with the query in proposed scheme. As a result, it can decrease the access latency. As the results of performance evaluation, the proposed scheme showed that the tuning time was reduced by about 13% and the access latency was reduced by about 8% on average over the existing broadcast scheme.

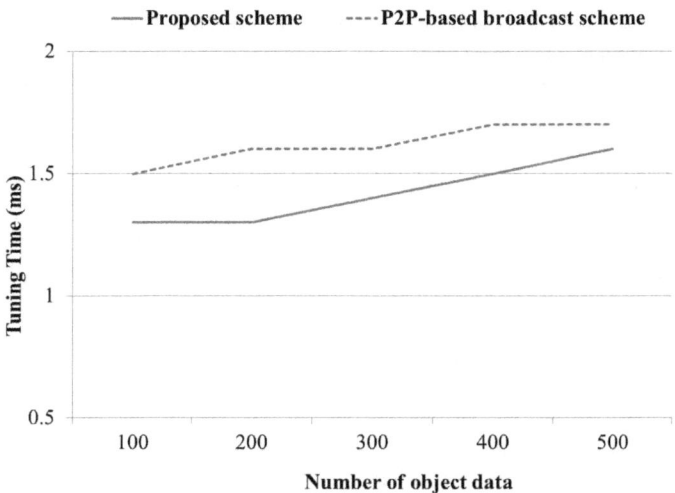

Fig. 4. Tuning Time According to the Amount of data

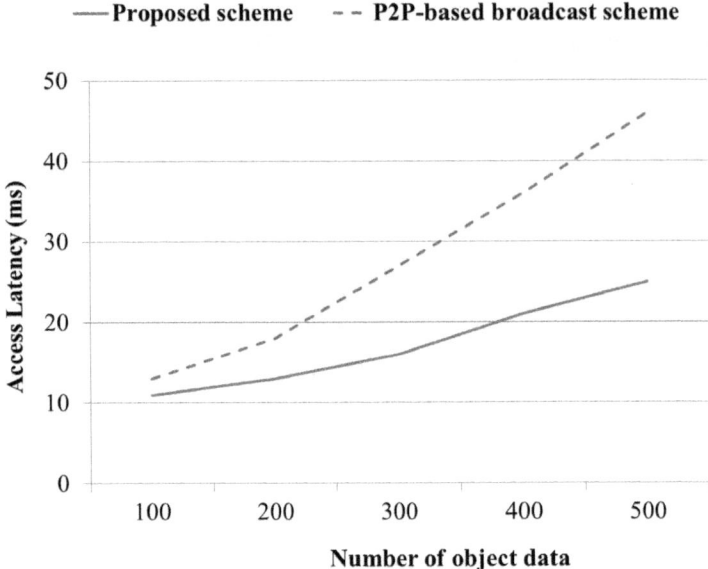

Fig. 5. Access Latency According to the Amount of data

4   Conclusions

In this paper, we have proposed a bidirectional linear broadcasting scheme that can recognize road environments based on the sensor networks. The proposed scheme splits a service area into the sensor cluster-based segments in order to process a query efficiently. The proposed scheme also performs the optimized data updates based on road information collected in sensor networks by considering the frequently changed moving object environments. As a result, our scheme reduces the tuning time by about 13% and access latency by about 22% over the conventional schemes. Moreover, our scheme improves about 8% query response rates over the existing schemes on average.

Acknowledgments. This work was supported by Technology Development Program for ('Agriculture and Forestry' or 'Food' or 'Fisheries'), Ministry for Food, Agriculture, Forestry and Fisheries, Republic of Korea and Basic Science Research Program through the National Research Foundation of Korea(NRF) grant funded by the Korea government(MEST)(No. 2009-0080279).

References

1. Sadoun, B., Al-Bayari, O.: Location Based Service Using Geographical Information Systems. Computer Communications 30(16), 3154–3160 (2007)
2. Abrougui, K., Boukerche, A., Pazzi, R.W.N.: An Efficient Fault Tolerant Location Based Service Discovery Protocol for Vehicular Networks. In: 2010 IEEE Global Telecommunications Conference (2010)

3. Zheng, B., Lee, D.L.: Information Dissemination via Wireless Broadcast. Magazine Communications of the ACM 48(5), 105–110 (2005)
4. Xu, J., Lee, W., Tang, X., Gao, Q., Li, S.: An Error-Resilient and Tunable Distributed Indexing Scheme for Wireless Data Broadcast. IEEE Transactions on Knowledge and Data Engineering 18(3), 392–404 (2006)
5. Ku, W.S., Zimmermann, R., Wang, H.: Location-Based Spatial Query Processing in Wireless Broadcast Environments. IEEE Transactions on Mobile Computing 7(6), 778–792 (2008)
6. Culler, D., Estrin, D., Srivastava, M.: Guest Editors' Introduction: Overview of Sensor Networks. IEEE Computer 37(8), 41–49 (2004)
7. Knight, A., Yu, Q., Rege, M.: Efficient range query processing on uncertain data. In: IEEE International Conference on, Information Reuse and Integration, IRI (2011)

Spatial Keyword Queries

Gao Cong

School of Computer Engineering,
Nanyang Technological University, Singapore 639798
gaocong@ntu.edu.sg

Abstract. Spatial-textual content is becoming increasingly prevalent:

— Location-based services from major commercial search engines. For example, in Google Maps many (geo-referenced) points of interest—e.g., clinics, stores, tourist attractions, hotels, entertainment services, public transport, and public services—are being associated with descriptive texts.
— Websites with location content. For example, online yellow pages, documents of Wikipedia, Tweets in Twitter, photos in Flickr, points of interest in Foursquare, etc.
— Moving objects associated with texts. An example scenario is that each healthcare worker has certain skills, described in keywords, and moves around in a large hospital.

These call for spatial-keyword search from the perspectives of both the users and the service providers. From the user's perspective, users may want to issue queries such as "health screening clinics near NTU, Singapore", which has a location component "NTU, Singapore" and a keyword component "health screening clinics". Indeed, location-based services (e.g., Google Maps) and Twitter already support such types of queries. From the perspective of service providers, they want to know the number of customers who are interested in their services compared with competitors. For example, a nutrition store may want to find potential customers whose profiles are relevant to the products of the store and whose locations are close to this store. The talk covers recent results [1–6] on spatial keyword querying obtained by the speaker and his colleagues.

References

1. Cao, X., Cong, G., Jensen, C.S.: Mining significant semantic locations from gps data. PVLDB 3(1), 1009–1020 (2010)
2. Cao, X., Cong, G., Jensen, C.S.: Retrieving top-k prestige-based relevant spatial web objects. PVLDB 3(1), 373–384 (2010)
3. Cao, X., Cong, G., Jensen, C.S., Ooi, B.C.: Collective spatial keyword querying. In: SIGMOD Conference, pp. 373–384 (2011)
4. Cong, G., Jensen, C.S., Wu, D.: Efficient retrieval of the top-k most relevant spatial web objects. In: PVLDB, pp. 337–348 (2009)
5. Lu, J., Lu, Y., Cong, G.: Reverse spatial and textual k nearest neighbor search. In: SIGMOD, pp. 349–360 (2011)
6. Wu, D., Yiu, M.L., Jensen, C.S., Cong, G.: Efficient continuously moving top-k spatial keyword query processing. In: ICDE, pp. 541–552 (2011)

H. Yu et al. (Eds.): DASFAA Workshops 2012, LNCS 7240, p. 250, 2012.
© Springer-Verlag Berlin Heidelberg 2012

Evaluating Spatial Keyword Queries under the MapReduce Framework⋆

Wengen Li[1], Weili Wang[1], and Ting Jin[2]

[1] Dept. of Computer Science and Technology, Tongji University, Shanghai, China
{lwengen,ken.wlwang}@gmail.com
[2] School of Computer Science, Fudan University, Shanghai, China
tingj@fudan.edu.cn

Abstract. Spatial keyword queries, finding objects closest to a specified location that contains a set of keywords, are a kind of pervasive operations in spatial databases. In reality, there is some spatial data that is not stored in databases, instead in files. And generally this kind of spatial data is textual, noisy, large-scale and used now and then, which makes it be quite costly to conduct spatial keyword querying on such spatial data. To solve this problem, in this paper we propose an efficient method by using a distributed system based on MapReduce. The algorithm for spatial keyword query evaluation under the MapReduce model is developed and implemented. Experimental results demonstrate that this method can process spatial keyword queries effectively and efficiently.

Keywords: Spatial keyword query, MapReduce, Hadoop, HDFS.

1   Introduction

Spatial keyword queries [1] are a widely used in spatial databases. The goal is to find some objects that are closest to the query location and contain the specified set of keywords. With the proliferation of global positioning systems (GPSs) and their applications, more and more objects on the Web are associated with geographical locations, in addition to textual labels. This development gives significance to spatial keyword queries.

In practice, there are application scenarios like this. We have large scale spatial datasets that are not stored in databases, and we want to find some specific objects from the datasets within a limited time. In addition, every object has spatial attributes and textual attributes. For example, the geographic objects recorded by the U.S. Board on Geographic Names are just stored as text files. We call this kind of spatial data none-in-database spatial data.

The existing methods for spatial keyword queries have limitations to deal with the none-in-database data. First, it is highly expensive to put the data into databases. Because the data can be dirty, type mixed and unstructured. So a

---

⋆ This work was supported by National Natural Science Foundation of China under grant No. 60873040.

H. Yu et al. (Eds.): DASFAA Workshops 2012, LNCS 7240, pp. 251–261, 2012.

lot of work has to be done before we can put the data into databases, including extracting useful attributes, deleting meaningless objects, checking the attributes with null value etc. Second, queries are issued now and then, which means that most time we just leave the data there without operations. So there is no need to spend a lot of time and other resources to maintain it. What is more, the index for the data will be quite large and a lot disk space is needed to store the data and indices. Last but not least, the more result objects are required, the more cost has to pay. So it is important to find new techniques to deal with big spatial data over which spatial keyword queries can be conducted efficiently without index support. The MapReduce [2] is a good choice.

MapReduce is a distributed programming model put forward by Google, which is one of google's core technologies. MapReduce is designed to process large-scale datasets by dividing the data into many small blocks and dealing with them in parallel, which can reduce the run time substantially. In this work, we explore effective methods to efficiently support spatial keyword queries on none-in-database spatial data by using MapReduce.

The rest of this paper is organized as follows. Section 2 gives a description of the MapReduce programming model and introduces spatial keyword queries. Section 3 introduces how to evaluate spatial keyword queries based on MapReduce. Section 4 is experimental evaluation. Finally, Section 5 concludes the paper.

2   Preliminaries

2.1   The MapReduce Programming Model

MapReduce is a widely used programming model designed to process large scale of data on distributed system. The main idea of MapReduce is to divide the original big data into many small blocks and then process them on different machines at the same time. Usually the size of a block can be up to 64M which is much larger than the disk page. MapReduce can implement this with the help of HDFS which is the open source realization of GFS [3], a distributed file system.

MapReduce processes the data through two phases, i.e., map phase and reduce phase. In map phase the main task is to obtain useful information from the data file and form the key/value pair which is an important feature in MapReduce. In general, one map task takes care of one block and the output of the map task is divided into R pieces. Here the value of R can be defined by the user. In the second phase a reduce task takes a piece of data from every map task output and then put them together. In the reduce phase the objects with the same key are going to be combined together. Between the two phases there are some other important operations, like remote read, shuffle and sort. At last R final files will be produced by R reduce tasks. Fig. 1 shows the implementation of MapReduce [4].

Like the figure describes it is quite clear for programmers to program their own distributed programs by MapReduce. But in practice there is still a lot of work to do, like the data division design, the map operation, the shuffle operation, the sort operation and the reduce operation.

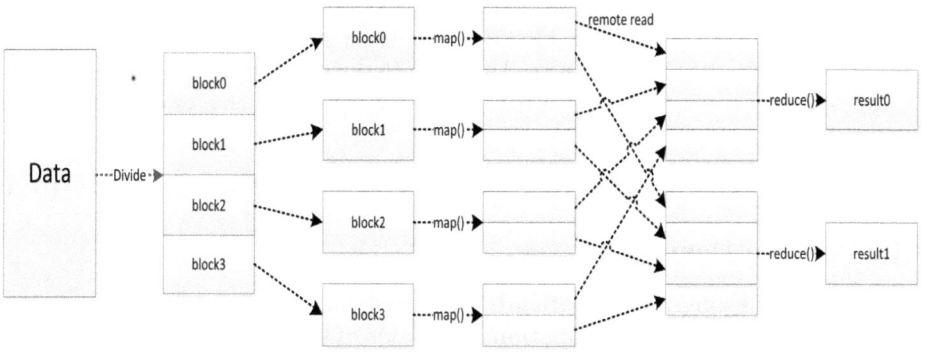

Fig. 1. The workflow of MapReduce

2.2   Spatial Keyword Queries

Spatial keyword query is a widely used query in spatial database. Many applications require spatial keyword query operation. For example, one may want to find the nearest supermarket which sells shoes from his or her home.

Assume that D is a spatial database. Each object o in D has the form $o(o.ID, o.s, o.t)$, where $o.ID$ is the identifier, $o.s$ is the spatial attributes and $o.t$ is the keywords list. The query condition is $q(k, q.s, q.t)$, where the $q.s$ and $q.t$ are the same with $o.s$ and $o.t$. The value of k in the query condition is based on the user's requirement for the number of final objects.

Specially, when k is equal to 1 spatial keyword query can be defined like this. The result object should have the keywords in the $q.t$, which means $q.t$ should be a subset of $o.t$. And also the object should have *Min (dist (o.s, q.s))* in the database D. So the spatial keyword query can be defined as follows.

$$(r.t \subseteq q.t) \wedge dist(r.s, q.s) = min(dist(o.s, q.s), \forall o \in D). \tag{1}$$

The object r is the result.

If more objects are wanted, we can just do a kNN query [5,6] based on $dist(o.s, q.s)$.

The straightforward method for spatial keyword queries is to check all the objects one by one until the right answer is found, which will lead to very high cost. The practical method is to construct some kind of index structure for spatial database. The common index structure for spatial data is R-tree [7] and its variations, like R*-tree [8], packed R-tree [9]. And the common index structure for keywords is inverted file and signature file [10]. The original way for spatial keyword query is to search the spatial attributes and keywords respectively and then combine them together. This can results in some unexpected problems like how to combine them and how to make sure the final result is right. Then researchers come up with IR-tree [11], IR2-tree [12] and BR*-tree [13], and the main idea is to create hybrid index [14] for both spatial and textual attributes.

But these methods have some disadvantages to deal with none-in-database data which is described in the introduction. So we try to combine the spatial keywords query with MapReduce. And MapReduce can deal with these disadvantages well.

3   Spatial Keyword Queries Processing under MapReduce

3.1   Problem Definition

In the real world some spatial data are not stored in the database for some reason. For example, the location information of millions of buildings in the past 30 years is just stored as textual data. Another example is the temperature recording which has spatial feature. The temperature recording has a long history, may be from the early 20th century or even earlier. For many countries the temperature recording is stored as text file.

Both of the above two kinds of data have the following features: 1) Dirty: The data can be in a mess. In general the history of the data may be quite long. During the period some information could have been lost. And another possibility is that the data format at different time can be different. 2) Textual: The data is stored as a text file and has no existed standard structure to maintain it. 3) Big: The data can be quite large. We need to store all the information for the buildings or the temperature recording. So the data can be dozens of GB or even TB. 4) Not in database: The data is just in the text file. It is not stored in some kind of database. 5) Used now and then.

As history data it is used only when someone wants to have a study on it. At most of the time, the data is just stored there. We may have the following application requirements on this kind of data: 1) Obtain one spatial keyword query result object; 2) Obtain k spatial keyword query result objects; 3) Obtain all the objects which have the given keywords.

We can define them formally. Every object has the form $o(o.ID, o.s, o.t, o.m)$, where $o.ID$, $o.t$ and $o.t$ are the same with the section 3. Here o.m represents some other unconcerned information. The query is $q(k, q.s, q.t)$.

For the above three application requirements there may be some requirements. For example, the result should be returned in a tolerant time. And the cost should be reasonable. This makes it impossible for us to analyze the data and put them in the database and create an efficient index structure for it. So here the distributed model MapReduce is employed to solve these problems.

We divide the whole process into six steps, which are respectively as follows: 1) Data division; 2) Key/value pair extraction; 3) Keyword check and spatial attributes extraction; 4) Distance computation; 5) Partition and sorting; 6) Obtain the result.

The Step 1) is responsible for dividing the input data set into many small blocks. Steps 2)–4) belong to the map phase and Step 6) is the reduce phase, and Step 5 is the middle operation between the map phase and the reduce phase.

3.2   Data Division

The original data need to be divided into blocks before the map operation. Because the data can be quite large, may be terabytes, or even petabytes. So in practice it is extremely difficult for one computer to process it. We need to divide the data into many blocks and put them in different machines to process. In MapReduce the size of each block can be 16M to 64M, or bigger. Here we choose 64M.

If the input file is just text type then we just divide it into blocks from the preliminary to the end. This is quite simple. For example, if the file is 200M we divide it into four blocks and the first three are 64M while the last one is only 8block. One thing need to notice is that two blocks may share a line. It means some objects may belong to two blocks. In this situation we have to read the two blocks when the object is needed.

If the input file has been compressed we need to consider whether the compression format support division. If the compression format is ZIP or bzip2 then we can just divide it like the text type file. But some compression formats, like DEFLATE, Gzip, LZO, do not support division. In this situation we have two ways to process it. The first one is to extract it. After that we can process it like the common file. Another way is just give it to a map task which means the map task has to process a big file. The second way goes away from our original intension. So it is used now and then.

3.3   Key/Value Pair Extraction

Before we begin to extracting the key/value pair we need to describe the format of our data. It will be more convenient for us to show our idea if the data format can be fixed although the data can be of all kinds of formats.

We assume the data is comprised of many objects and each object is a line. Every object consists of four parts. One is the *ID* which identifies the object. The second part is the spatial attributes which can be any dimensional. For easy to present the process procedure we just consider two dimensions. So the spatial attributes can be just represented with a two-tuples (x, y). In practice the spatial can be hid in the object and we need more effort to find it out. The third part of the related keywords and the fourth pare is some other information.

Here we assume the every line is a object. If the data is not copperplate we can just use additional operation to clear it.

So the key/value pair can be defined like this. The key value is the offset of each line and the value is the content of the line or the object.

3.4   Keyword Check and Spatial Attributes Extraction

In spatial key word query it is quite difficult to take both spatial attributes and textual attributes into account. The main challenge is that they are two different kinds of data and they have different features. The traditional way to is to process the spatial attributes and textual attributes separately and combine the two results together. But it is quite difficult to make the combination. Another way is create hybrid index which has both the spatial attributes and the keywords. Here we do not have the index and we do not have the necessary to create it. So the original way is used to process the data. By using MapReduce the disadvantage of the original way can be avoided. Because every block is just 64M and it will be quite easy to traverse it.

If we just want to obtain the object which has all the required keywords we can scan the block to check it. The object can be a candidate if and only if all the required keywords are included by it. If we do not need the object to include all the keywords or we have other requirements on the keywords we can just define a function and use it to choose the candidates.

Then it is needed to obtain the spatial attributes from the candidates and turn them into standard format. Every candidate is a line. It can consist of all kinds of attributes. So we need to spend much effort to analyze it and find out the spatial data.

After that we can calculate the spatial distance between the candidate and the query point.

3.5   Distance Computation

The distance between the candidate and the query point can be the simple Euclidean distance, or some other distance like Manhattan distance. User only needs to describe the way to calculate the distance. MapReduce can do it on every candidate for you.

After that it is needed to point out the output format which is another format of key/value pair. Here the distance is used to be the key because we need to obtain result by distance, and the content of the object is the value.

3.6   Partition and Sorting

A big part of the objects which do not have the keywords can be pruned through the map phase but there may be still a lot of candidates. So we need to have an efficient way to deal with this problem. In MapReduce we can use more than one reduce task to process the output from map tasks. The key point is how to distribute the output to the reduce tasks.

There are two challenges for partition. First we should make sure that the output of each reduce work can combine to be a sorted file. This requires the partition strategy to put the data into reasonable groups. Secondly, we should make sure the partition is fair for each reduce work. Because it is unreasonable to ask one reduce task to do the 90% of the work and the others only need to process the rest. This can reduce in bad load balance.

Hash function is the general method for partition. By doing this pairs which have the same key can be put into the same group. Here we need to define a new partition function because we have to make sure that the candidates in one reduce task are close to each other by the key. For example, if the number of reduce tasks is two, then we need to divide the output of map phase into two parts. Assume the output is as follows (just consider the key):

5, 1, 9, 6, 2, 4,

then we can define a partition function like this:

```
if key >= 5
    put it into part1;
else
    put it into part2;
```

So the six keys are put into two groups, one includes 5, 9 and 6, and the other has 1, 2 and 4, which makes sure that the number in part2 is always smaller than part1. But how to decide the partition value, here is 5, is a challenge. We need first to assess the data and obtain an approximate value. The general method is to take a random sample from the big data and then find out the distribution. After partition, every reduce task have a temp file which consists of the parts from every map task. The next work is to combine them together and sort them. This is quite simple because any efficient sort algorithm can be used to do the work.

3.7   Getting the Result

Different applications may have different requirements for the final result. For example, we just want to find the nearest object from query point and the object should have all the required keywords. This is the simplest application. More general situation is to find more than one object that satisfies the requirements. In this situation, the cost of traditional spatial keyword query will increase by the size of results. Fortunately, we do not need additional cost to do this by MapReduce. We can just open the output file by reduce task and fetch the *top k* objects. Extremely, if all the objects including the keywords are required the traditional methods will be highly costly. But in MapReduce the output of each reduce task is the final answer.

4   Experiments

4.1   Datasets

There are two sources of data, i.e., synthetic data and real data. The synthetic data is randomly generated and contains about 50,000,000 objects. Each object consists of three parts, the ID, the spatial attributes and the keyword list. The ID is a positive integer; the spatial attributes are in two dimensional spaces which have the format of (x, y); the keyword list consists of one or more keywords. We first choose 100 keywords and then pick a random number of keywords from the set for each object. We use S-data to represent the synthetic data.

The real data is from the U.S. Board on geographic names (geonames.usgs.gov). We use R-data to represent this kind of data. Every object in R-data is a geographic location and it has a large number of words to describe some related information. The real data needs to expand because its size is too small, only about 200MB. Here we transform the spatial attributes by adding a random number to each of the spatial attribute. Finally we obtain the same scale of data with S-data.

4.2   Query Design

First we define three kinds of queries. They are Single Spatial Keyword Query (SSKQ), k-Spatial Keyword Query (kSKQ), and Complete Spatial Keyword Query (CSKQ). They respectively return one record, k records and all the required records.

These queries are executed on a single computer and the distributed system which is implemented by the MapReduce. And the size of data varies from 100MB to 3.4GB.

4.3   Experiment Setup and Performance Metrics

We implemented all the algorithms with JDK1.6 and the configuration for each computer is an Intel(R) Core(TM)i3 CPU @2.6GHz with 2GB of RAM. The systems consists of 32 computers and its topological structure is a typical two-level network architecture which is described as Fig. 2.

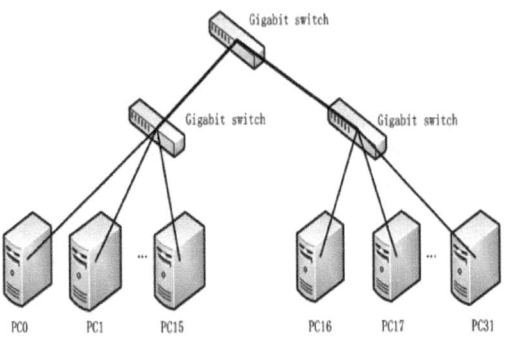

Fig. 2. The topological structure of MapReduce cluster System

The Hadoop[4] is used to build the distributed system. Hadoop consists of MapReduce and HDFS. When running the program we start half of the 32 machines. One of the 16 machines is used as master node which works as the namenode and jobtracker. And the other 15 machines are used to serve as datanodes and tasktrackers. Fig. 3 lists the 15 live datanodes.

When running the program we can set different number of map tasks and reduce tasks to test our solutions. And the running time is used to assess the efficiency of each query.

Live Datanodes : 15

Node	Last Contact	Admin State	Configured Capacity (GB)	Used (GB)	Non DFS Used (GB)	Remaining (GB)	Used (%)	Used (%)	Remaining (%)	Blocks
hadoop-10	2	In Service	91.67	0	12.82	78.85	0		86.02	1
hadoop-11	2	In Service	91.67	0	13.63	78.04	0		85.13	0
hadoop-12	1	In Service	91.67	0	13.19	78.48	0		85.61	0
hadoop-13	0	In Service	91.67	0	13.4	78.26	0		85.38	0
hadoop-14	0	In Service	91.67	0	12.7	78.96	0		86.14	0
hadoop-15	0	In Service	91.67	0	14.47	77.19	0		84.21	0
hadoop-16	0	In Service	91.67	0	14.06	77.61	0		84.66	0
hadoop-2	1	In Service	91.67	0	13.77	77.9	0		84.98	0
hadoop-3	0	In Service	91.67	0	13.12	78.55	0		85.69	0
hadoop-4	0	In Service	91.67	0	14.2	77.47	0		84.51	0
hadoop-5	2	In Service	91.67	0	13.92	77.75	0		84.82	0
hadoop-6	2	In Service	91.67	0	13.34	78.33	0		85.45	0
hadoop-7	2	In Service	91.67	0	13.28	78.38	0		85.51	0
hadoop-8	2	In Service	91.67	0	13.95	77.71	0		84.78	1
hadoop-9	2	In Service	91.67	0	13.85	77.81	0		84.89	1

Hadoop, 2011

Fig. 3. The live data nodes

4.4   Results and Analysis

Fig. 4 shows the running time with different k value on single computer on the S-data. As can be seen from the Fig. 4 the running time grows with the k value almost linearly. It means the running time will be bigger if we want to find more records. For example if we want to find 50000 records the time cost will be approximately 280 seconds. And we can also see from the Fig. 4 that the running time grows more rapidly if the data size is larger.

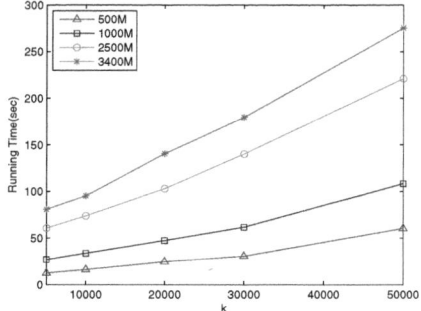

Fig. 4. Running time *vs.* k value

The following two figures show the results of the three queries on the two kinds of data. The k value is 10 for the kSKQ. As for the CSKQ we set k to 50000 which is smaller than the required value. We do this because the running time will be too big to show if we ask for all the required records.

From Fig. 5(a) we can see that the running time for three queries on single compute increases with the data size linearly. The SSKQ and kSKQ almost have the same polyline because the difference between 1 and 10 is not outstanding. But the running time for CSKQ grows much more rapidly because it has quite large k value. The running time on MapReduce is almost the same when the data size changes. And the running time is much bigger than query on single computer when the dataset is small. The reason is that MapReduce needs about 30 seconds to initialize. But after that the data is divided into many small blocks and these blocks are processed in parallel. So the running time is rather stable. In addition, the running time on MapReduce has nothing to with the k value. That is why the three polylines for SSKQM, kSKQM and CSKQM are overlaid with each other. This is a big advantage compared with the single computer query.

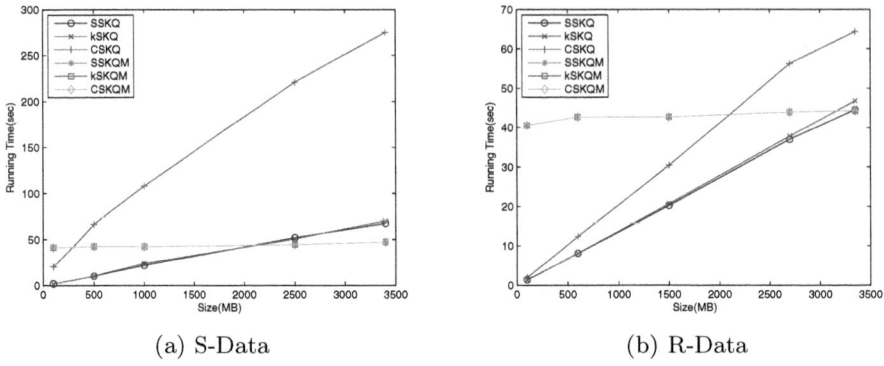

(a) S-Data   (b) R-Data

Fig. 5. Single machine *vs.* MapReduce cluster system

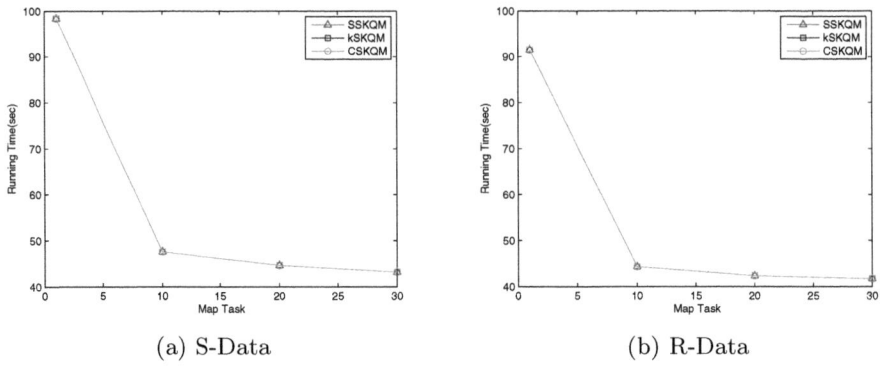

(a) S-Data   (b) R-Data

Fig. 6. Running time vs. the number of Map tasks

Fig. 5(b) shows the result on R-data. We can see that the running time is smaller than Fig. 5(a) totally. This is because a big part of the real data has been in order before processing. So a considerable part of sort cost is saved. The other features in this figure are the same with Fig. 5(a).

Fig. 6 describes the effect of number of various map task. We set four values of the number of map task, i.e., 1,10,20,30. We can see that the running time decreases with the increase of map task. Both datasets are about 3.4GB and each block is 64MB. So we have about 50 blocks. More map tasks means more blocks can be processed at the same time, which reduces the running time. As shown in Fig. 5 and Fig. 6, if there is only one map task, the running time of distributed system is bigger than the single computer. Because in distributed system the data is not stored locally and map task needs to read remote data, which leads to high cost.

5   Conclusion and Future Work

In this paper, we use MapReduce to perform spatial keyword queries. Experimental results show that it is efficient to deal with spatial keyword queries over big, seldom used, type mixed spatial data with MapReduce. As for future work, how to employ index mechanism with MapReduce is a promising direction.

References

1. Ramaswamy, H., Bijit, H., Chen, L., et al.: Processing spatial-keyword (SK) queries in GIR systems. In: SSDBM (2007)
2. Dean, J., Ghemawat, S.: MapReduce Simplified Data Processing on Large Clusters. In: OSDI, pp. 137–150 (2004)
3. Ghemawat, S., Gobioff, H., Leung, S.-T., et al.: The Google File System. In: ACM/SOSP, pp. 29–43 (2003)
4. White, T.: Hadoop: The Definitive Guide. O'Reilly, CA (2009)
5. Roussopoulos, N., Kelley, S., Vincent, F.: Nearest neighbor queries. In: SIGMOD, pp. 71–79 (1995)
6. Cong, G., Jensen, C.S., Wu, D.: Efficient retrieval of the top-k most relevant spatial web objects. PVLDB 2(1), 337–348 (2009)
7. Guttman, A.: R-Trees-a dynamic index structure for spatial searching. In: SIGMOD, pp. 47–57 (1984)
8. Timos, K.S.: The R*-Tree: An Efficient and Robust Access Method for points and Rectangles. In: SIGMOD, pp. 322–331 (2000)
9. Roussopoulos, N., Leifker, D.: Direct spatial search on pictorial databases using packed R-trees. In: SIGMOD, pp. 17–31 (1985)
10. Zobel, J., Moffat, A., Ramamonhanarao, K.: Inverted files versus signature files for text indexing. ACM Transactions on Database System 23(4), 453–490 (1998)
11. Li, Z., Lee, K.C.K., Zheng, B., et al.: IR-tree-An efficient index for geographic document search. IEEE Transactions on Knowledge and Data Engineering 23(4), 585–599 (2011)
12. Felipe, I.D., Hristidis, V., Rishe, N.: Keyword search on spatial databases. In: ICDE, pp. 656–665 (2008)
13. Zhang, D., Chee, Y.M., Mondal, A., et al.: Keyword search in spatial databases: Towards searching by document. In: ICDE, pp. 688–699 (2009)
14. Zhou, Y., Xie, X., Wang, C., et al.: Hybrid index structures for location-based web search. In: CIKM, pp. 155–162 (2005)

Detection of High-Risk Zones and Potential Infected Neighbors from Infectious Disease Monitoring Data[*]

Biying Tan[1], Lei Duan[1,**], Chi Gou[1], Shuyang Huang[1], Yuhao Fang[1],
Xing Zhao[2], and Changjie Tang[1]

[1] School of Computer Science, Sichuan University,
Chengdu 610065, China
[2] West China School of Public Health, Sichuan University,
Chengdu 610041, China
leiduan@scu.edu.cn

Abstract. Detecting the high-risk zones as well as potential infected geographical neighbor is necessary and important to reduce the loss caused by infectious disease. However, it is a challenging work, since the outbreak of infectious disease is uncertain and unclear. Moreover, the detection should be efficient otherwise the best control and prevention time may be missed. To deal with this problem, we propose a geography high-risk zones detection method by capturing the significant change in the infectious disease monitoring data. The main contribution of this paper includes: (1) Analyzing the challenges of the early warning and detection of infectious disease outbreak; (2) Proposing a method to detect the zone that the number of monitoring cases changes significantly; (3) Defining the infection perturbation to describe the infection probability between two zones; (4) Designing an algorithm to measure the infection perturbation of infectious disease between adjacent zones; (5) Performing extensive experiments on both real-world data and synthetic data to demonstrate the effectiveness and efficiency of the proposed methods.

Keywords: Data Mining, Change Mining, Spatial Mining, Time Series.

1 Introduction

The issue of infectious disease prevention and control is important in public health and humans' daily life. For example, the acute piratory syndrome (SARS) outbreak in 2002 caused huge losses. According to the report from WHO, 8069 people infected SARS and 775 people lost lives in 32 main countries of the world since its outbreak to July.11th ,2003. Not only the direct loss but also the indirect loss that caused by the infectious disease bring heavy burden to the society. Take SARS event for example,

---

[*] This work was supported by the National Natural Science Foundation of China under grant No. 61103042, the National Research Foundation for the Doctoral Program by the Chinese Ministry of Education under grant No.20100181120029, and the Young Faculty Foundation of Sichuan University under grant No. 2009SCU11030.
[**] Corresponding author.

H. Yu et al. (Eds.): DASFAA Workshops 2012, LNCS 7240, pp. 262–273, 2012.

China mainland lost 2.2 billion US dollars, Korea lost 2 billion US dollars, Hong Kong lost 1.7 billion US dollars, Japan lost 1.1 billion US dollars.

As a result, taking effective early warning and detection for the infectious disease is urgent when the epidemic outbreaks. Specifically, it's required to detect the outbreak zones, and take measures to control the disease happened there. Moreover, it is necessary to predict the future potential outbreak zone and adopt prevention measures. Thus, applying statistics and data mining techniques to help experts detect the disease outbreak, as well as analyze the infection impact to the other zones is a desirable work, which has got many attentions, such as [1-9].

Many countries have set up the public health surveillance system, which can collect the number of cases of infectious disease in certain zones. The monitoring data, collected by the system, offers data sources for the researches on outbreak detection and early warning [10-12]. The goal of our work is to detect the high-risk zones and potential infected zones based on the data supplied by the surveillance system of China. It is a challenging work, since the spread of infectious disease has following features.

- *Uncertain*: For most infectious diseases, the spread is affected by various factors, such as daily communication, atmospheric flowing, living habit, air travel and climatic change.
- *Unclear*: For unknown infectious diseases and some new subtypes of an infectious disease, the transmission media of the related pathogens (cause of disease) is unclear at the beginning of the outbreak.
- *Timeliness:* For some urgent infectious diseases, the affected population and range may be expanded in a short time. As a result, fast prediction is expected. Otherwise, the best intervention time may be missed, and the loss caused by the infectious disease will be increased.

Moreover, once the infectious disease outbreaks, more attention should be paid to the neighboring zones of the outbreak zone. Observation 1 is a real-world case to illustrate this fact.

Observation 1. In 2002, SARS firstly broke out at Guangdong Province in China. Several days later, SARS cases were discovered at Hong Kong. Several weeks later, Southeast Asia countries (regions) discovered SARS cases while the epidemic of SARS in Guangdong was more serious.

From Observation 1, we can see that when SARS broke out in Guangdong Province, China, the infection probability for Hong Kong was larger than other countries (regions) far away from Guangdong, since Hong Kong is closer to Guangdong geographically. Intuitively, the neighboring zones of the outbreak zone are more vulnerable to be infected by the infectious disease. We think the reasons are: (i) the communications between two neighboring zones are frequent and intensive. So, there exists many transmission media between them. (ii) The environment, including living habits, climate, etc., between two neighboring zones is similar.

Naturally, the influence on the neighbor that has more communications and similar environment to the outbreak zone is stronger, vice verse. In other words, the neighbor has more probability to be infected in a short time. Thus, prevention measures should be done immediately at this neighbor.

There is little previous work on geographical infection analysis between zones. We detect the high-risk zones and the infection perturbation between adjacent zones based on the monitoring data and geographical information. The main contributions include: (1) Analyzing the challenges of the early warning and detection of infectious disease outbreak; (2) Proposing a method to detect the zone that the number of monitoring cases changes significantly; (3) Defining the infection perturbation to describe the infection probability between two zones; (4) Designing an algorithm to measure the infection perturbation of infectious disease between adjacent zones; (5) Performing extensive experiments on both real-world data and synthetic data to demonstrate the effectiveness and efficiency of the proposed methods.

2   Related Works

Previous studies on outbreak detection and early warning for infectious disease can be divided into two groups: space-time scan statistic and anomaly pattern detection.

2.1   Space-Time Scan Statistic

The main feature of this method is applying statistic methods on analyzing the variety of the numbers of infectious disease cases along with the space and time changes. The spatial scan statistic method that mentioned in [1] extended the scan statistic in three directions: clusters' detection, the scanning window and the baseline.

The authors in [2] proposed an elliptic spatial scan statistic with a scanning window of variable location, shape, angle and size, since circular scanning window was not suitable for the non-circluar zones. Furthermore, Reference [3] raised a flexibly shaped space-time scan statistic method to satisfy the requirement of detecting and monitoring in irregular shaped areas.

The monitoring data in the real-world is ordinal or continuous. Any discretization method may cause the loss of information. Reference [4] proposed a spatial scan statistic method to analyze the data incorporating the ordinal structure. Furthermore, Reference [5] and Reference [6] proposed methods to deal with the survival data and the heterogeneous population data, respectively.

However, the methods of space-time scan statistic are not suitable for our problem, since the main concern of these methods is beyond the infection impact analysis between zones.

2.2   Anomaly Pattern Detection

The main idea of anomaly pattern detection includes two aspects: (i) defining an anomaly; (ii) comparing recent dataset with baseline distribution to discover significant patterns of anomalies. Wong et al. gave a rule-based anomaly pattern detection algorithm in [7]. This algorithm can overcome the limit of some early methods adopting individual data points without particular combinations of features. It is difficult to define baseline, since there are various trends in disease spread. The authors in [8] proposed a Bayesian network based method to define the baseline.

The authors in [9] proposed a pattern based method, named WSARE, for disease outbreak early warning. WSARE can find out the anomalous patterns from discrete dataset. The experimental study shows that it is efficient and accurate on disease detection.

3   High-Risk Zones Detection and Potential Neighbors Prediction

This section mainly discusses the details of our proposed detection and prediction method, named as HZD-PNP, for detecting high-risk zones and predicting potential infected neighbors.

3.1   High-Risk Zone Detection

For the analysis on the infectious disease monitoring data, one important issue is infectious disease outbreak detection. However, it is hard to exactly define the outbreak, the reasons include: firstly, different infectious diseases have different outbreak standards; secondly, for most infectious diseases, the standards of outbreak are still unspecified in the domain of public healthy.

To deal with this problem, most researches focus on the change of number of disease cases. Generally, HZD-PNP follows the same idea of previous work on outbreak detection. The initial step of HZD-PNP is to detect the zones, where the number of cases increases significantly in recent time window. Formally, given a recent time window T_r, a historical time window T_h. Let $CE(T_r)$, $CE(T_h)$ be the numbers of cases in T_r and T_h at zone Z in the same time unit, respectively. If the difference between $CE(T_r)$ and $CE(T_h)$ is larger than a predetermined threshold, zone Z is called as a **high-risk zone**. In HZD-PNP, the recent time window T_r is a time period determined by the user, such as current day or current week. The historical time window T_h is also a user determined time period which is prior to T_r.

The high-risk zones detection in HZD-PNP includes three steps as follows. Step 1: define the recent time window and historical time window; Step 2: for each zone, count the numbers of cases in recent and historical time windows, respectively. Step 3: for each zone, apply the statistic method to examine the significance of difference between the numbers of cases in recent and historical time windows.

As stated in [9], determining the recent time window T_r and the historical time window T_h is important for high-risk zones detection. The reason lies that, firstly, if the historical time window is so close to the recent time window, the changes between the recent and history may be too small to be captured. On the other hand, if the historical time window is long before the recent one, the difference may not be caused by the infectious disease outbreak. As a result, the selection of historical time window is critical, it should not be near, nor be long before the recent time window.

Similar to using *Fisher's Exact Test* [13] to evaluate each anomaly pattern detection rule in [9], we try to adopt *Fisher's Exact Test* in HZD-PNP to examine the significance of the difference for high-risk zones detection.

Example 1. Given zone Z. Suppose in recent time, the number of sick persons is a, the number of non-sick persons is b. And suppose in historical time, the number of sick persons is c, the number of non-sick persons is d. Then the contingency table is constructed as follows.

Table 1. An example of a contingency table

Group	$CE(T_r)$	$CE(T_h)$
Sick	a	c
Non-sick	b	d

In Table 1, the values of $(a + b)$ and $(c + d)$ equal to the numbers of population in recent and historical time windows at zone Z, respectively. That is, the sum of a, b, c and d is a big number. Thus, *Fisher's Exact Test* may be not suitable for this situation. In this case, we use *Chi-Square test* instead of *Fisher's Exact Test* for high-risk zones detection, since *Chi-Square test* is an approximation to *Fisher's exact test* when counts are large. In this study, we set the threshold to be 0.05. That is, if the result of *Chi-Square test* is less than 0.05, the zone is detected as a high-risk zone. For all detected high-risk zones, the adjacent ones are merged together to form a high-risk clustering region. Thus, the numbers of population and cases of a clustering region are the sum of the population and cases of each zone in the region, respectively. Then we can treat the clustering region as a big zone with larger size and population.

3.2   Potential Infected Neighbors Prediction

As stated in Section 1, the factors that determine the spreading direction of infectious disease are complex, various and uncertain. The spread of infectious disease is prone to the zones which have frequent and intensive communications and the similar environment to the outbreak zone.

In this work, we predict the potential infected zones, which may break out the same infectious disease as the high-risk zones. Specifically, we focus on the zones that are adjacent to the high-risk zones (clustering regions), since, in most cases, these zones easily meet the spread conditions of infectious disease, such as the frequent and intensive communication and the similar environment. As a result, we can predict the potential infected neighbors as an early warning by listing the adjacent zones that have high risk of disease outbreak. We take one day as the basic time unit, and convert the numbers of daily cases, supplied by the surveillance system, into a time series data, which is composed of basic case ratios.

Definition 1 (Basic Case Ratio). Let $NC(C) = \{c_1, c_2,..., c_k\}$ be the time series of monitoring cases at a zone in basic time unit. The basic case ratio of c_i is c_i / $(c_1+c_2+...+c_k)$, denoted as $BCR(C) = \{BCR(c_1), BCR(c_2),..., BCR(c_k)\}$.

We analyze the infection influence between two zones based on the basic case ratios. Let $BCR(p_i)$ and $BCR(q_i)$ be the basic case ratios of zone P and zone Q on day i, respectively. Then the closer the relationship between zone P and zone Q is, the value of $BCR(p_{i+j})$ ($j \geq 1$) will be more markedly affected by the value of $BCR(q_i)$, and vice versa. In this study, we call this kind of change of cases number caused by the influence from other zones, as **infection perturbation**. Infection perturbation reflects the degree of a zone being affected when the other zone breaks out infectious disease. We use the relative entropy [14], which is commonly used to measure the influence and similarity between two models [15-17], to measure the infection perturbation between two zones based on their basic case ratios. The infection perturbation is defined in Definition 2.

In our practice, to avoid the relative entropy being infinite that is caused by the zero basic case ratio, data preprocessing is necessary. Let the number of population at zone P be $Pop(P)$. Then, for each value in $BCR(P)$, $BCR(P_i)$, is added by $1/Pop(P)$.

Definition 2 (Infection perturbation). Suppose $BCR(P)$ and $BCR(Q)$ be the time series of basic case ratio at zone P and zone Q, respectively. The infection perturbation between P and Q, denoted as $IP(P, Q)$, is calculated as follows.

$$IP(P, Q) = 1 / (\sum_i BCR(p_i) \log \frac{BCR(p_i)}{BCR(q_i)} + 1) \tag{1}$$

Equation (1) shows that for two zones, the less the relative entropy between them is, the stronger the infection perturbation between them exists. That is, one zone has higher probability to be infected if it has greater infection perturbation with adjacent places where breaks out infectious disease.

3.3   Workflow of HZD-PNP

In this subsection, we present the details of HZD-PNP for the high-risk zones detection and potential infected neighboring zones prediction. There are two start ways of HZD-PNP: (i) HZD-PNP detects the monitoring data, and selects the zones, in which the number of cases increases significantly in recent time window. (ii) If an infectious disease outbreaks, HZD-PNP can start the detection of potential infected zones that are adjacent to the predetermined zones.

The next step in HZD-PNP is to evaluate the infection perturbation between each high-risk zone (clustering region) and its neighboring zones. This step consists of two main procedures: selecting neighboring zones and measuring the infection perturbation for each neighboring zones.

The strategy of HZD-PNP to select neighboring zones is stated as follows. Given a high-risk zone Z, there are three opinions to select the neighbors of Z in HZD-PNP: (i) The neighbors are the zones that are geographically adjacent to Z; (ii) The neighbors are the zones that are the nearest k ($k \geq 1$) zones to Z. The distance between two zones is evaluated by the Euclidean distance between the administrative centers of these two zones; (iii) The neighbors are the zones that are the nearest k ($k \geq 1$) zones and geographically adjacent to Z.

Secondly, for each neighboring zone, the infection perturbation between it and the high-risk zones is measured by Equation (1). However, there is another issue should be taken into account. When a zone is a neighbor to several high-risk zones, how to measure the infection perturbation? In this case, HZD-PNP conceives the related high-risk zones as a whole in logical. That is, given a zone Y, let $RN(Y)$ be the set of high-risk zones that takes Y as a neighboring zone. For $RN(Y)$, the number of population is $\Sigma\ Pop(rn)$, $rn \in RN(Y)$, and the time series of monitoring cases is the sum of cases in the related high-risk zones. Then the infection perturbation between Y and $RN(Y)$ is measured by Equation (1).

The last step of HZD-PNP is to detect the potential infected zones that are adjacent to high-risk zones (clustering regions) based on the infection perturbation evaluation. To this end, HZD-PNP lists the zones whose infection perturbation is large with the high-risk zones, which means the zone may be easily affected by the high-risk zones; therefore the cases distribution of the neighboring zone may be similar with that of the high-risk zones. Then, HZD-PNP predicts the zones may be the next infectious disease outbreak regions.

There are two ways to predict the next infected zones: (1) HZD-PNP outputs all zones whose infection perturbation is greater than the given threshold; (2) outputting zones with top k largest infection perturbation measures. In this work, we predict the most k potential infected neighbors from current high-risk zones.

Algorithm 1 describes the pseudo code of HZD-PNP. In Algorithm 1, *HRZones* is the set of high-risk zones detected by HZD-PNP or predetermined outbreak ones. Function *testRisk(Tr, Th, P)* evaluates the change of cases between the recent period and the historical period for zones, and returns the detected high-risk zones in Step 3. Function *Merge(HRZones)* merges adjacent high-risk zones to form a clustering region. Step 5 is an interactive procedure that the experts can identify the zones found in Step 4. Function *nbrSelect(HRZones, G)* returns all neighbors to zones in *HRZones* in Step 6. From Step7 to Step 9, the zones which are neighboring to several high-risk zones are marked. The related high-risk zones are conceived as a whole in logical in the next step. The infection perturbation between each high-risk zone and its neighboring zones are evaluated in Step 12. The zones that have the largest infection perturbation with the high-risk zones will be output.

Algorithm 1: HZD-PNP (Tr, Th, P, G, HRZones)
Input: (1) monitoring dataset of recent period: Tr; (2) monitoring dataset of historical period: Th; (3) population dataset: P; (4) geographic dataset: G; (5) high-risk zones: HRZones
Output: possible outbreak zones: *POZ*.

```
begin
    1. if HRZones ≠ null go to Step 5
    2. For each monitoring zone in Tr
    3.     HRZones ← HRZones + testRisk(Tr, Th, P)
    4. HRClusters ← Merge(HRZones)
    5. interactiveIdentify(HRClusters)
    6. NbrZones ← nbrSelect(HRClusters,G)
    7. For each n in NbrZones
    8.     if n is a neighbor to several z in HRZones
    9.         mark n is affected by several high-risk zones
   10. POZ ← null
```

```
  11. For each z in HRZones
  12.     POZ ← POZ + ipEval(Th,NbrZones)
  13. return POZ
end.
```

4   Performance Study

To evaluate the performance of HZD-PNP, we implement HZD-PNP in Java. The experiments are performed on an Intel Pentium Dual 1.80 GHz (2 Cores) PC with 2G memory running Windows XP operating system.

4.1   Effective Study on the Real-World Monitoring Data

4.1.1   Dataset

We perform the effective study of HZD-PNP on the real-world disease monitoring data provided by the Department of Health Statistics, Sichuan University. Since the data is sensitive, we skip over the semantic details and formulate the data formally as follows. Let MD be the monitoring dataset of a certain infectious disease at region CQ from June 4th to December 30th in 2009, $MD = \{D_i \mid 1 \leq i \leq 210\}$, where D_i is the set of daily number of new cases at all 40 monitoring zones in region CQ. Suppose $ZONE = \{z_i \mid 1 \leq i \leq 40\}$ is the set of monitoring zones, then each $d_j \in D_i (1 \leq j \leq 40)$ represents the number of new cases at zone z_j on day i. The total number of cases is 17,478 in MD. We take the data in one week as the basic analysis time unit. In other words, HZD-PNP predicts the zones that are possible to break out infectious disease in the next week. Based on MD, we convert it into MW, which contains the average daily number of cases in each week, $MW = \{W_i \mid 1 \leq i \leq 30\}$.

In this experimental study, we set the historical time window (T_h) to contain 5-week data, the recent time window (T_r) to contain 1-week data. T_h is one week prior to T_r. We use HZD-PNP to detect high-risk zones and potential infected adjacent zones of the infectious disease in the following week next to T_r. By this means, we divide MW into 23 training and test data groups, denoted as MW' (there are 30 weeks during June 4th to December 30th, each consecutive 8 weeks compose a group). For example, in the first group, the training data consists of T_h and T_r, where $T_h = \{W_1, W_2, W_3, W_4, W_5\}$, $T_r = \{W_7\}$; the test data is $\{W_8\}$.

4.1.2   Sensitivity Test on High-Risk Zones Detection

As stated before, one of the start ways of HZD-PNP is high-risk zone detection. The destination of this experiment is demonstrating the detection sensitivity under different time units. To this end, we apply HZD-PNP to dataset MW', the detected high-risk zones is denoted as HRZ_w. Furthermore, we apply HZD-PNP to each daily monitoring data, and record the union of each week day's detected zones. The result is denoted as HRZ_d. Table 2 lists the experimental results. In Table 2, the zones in a high-risk clustering region are enclosed in parentheses. For example, a high-risk clustering that composed by z_i and z_j is denoted as (z_i, z_j). Due to the space limitation, we list the result in September $(T_r \in \{W_{13}, W_{14}, W_{15}, W_{16}, W_{17}\})$ in Table 2.

Table 2. The experimental results on high-risk zones detection test

T_r	HRZ_w	HRZ_d
W_{13}	$\{z_{39}\}$	$\{z_1, (z_6,z_{16}), z_{10}, z_{22}, z_{39}\}$
W_{14}	$\{z_7, (z_{10},z_{13},z_{14},z_{17},z_{18},z_{19},z_{20},z_{26}, z_{28},z_{30}), (z_{39},z_{40})\}$	$\{z_1,z_5,z_7,(z_{10},z_{11},z_{12},z_{13},z_{14},z_{15},z_{17}, z_{18},z_{19},z_{20},z_{26},z_{27},z_{28},z_{30},z_{31}), z_{38}, (z_{39},z_{40})\}$
W_{15}	$\{z_1, (z_{10},z_{11},z_{14},z_{19},z_{28},z_{40}), z_{30}\}$	$\{(z_1,z_3), (z_7,z_8,z_9,z_{22},z_{25},z_{33},z_{36}), (z_{10},z_{11},z_{12},z_{13},z_{14},z_{17},z_{18},z_{19},z_{20},z_{26},z_{28},z_{39},z_{40}), (z_{29},z_{30})\}$
W_{16}	$\{(z_1,z_2), (z_{11},z_{14}), z_{15}, (z_{23}z_{31})\}$	$\{(z_1,z_2,z_3), z_8, (z_{10},z_{11},z_{12},z_{14},z_{40}), (z_{15},z_{36}), (z_{18},z_{19},z_{23},z_{31}), z_{29}\}$
W_{17}	$\{z_1,z_2,z_3,z_{10},z_{15},z_{23},z_{27},z_{30},z_{31}\}$	$\{(z_1,z_2,z_3,z_5,z_6,z_7,z_8z_{15},z_{21}, z_{22},z_{33},z_{36}), (z_{10},z_{12},z_{13},z_{14},z_{20}), z_{18}, z_{27}, (z_{23},z_{29},z_{30},z_{31})\}$

From Table 2, we can see that HZD-PNP is effective to detect high-risk zones. Compared with HRZ_d, the number of detected high-risk zones in HRZ_w is decreased, and all zones detected in HRZ_w are also detected in HRZ_d. The reason lies that the number of cases may change significantly in some day, while the change may not result in the number of the week changes remarkably. So to get accurate detection, the basic time unit should not be too long. In the other hand, to avoid the false alarms caused by the high sensitivity detection, it is better to set the basic time unit according to the specific situation.

4.1.3   Accuracy Test on the Detection and Prediction Results of HZD-PNP

The aim of HZD-PNP includes: high-risk zones detection and potential next-infected neighbors prediction. In this Subsection, we perform experiments to demonstrate the accuracy of HZD-PNP. Firstly, we apply HZD-PNP to MW' to evaluate the infection perturbations from the high-risk zones (list in Table 2) to its neighbors. Then HZD-PNP performs early warning for the neighbors with the largest infection perturbations.

Next, as WSARE, proposed in [9], can find the most significant changes between the recent period and historical period, we apply it to the test data to find the zones which should be early warned for disease outbreak. Then, we compare the possible outbreak zones that are predicted by HZD-PNP with the results of WSARE, followed by excluding the zones, where the number of cases decreases dramatically, found by WSARE. There are three versions of WSARE (2.0, 2.5, 3.0) based on different selections of baseline. WSARE 2.5 selects all historical data as the baseline. We conduct WSARE 2.5 in the experiment, since the way of it to select baseline is consistent with historical time window selection in HZD-PNP. Moreover, since HZD-PNP merely focuses on the zones that are geographically adjacent to the high-risk zones, we apply WSARE 2.5 to the same zones to find out the zones to be warned for disease outbreak. Due to the space limitation, we list the predicted zones by HZD-PNP and related neighboring high-risk zones, as well as the zones found by WSARE 2.5 in September ($T_r \in \{W_{13}, W_{14}, W_{15}, W_{16}, W_{17}\}$) in Table 3.

Table 3. The predicted zones by **HZD-PNP** and the zones found by WSARE 2.5

T_r	predicted zones by HZD-PNP	neighboring high-risk zones	zones found by WSARE 2.5
W_{13}	$\{z_{40}\}$	$\{z_{39}\}$	$\{z_{40}\}$
W_{14}	$\{z_9\}$	$\{z_7\}$	$\{z_{11}\}$
W_{14}	$\{z_{11}\}$	$\{z_{10},z_{13},z_{14},z_{17},z_{18},z_{19},z_{20}\}$	$\{z_{11}\}$
W_{14}	$\{z_{11}\}$	$\{z_{26}\}$	$\{z_{11}\}$
W_{14}	$\{z_{31}\}$	$\{z_{10},z_{13},z_{14},z_{17},z_{28}\}$	$\{z_{11}\}$
W_{14}	$\{z_{31}\}$	$\{z_{30}\}$	$\{z_{11}\}$
W_{14}	$\{z_{11}\}$	$\{z_{14},z_{26},z_{39},z_{40}\}$	$\{z_{11}\}$
W_{15}	$\{z_2\}$	$\{z_1\}$	$\{z_2,z_{12},z_{18},z_{19},z_{31},z_{39},z_{40}\}$
W_{15}	$\{z_{12}\}$	$\{z_{10},z_{11},z_{14}\}$	$\{z_2,z_{12},z_{18},z_{19},z_{31},z_{39},z_{40}\}$
W_{15}	$\{z_{18}\}$	$\{z_{19}\}$	$\{z_2,z_{12},z_{18},z_{19},z_{31},z_{39},z_{40}\}$
W_{15}	$\{z_{31}\}$	$\{z_{10},z_{28}\}$	$\{z_2,z_{12},z_{18},z_{19},z_{31},z_{39},z_{40}\}$
W_{15}	$\{z_{31}\}$	$\{z_{30}\}$	$\{z_2,z_{12},z_{18},z_{19},z_{31},z_{39},z_{40}\}$
W_{15}	$\{z_{39}\}$	$\{z_{11},z_{14},z_{40}\}$	$\{z_2,z_{12},z_{18},z_{19},z_{31},z_{39},z_{40}\}$
W_{16}	$\{z_3\}$	$\{z_1,z_2\}$	$\{z_3,z_{10},z_{18},z_{20}\}$
W_{16}	$\{z_{10}\}$	$\{z_{11}\}$	$\{z_3,z_{10},z_{18},z_{20}\}$
W_{16}	$\{z_{10}\}$	$\{z_{14}\}$	$\{z_3,z_{10},z_{18},z_{20}\}$
W_{16}	$\{z_{20}\}$	$\{z_{15}\}$	$\{z_3,z_{10},z_{18},z_{20}\}$
W_{16}	$\{z_{30}\}$	$\{z_{23},z_{31}\}$	$\{z_3,z_{10},z_{18},z_{20}\}$
W_{17}	$\{z_{32}\}$	$\{z_1,z_2\}$	$\varnothing$
W_{17}	$\{z_5\}$	$\{z_3\}$	$\varnothing$
W_{17}	$\{z_{11}\}$	$\{z_{10}\}$	$\varnothing$
W_{17}	$\{z_{16}\}$	$\{z_{15}\}$	$\varnothing$

From Table 3, we can see that some predicted zones are neighbors of several high-risk zones, such as z_3 in W_{16} is a neighbor of both z_1 and z_2. Furthermore, we can see that except for during W_{17}, some predicted zones in W_{13}, W_{14}, W_{15}, W_{16} can also be warned by WSARE 2.5. We think the differences of the prediction results between HZD-PNP and WSARE 2.5 lies that, WSARE 2.5 aims at finding zones whose number of cases changes most dramatically, while HZD-PNP sets off alarms once the numbers of cases in some zones have an increasing trend. Thus, HZD-PNP finds more potential outbreak zones than WSARE 2.5; On the other hand, the results of WSARE 2.5 are based on the predicted week. If some disease intervention measures have already been conducted, the number of cases may be decreased. As a result, WSARE 2.5 may not find those zones. Based on above comparison and analysis, we believe that applying HZD-PNP for the real-world infectious disease surveillance is desirable.

4.2   Scalability Test on the Synthetic Data

As stated in Introduction section, the detection and early warning problem of infectious disease is a time-sensitive work, which should be done in a short time. In order to demonstrate HZD-PNP is effective for long-term monitoring dataset, we extend the scale of the dataset by copying the available monitoring dataset $50 \times n$ ($1 \leq n \leq 10$) times, respectively. As the result, the numbers of both the cases and monitoring sites are increased. The 500 times means that the scale of dataset is the same as the 5-years monitoring dataset of China. Figure 1 illustrates the running time of HZD-PNP on each dataset.

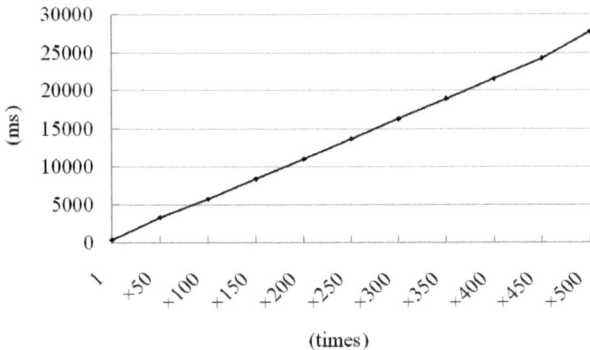

Fig. 1. The running time of HZD-PNP under datasets with different scales

From Figure 1, we can see that the running time of HZD-PNP grows linearly as long as the scale of dataset (numbers of both the cases and monitoring sites) is increased. As a result, it is desirable to use HZD-PNP in a large dataset.

5   Conclusions

Accurately detecting high-risk zones and timely warning the next infected zones can reduce the loss caused by the infectious disease. Meanwhile, it is a challenging work, since the spread of an infectious disease may be uncertain and unclear to human. To supply the decision support on implementing the control and prevention measures in the best intervention time, we propose a method, named HZD-PNP. HZD-PNP can detect the zones, where the number of cases increases dramatically during recent time, and predicts potential next-infected neighboring zones by evaluating the infection perturbation between each neighbor and the detected high-risk zone. The performance study of HZD-PNP on both real-world data and synthetic data demonstrates that HZD-PNP is effective and efficient for predicting potential next-infected zones.

There are several possible directions for future work. (i) Consider the zones that are not neighboring to high-risk zones in the procedure of infection perturbation evaluation. In this case, the efficiency of HZD-PNP may be decreased, since the number of zones to be evaluated is large. (ii) Different infectious diseases may be affected by different factors. How to take other factors that are related to the disease spread, such as the climate, temperature, air travel, into consideration is a challenging and meaningful work. (iii) Introduce the transmission dynamics of infectious diseases into the spread direction prediction. It is challenging to implement the transmission dynamics, which are from the domain of epidemiology, in HZD-PNP. (iv) Study the spread of infectious disease in temporal aspect. As the condition of infectious disease varies over time, we believe that more accurate prediction results can be got by considering the transmission dynamics in infection perturbation measuring.

References

1. Kulldorff, M.: A Spatial Scan Statistic. Communications in Statistics: Theory and Methods 26(6), 1481–1496 (1997)
2. Kulldorff, M., Huang, L., Pickle, L., Duczmal, L.: An Elliptic Spatial Scan Statistic. Statistics in Medicine 25(22), 3929–3943 (2006)
3. Takahashi, K., Kulldorff, M., Tango, T., Yih, K.: Flexibly Shaped Space-time Scan Statistic for Disease Outbreak Detection and Monitoring. International Journal of Health Geographics 7, 14 (2008)
4. Jung, I., Kulldorff, M., Klassen, A.: A Spatial Scan Statistic for Ordinal Data. Statistics in Medicine 26(7), 1594–1607 (2007)
5. Huang, L., Kulldorff, M., Gregorio, D.: A Spatial Scan Statistic for Survival Data. Biometrics 63(1), 109–118 (2007)
6. Huang, L., Tiwari, R.C., Zou, Z., Kulldorff, M., Feuer, E.J.: Weighted Normal Spatial Scan Statistic for Heterogeneous Population Data. Journal of the American Statistical Association 104(487), 886–898 (2009)
7. Wong, W.K., Moore, A., Cooper, G., Wagner, M.: Rule-based Anomaly Pattern Detection for Detecting Disease Outbreaks. In: Proc. the 18th National Conference on Artificial Intelligence (AAAI 2002), pp. 217–223. MIT Press (2002)
8. Wong, W.K., Moore, A., Cooper, G., Wagner, M.: Bayesian Network Anomaly Pattern Detection for Disease Outbreaks. In: Proc. the 20th International Conference on Machine Learning (ICML 2003), pp. 808–815. AAAI Press (2003)
9. Wong, W.K., Moore, A., Cooper, G., Wagner, M.: What's Strange About Recent Events (WSARE): An Algorithm for the Early Detection of Disease Outbreaks. Journal of Machine Learning Research 6, 1961–1998 (2005)
10. Wagner, M.M., Tsui, F.C., Espino, J.U., Dato, V.M., Sittig, D.F., Caruana, R.A., McGinnis, L.F., Deerfield, D.W., Druzdzel, M.J., Fridsma, D.B.: The Emerging Science of Very Early Detection of Disease Outbreaks. Journal of Public Health Manag. Pract. 7(6), 51–59 (2001)
11. Centers for Disease Control and Prevention. Updated Guidelines for Evaluating Public Health Surveillance Systems: Recommendations from the Guidelines Working Group. MMWR 50(RR-13): 1-35 (2001)
12. Russell, K.L., Rubenstein, J., Burke, R.L., Vest, K.G., Johns, M.C., Sanchez, J.L., Meyer, W., Fukuda, M.M., Blazes, D.L.: The Global Emerging Infection Surveillance and Response System (GEIS), a U.S. Government Tool for Improved Global Biosurveillance: a review of 2009. BMC Public Health 11(suppl. 2), S2 (2011)
13. Agresti, A.: An Introduction to Categorical Data Analysis, 2nd edn., New Jersey. Wiley Series in Probability and Statistics (2007)
14. Kullback, S., Leibler, R.A.: On Information and Sufficiency. The Annals of Mathematical Statistics 22(1), 79–86 (1951)
15. Cover, T., Thomas, J.: Elements of Information Theory. Wiley Series in Telecommunications. John Wiley and Sons, New-York (1991)
16. Hershey, J., Olsen, P.: Approximating the Kullback Leibler Divergence between Gaussian Mixture Models. In: Proc. of ICASSP, Honolulu, USA (2007)
17. Minka, T.: Divergence Measures and Message Passing. Technical Report MSR-TR-2005-173, Microsoft research, Cambridge (2005)

Bayesian Network-Based Probabilistic XML Keywords Filtering

Chenjing Zhang[1,2], Kun Yue[3], Jinghua Zhu[2],
Xiaoling Wang[4], and Aoying Zhou[4]

[1] College of Information Technology, Shanghai Ocean University, China
cjzhang@shou.edu.cn
[2] School of Computer Science and Technology, Fudan University, China
{cjzhang,jhzhu}@fudan.edu.cn
[3] School of Information Science and Engineering, Yunnan University, China
kyue@ynu.edu.cn
[4] Shanghai Key Laboratory of Trustworthy Computing,
Software Engineering Institute, East China Normal University
{xlwang,ayzhou}@sei.ecnu.edu.cn

Abstract. Data uncertainty appears in many important XML applications. Recent probabilistic XML models represent different dependency correlations of sibling nodes by adding various kinds of *distributional* nodes, while there does not exist a uniform probability calculation method for different dependency correlations. Since Bayesian Networks can denote various dependency correlations among nodes just by conditional probability table(CPT), this paper proposes the Bayesian Networks based probabilistic XML model PrXML-BN, and combines SLCA semantic meaning of keyword query into Bayesian Networks, then implements keywords filtering on SLCA semantic meaning. To optimize the performance of keywords filtering, two optimization strategies are proposed in this paper. In the end, experiments verify the performance of keywords filtering algorithm based on SLCA in model PrXML-BN.

Keywords: Probabilistic XML, Bayesian Networks, Keywords Filtering, SLCA.

1 Introduction

Data uncertainty appears in various XML applications, such as data integration, automatic information extraction and so on. Many developed works are focused on how to model probabilistic XML [1–4]. They denote probabilistic XML documents as *p-document*, which include two kinds of nodes, *distributional* node and *ordinary* node. *Distributional* node denotes the correlations between sibling nodes in original XML document and has 5 types: $\{IND, MUX, DET, EXP, CIE\}$. According to types of *distributional* node, [4] defines the family of *p-document* as $\mathrm{PrXML}^C, C \subseteq \{IND, MUX, DET, EXP, CIE\}$. The research of [5–8] is about structural query in probabilistic XML. There are fewer works dealing with the

H. Yu et al. (Eds.): DASFAA Workshops 2012, LNCS 7240, pp. 274–285, 2012.

keywords query in probabilistic XML except for [9], which pays attention to IND and MUX *distributional* nodes.

There are still some other popular and useful data dependency which should be studied. For example, let Hung and Jonson be the authors of a document. If Hung appears on another document, then Jonson will more likely (probabilistically) appear on the same document in co-authorship. This kind of data dependency relationship is valuable for the data maintenance and consistency, and needs to be considered uniformly with other dependencies.

Bayesian Networks support more general dependency correlations among data and to perform the uncertain computation works. And SLCA semantic meaning [10, 11] is one of the widely accepted semantic meanings in keywords query. This paper will use Bayesian Networks to model probabilistic XML and discuses SLCA-based keyword filtering.

The main contributions of this paper are listed below:

− This paper proposes the Bayesian Networks based probabilistic XML model PrXML-BN, which can represent sibling nodes dependency relationship uniformly.
− This paper gives the probability presentation of SLCA semantic meaning in PrXML-BN and implements keywords filtering based on SLCA.
− Extensive experiments verify the performance of keywords filtering based on SLCA semantic meaning in PrXML-BN.

The rest of this paper is organized as follows: Section 2 will define SLCA semantic meaning in probabilistic XML and propose the method of modeling PrXML-BN. How to represent and compute the probability of SLCA node semantic meaning in PrXML-BN will be discussed in Section 3. Section 4 gives the system architecture and its implementation. Two strategies of optimizing keywords filtering based on SLCA will also be proposed in Section 4. Section 5 is the evaluation of experiments. Conclusions and future work are in the end.

2   Preliminary and Model

The SLCA node is a node that should satisfy: (1) all the keywords appear in the subtree rooted at the node, and (2) the node have no descendent nodes whose subtrees also contain all the keywords. [9] uses a binary tuple $< v, p >$ to denote the SLCA node in probabilistic XML, where v is a SLCA node and p is the probability of v to be a SLCA node. So our work is to obtain p. If no descendants are SLCA nodes, to compute the probability of v to be a SLCA node can be transformed into computing the probability of the possible world, which contains v and all keywords appear in $SubT(v)$'s current possible world.

2.1   PrXML-BN Model

In probabilistic XML studies, some important assumptions are stated in [1]. If parent nodes exist, children nodes exist with definite probability; otherwise, the probability of the existence of children nodes is "0". Except for these nodes, the rest nodes are independent in default. The probabilistic XML model, PrXML-BN, considers the dependent relationships not only between father-children nodes but also between sibling nodes. All dependencies are treated as father-children relationship. The quantities uncertain dependency between all nodes and their parent nodes will be represented in CPT. The constraint is no circles are among sibling nodes' dependency correlations.

Definition 1 maps probabilistic XML to Bayesian Networks X-BN.

Definition 1. *For a given probabilistic XML document p-document with CPT, it is denoted as a binary tuple $< \mathbb{N}, \mathbb{F} >$. $\mathbb{N}$ is node set of the document. The conditional probability of all the nodes in p-document is denoted as $\mathbb{F}$ which includes the quantities uncertain dependency between node v and its parent nodes in p-document. The triple $< \mathbb{X}, \mathbb{E}, \mathbb{P} >$ denotes the mapping from p-document to Bayesian Networks X-BN. $\mathbb{X}$ is node set, $\mathbb{E}$ is edge set and $\mathbb{P}$ is CPT set of nodes in Bayesian Networks. The mapping process is according to following rules:*

1. *The nodes in $\mathbb{N}$ and $\mathbb{X}$ are one-to-one correspondence. The value of nodes in $\mathbb{X}$ is "1" denoting the node exists and "0" for the opposite case.*
2. *When the corresponding relationship between the nodes in $\mathbb{N}$ and $\mathbb{X}$ is given, the conditional probabilities in $\mathbb{F}$ and $\mathbb{P}$ are one-to-one correspondence. The uncertain quantitative relationships between any given node X_i and its parent node set $\pi(X_i)$ are included in $\mathbb{P}$.*
3. *For any given X_i in $\mathbb{X}$, a directed edge from each node in $\pi(X_i)$ to X_i will be appended. $\mathbb{E}$ is consisted of these directed edges.*

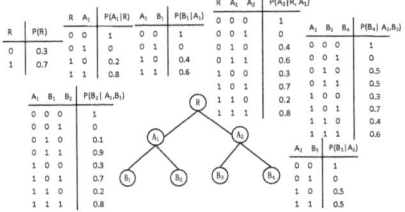 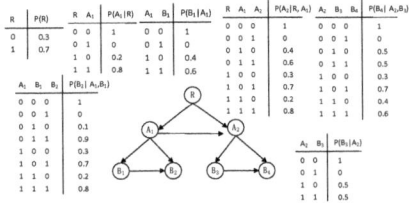

Fig. 1. a Probabilistic XML Document Tree

Fig. 2. Bayesian Networks X-BN for Probabilistic XML

Fig. 1 is a probabilistic XML document with CPT. According to Definition 1, Fig. 1 can be transformed into Bayesian Networks shown in Fig. 2. Therefore, the probability computation works of nodes in Fig. 1 can use the probability computation method of Bayesian Networks(X-BN, shown in Fig. 2).

2.2   XK-BN for SLCA Computation in PrXML-BN

Now the SLCA semantic meaning will be added into X-BN for filtering keywords.

Example 1. *keyword query:* K_1, K_2.

For the given keyword query in Example 1, Let keywords be denoted as nodes K_1, K_2. Then the value of keyword nodes is "1" or "0" (which means the existence of keyword nodes). The values of different keywords are independent.

Definition 2. *The dependency between document node n and single keyword node K:*

1. *If n includes keyword K, n and K are dependent. And n is the parent node and K is the child node. It is denoted as $n \to K$.*
2. *If n does not include any keywords, n and K are independent.*

Definition 2 gives the dependency relationship between document node and keyword node. Keyword nodes' $CPTs$ follow the rules: if the value of its parent node is "1", the probability of the keyword node's value to be "1" is "1" and "0" for it's value to be "0"; if there does not exist any parent node's value to be "1", the probability of the keyword node's value to be "1" is "0" and "1" for its value to be "0".

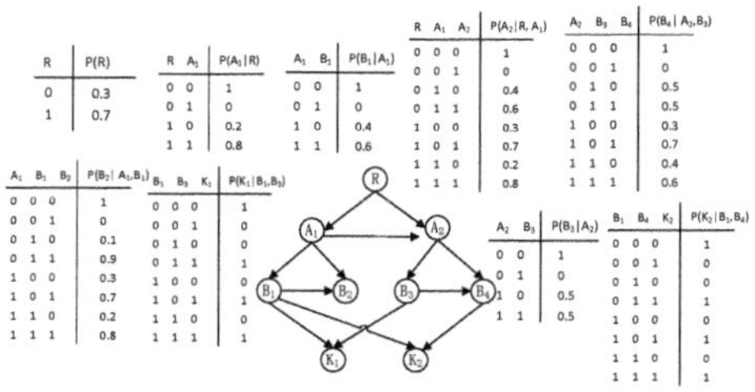

Fig. 3. Bayesian Networks XK-BN Which Include Keywords

Insert keyword nodes $\{K_1, K_2\}$ into node set $\mathbb{X}$; according to Definition 2, append the directed edges from document nodes to keyword nodes into $\mathbb{E}$; and add conditional probabilities of keyword nodes into $\mathbb{P}$. Finally, the newly triple $< \mathbb{X}, \mathbb{E}, \mathbb{P} >$ denotes the Bayesian Networks XK-BN which includes keyword nodes. Fig. 3 shows an example of XK-BN. Node B_1 includes both keywords. Node B_3 and B_4 include keyword K_1 and K_2 separately. The $CPTs$ of node K_1 and K_2 can be generated by preprocessing given XML document and query.

3 Probability Computing in PrXML-BN

3.1 Main Idea

In model PrXML-BN, the probability of node n to be a SLCA node is the total probability of all possible worlds in which n exists and all keywords appear in $SubT(n)$'s current possible world. The main idea of computing the probability of a node to be a SLCA node is briefed in Algorithm 1.

Algorithm 1. ComputeProb

Input:
   current node,n;
Output:
   probability of current node n to be a SLCA node, *prob*;
1: *prob* = 0;
2: **if** no SLCA nodes in n's descendants **then**
3:    **for** each possible world of XK-BN **do**
4:       **if** SubT(n) includes all keywords **then**
5:          add current possible world's probability to *prob*;
6:       **end if**
7:    **end for**
8: **else**
9:    $n.flag$ = true;
10: **end if**
11: return *prob*;

Scanning document node bottom-up, algorithm 1 computes current node n's probability to be SLCA. If n's descendants aren't SLCA nodes(line 2) and XK-BN's current possible world satisfies that $SubT(n)$ include all keywords, add the possible world's probability to *prob*, which is n's probability to be SLCA. If any n's descendant is SLCA, $n.flag$ records weather there are SLCAs in $SubT(n)$ and will be transferred to upper nodes. Following will discuss how to find the possible worlds in which $SubT(n)$ includes all keywords(line 3 and line 4).

3.2   The Correct Possible World

The subsection will discuss how to depict the possible worlds in which node n exists and all keywords appear in $SubT(n)$.

Definition 3. *Node Set Partition.*

For any non-keyword node n in XK-BN, it will separate all nodes in XK-BN into four parts: the keyword node set $\mathcal{K}$; the node set of subtree $SubT(n)$ in original XML tree, $\mathcal{D}_n$; the ancestor node set of n, $\mathcal{A}_n$; the rest nodes set, $\mathcal{O}_n$.

$\mathcal{A}_n$ includes all the nodes on the paths from the root node (which is the root node of the original XML tree) to current node n, except for n itself. Take the node B_1 in Fig. 3 as an example. $\mathcal{K} = \{K_1, K_2\}$, $\mathcal{D}_{B_1} = \{B_1\}$, $\mathcal{A}_{B_1} = \{R, A_1\}$, $\mathcal{O}_{B_1} = \{B_2, A_2, B_3, B_4\}$.

If node set $\mathcal{D}_n$ contains m nodes $D_{n,1}, D_{n,2}, \cdots, D_{n,m}$, where $D_{n,1}$ is the root node n of subtree $SubT(n)$, $\mathcal{D}_n$ can be denoted as vector $(D_{n,1}, D_{n,2}, \cdots, D_{n,m})$. The value of vector $\mathcal{D}_n$ can be denoted by vector $(v_{D_{n,1}}, v_{D_{n,2}}, \cdots, v_{D_{n,m}})$. The value of $D_{n,i} = v_{D_{n,i}}(1 \leq i \leq m)$ will be set as "0" or "1".

Definition 4. *For a given non-keyword node n and $\mathcal{D}_n$ in XK-BN, $D_{n,1}$ which is the first node of $\mathcal{D}_n$ is node n, the value set $V_{\mathcal{D}_n}$ of $\mathcal{D}_n$ is defined as follows:*

$$V_{\mathcal{D}_n} = \{(v_{D_{n,1}}, v_{D_{n,2}}, \cdots, v_{D_{n,m}}) \mid v_{D_{n,1}} = 1 \wedge \{\forall i, K_i \in \mathcal{K},$$
$$\exists j, 1 \leq j \leq m, D_{n,j} \in \mathcal{D}_n \wedge D_{n,j} \rightarrow K_i \wedge v_{D_{n,j}} = 1\}\}. \tag{1}$$

Definition 4 gives the node value vector set $V_{\mathcal{D}_n}$ of $\mathcal{D}_n$, and $V_{\mathcal{D}_n}$ satisfies following condition: such nodes in $\mathcal{D}_n$ whose value is "1" contains all keywords (which indicates that all the keyword nodes' value must be "1" in $XK\text{-}BN$). $\mathcal{D}_n$ may be empty set. If node vector $\mathcal{D}_n$ is set to be any value in $V_{\mathcal{D}_n}$, n satisfies the first condition of SLCA. Any node value vector, in which there exist nodes (whose value is "1") contain all keywords, should be in $V_{\mathcal{D}_n}$.

Take node B_1 in Fig. 3 as an example. $\mathcal{D}_{B_1}$ only has one node B_1, $V_{\mathcal{D}_{B_1}} = \{(1)\}$, which indicates that to set B_1 to be "1" can guarantee current possible world contains all keywords.

Definition 5. *Set's Value.*

Given set $\mathcal{E}$, constant c whose value is "0" or "1", value vector set C whose vector is consisted of "0" and "1". $E = c$ means the value of each node in E is c. $E = C$ stands for that E can be treated as a node vector and its value can be each value in C. The value of each node in node vector E is the corresponding value of current value vector of C.

For example, keyword node set $\mathcal{K} = 1$ denotes that all the keyword nodes' value are "1". The nodes in $\mathcal{D}_{A_1}$ constitute a vector (A_1, B_1, B_2) and the value vector set C is $\{(1,1,1),(1,1,0)\}$. $\mathcal{D}_{A_1} = C$ means (A_1, B_1, B_2) can equal to vector $(1,1,1)$ ($A_1=1$, $B_1=1$, $B_2=1$) or vector $(1,1,0)$ ($A_1=1$, $B_1=1$, $B_2=0$).

It can be found from Definition 4 and Definition 5 that $\mathcal{D}_n = V_{\mathcal{D}_n}$ can guarantee $SubT(n)$ contains all keywords for any non-keyword node n in XK-BN. If no n's descendants are SLCA nodes, the possible worlds presented by "$\mathcal{D}_n = V_{\mathcal{D}_n}, \mathcal{K} = 1$"(in fact, $\mathcal{K} = 1$ can be omitted.) are correct possible worlds in which n is a SLCA node. The total probability of these possible worlds is the probability of n to be a SLCA node, which is shown in formula 2.

$$P(\mathcal{K} = 1, \mathcal{D}_n = V_{D_n})$$
$$= \sum_{\substack{\mathcal{D}_n - \{n\} \\ \mathcal{D}_n = V_{D_n}}} \sum_{\mathcal{A}_n, \mathcal{O}_n} P(n = 1, \mathcal{K} = 1, \mathcal{A}_n, \mathcal{O}_n, \mathcal{D}_n - \{n\}). \tag{2}$$

4   Architecture and Implementation

4.1   Architecture

Fig. 4 shows the system architecture of SLCA-based probabilistic XML keywords filtering. In Fig. 4, after query submitted, system will firstly find whether the results are in the cache. If it does not find the results, it will call the parser and computation module to generate and return results. This paper does not discuss further which one will be cached, and only part of source queries (for example, half of the submitted queries) will be cached. It will randomly choose double of these cached queries to do query computation.

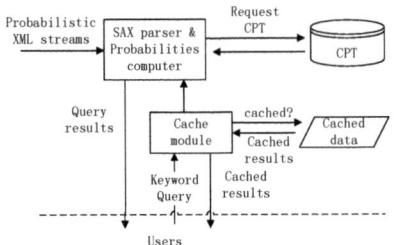

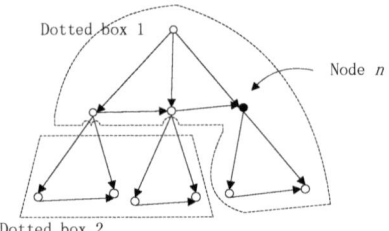

Fig. 4. system architecture using cache strategy

Fig. 5. node partition while computing probability

4.2   Parsing and Computing Module

When XML document is read by SAX parser, a record $(level, id, flag)$ will be saved in the stack for each XML node. $level$ denotes the depth of current node in document. id denotes the document node. $flag$ records whether current node and its descendants contain SLCA nodes. $flag = true$ means the subtree rooted at current node has contained SLCA nodes, which will be transformed to the up and prevent ancestors from being SLCA nodes.

Algorithm 2. EndElement

Input:
Output:
1: $info \leftarrow Stack$.pop();
2: **if** $info$.getLevel() $== level$ **then**
3:     $prob = $ ComputeProb($info$.getID());
4:     **if** $prob >0$ **then**
5:         $info$.setFlag(true);
6:         output $info$.getID() as SLCA node id, corresponding probability is $prob$;
7:     **end if**
8: **else**
9:     pop all children's records from $Stack$;
10:     $info \leftarrow$ pop current node's record from $Stack$;
11:     $prob = $ ComputeProb($info$.getID());
12:     **if** $prob >0$ **then**
13:         $info$.setFlag(true);
14:         output $info$.getID() as SLCA node id, corresponding probability is $prob$;
15:     **end if**
16: **end if**
17: $Stack$.push($info$);
18: $level = level$ - 1;

Before parsing documents, event *StartDocument* will initial some global variables, such as $level = 0$, $id = 0$.

StartElement. Event records current node's $(level, id, flag)$ and put it into stack.

EndElement. Event(Algorithm 2) computes the probability of the current node to be a SLCA node. Pop the current node's record and save it into *info* (line 1-2,line 9-10). Then Algorithm 1 will be called to compute probability of current node to be a SLCA node(line 3 and 11). If the probability is larger than "0", current node is a SLCA node and it will be output(line 4-7, line 12-15). At last, the current node's information will be push into stack.

4.3   Optimizations

Another optimization *Reducing Involved Nodes* is described in this subsection.

In XK-BN, it can be found that parent nodes of any non-keyword node n and of any node in $\mathcal{A}_n$ belong to $\mathcal{A}_n$; parent nodes of nodes in $\mathcal{D}_n - \{n\}$ are still in $\mathcal{D}_n - \{n\}$; parent nodes of $\mathcal{O}_n$'s nodes belong to $\mathcal{A}_n \cup \mathcal{O}_n \cup \{n\}$.

Definition 6. $P(\mathcal{E}|\mathcal{F})$ *denotes the continued multiplication of each node's conditional probability in* $\mathcal{E}$, *where the parent nodes of each node in* $\mathcal{E}$ *belongs to* $\mathcal{F}$. $P(\mathcal{E}|\mathcal{E})$ *has the same meaning but the parent nodes of each node* E_i *in* $\mathcal{E}$ *is in the difference set* $\mathcal{E} - \{E_i\}$.

$$P(\mathcal{E}|\mathcal{F}) = \prod_{\substack{E_i \in \mathcal{E} \\ 1 \le i \le m}} P(E_i|\mathcal{F}). \tag{3}$$

$$P(\mathcal{E}|\mathcal{E}) = \prod_{\substack{E_i \in \mathcal{E} \\ 1 \le i \le m}} P(E_i|\mathcal{E} - \{E_i\}). \tag{4}$$

For example $P(\mathcal{A}_n|\mathcal{A}_n) = \prod_{\substack{A_{n,i} \in \mathcal{A}_n \\ 1 \le i \le m}} P(A_{n,i}|\mathcal{A}_n - \{A_{n,i}\})$.

So formula 2 can be transformed to formula 5:

$$
\begin{aligned}
&P(\mathcal{K} = 1, \mathcal{D}_n = V_{D_n}) \\
&= \sum_{\substack{\mathcal{D}_n - \{n\} \\ \mathcal{D}_n = V_{D_n}}} \sum_{\mathcal{A}_n, \mathcal{O}_n} P(n = 1, \mathcal{K} = 1, \mathcal{A}_n, \mathcal{O}_n, \mathcal{D}_n - \{n\}) \\
&= \sum_{\mathcal{A}_n} P(\mathcal{A}_n|\mathcal{A}_n) P(n = 1|\mathcal{A}_n) \sum_{\substack{\mathcal{D}_n - \{n\} \\ \mathcal{D}_n = V_{D_n}}} P(\mathcal{D}_n - \{n\}|\mathcal{D}_n) \sum_{\mathcal{O}_n} P(\mathcal{O}_n|\mathcal{O}_n, \mathcal{A}_n, n = 1) P(\mathcal{K} = 1|\mathcal{O}_n, \mathcal{A}_n, \mathcal{D}_n) \\
&= \sum_{\mathcal{A}_n} P(\mathcal{A}_n|\mathcal{A}_n) P(n = 1|\mathcal{A}_n) \sum_{\substack{\mathcal{D}_n - \{n\} \\ \mathcal{D}_n = V_{D_n}}} P(\mathcal{D}_n - \{n\}|\mathcal{D}_n) \sum_{\mathcal{O}_n} P(\mathcal{O}_n|\mathcal{O}_n, \mathcal{A}_n, n = 1)
\end{aligned}
\tag{5}
$$

When formula 5 satisfies $\mathcal{D}_n = V_{D_n}$, whatever the value of nodes in $\mathcal{A}_n$ and $\mathcal{O}_n$ are, $P(\mathcal{K} = 1|\mathcal{O}_n, \mathcal{A}_n, \mathcal{D}_n) = 1$ is always true. Therefore, line 3 can be simplified to line 4 in formula 5.

According to Definition 6, it can get formula 6 as follows:

$$\sum_{\mathcal{O}_n} P(\mathcal{O}_n | \mathcal{O}_n, \mathcal{A}_n, n = 1)$$

$$= \sum_{\mathcal{O}_n} \prod_{\substack{1 \le i \le m \\ O_{n,i} \in \mathcal{O}_n}} P(O_{n,i} | \mathcal{O}_n - \{O_{n,i}\}, \mathcal{A}_n, n = 1) \qquad (6)$$

$$= \prod_{\substack{1 \le i \le m \\ O_{n,i} \in \mathcal{O}_n}} \sum_{O_{n,i}} P(O_{n,i} | \mathcal{O}_n - \{O_{n,i}\}, \mathcal{A}_n, n = 1)$$

In formula 6, for any node $O_{n,i} \in \mathcal{O}_n, (1 \le i \le m)$, if the values of nodes in $\mathcal{O}_n - \{O_{n,i}\}$ and $\mathcal{A}_n$ are given, $\sum_{O_{n,i}} P(O_{n,i} | \mathcal{O}_n - \{O_{n,i}\}, \mathcal{A}_n, n = 1) \equiv 1$. That is formula 6 equals "1". Then, formula 5 can be simplified to formula 7:

$$P(\mathcal{K} = 1, \mathcal{D}_n = V_{D_n})$$

$$= \sum_{\mathcal{A}_n} P(\mathcal{A}_n | \mathcal{A}_n) P(n = 1 | \mathcal{A}_n) \sum_{\substack{\mathcal{D}_n - \{n\} \\ \mathcal{D}_n = V_{D_n}}} P(\mathcal{D}_n - \{n\} | \mathcal{D}_n) \qquad (7)$$

Algorithm 3. ComputeProbOPT

Input:
　　current node,n;
Output:
　　the prob of which current node is a SLCA node,$prob$.
1: $prob=0$;
2: **if** no SLCA nodes in $SubT(n)$ **then**
3: 　$carelist$ = ConstructCareList(n);
4: 　$bt \in \{0, 1, 2, \ldots, 2^{carelist.size}\}$;
5: 　**for** each value of bt **do**
6: 　　**if** the current node's bit of bt is "1" **then**
7: 　　　**if** descendant nodes whose bits are"1" in bt include all keywords **then**
8: 　　　　$localprob = 1$;
9: 　　　　**for** each node m in $carelist$ **do**
10: 　　　　　$p(m|m's\ parents)$ = find node m's conditional probability from database;
11: 　　　　　$localprob = localprob \times p(m|m's\ parents)$;
12: 　　　　**end for**
13: 　　　　$prob = prob + localprob$;
14: 　　　**end if**
15: 　　**end if**
16: 　**end for**
17: **else**
18: 　$n.flag$ = true;
19: **end if**
20: **return** $prob$;

For any given non-keyword node n in XK-BN, formula 7 is the total probability of the possible worlds in which node n satisfying SLCA's first condition.

The optimization introduced above can eliminate some probability computation works. As is shown in Fig. 5, keyword nodes has been removed from Bayesian Networks. The black node in Fig. 5 is the node n await to compute probability. If condition $\mathcal{D}_n = V_{D_n}$ exists, nodes in $\mathcal{O}_n$ in dotted box 2 have no effect on computing probability of current node and only nodes in dotted box 1 are related to probability computation.

Algorithm 3 can replace Algorithm 1 to compute the probability of current node to be a SLCA node. If there are no SLCAs in $SubT(n)$(Line 2), counting n's referred nodes(Line 3). And using these nodes' probabilities to compute current node n's probability to be a SLCA node.

If there are n nodes in document and the sum of each node's $\mathcal{D}$ and $\mathcal{A}$ is m, $m \leq n$ in average, the time complexity of the whole method after optimization is $O(n \cdot 2^m)$.

5   Experiments

The experiment evaluations of SLCA based keywords filtering algorithms using Bayesian Networks are stated in this section, which include the studies for data sets features and query features.

The experiments are performed on an Intel dual-core with 2 GB memory computer and the programs are written in JAVA. The datasets are from DBLP dataset. Each XML packet tags a paper's information. The number of nodes in XML packet is from 7 to 16. Each kind of node contains about 1,000 records and its size is 500K in average. While building Bayesian Networks, for a given node v, except for its parent node, its sibling nodes which are in front of v will be choosed as v's parent nodes according to a certain percentage, such as 20%, 25% and 33%. Each node's probability will be generated randomly and saved in Berkeley DB, as well as the preprocessed keyword nodes' information. The structure of query is like [12], denoted as *kN-L-H*, where *kN* is the number of keywords, *H* and *L* is the highest and lowest keyword frequency, respectively. The keyword frequency is "100-100" in this paper. It will randomly generate a group of queries and repeat it five times. The average process time of each query will be recorded.

5.1   Date Set Feature

The query set is 2-100-100 and the number of data records is 1000. Fig. 6 and 7 show the average processing time of single query for 1,000 XML records. Fig. 6 records the algorithm without using cache category, whose processing time will increase with data dependency ration increased. While for the number of nodes in Bayesian Networks, the dependency ratio between sibling nodes has less effect on its processing time.

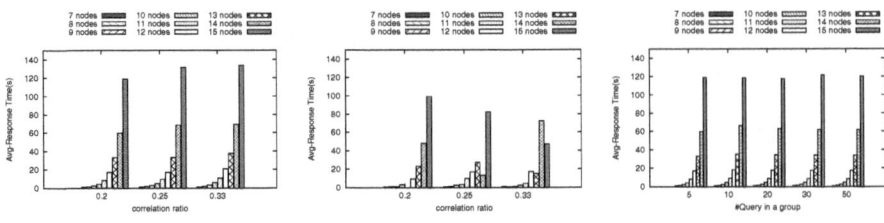

Fig. 6. Data Dependency Ratio's Effect

Fig. 7. Data Dependency Ratio's Effect Using Cache Strategy

Fig. 8. #Query's Effect

Fig. 7 shows the processing with cache strategy. Because of query's randomness, the ratio of results using cache strategy and the ration of reducing processing time is uncertain. Even then, compared with Fig. 6, the performance of using cache strategy is better than the other.

5.2 Query Feature

Fig. 8 and Fig. 9 study the influence of the query groups' size on average processing time. Both of them use nodes reducing optimization. And they also record the average processing time of each query on 1,000 XML packets before and after using cache strategy. Obviously, using cache strategy will reduce processing time remarkably.

Fig. 10 and Fig. 11 show the effects of the number of keyword nodes on average processing time. Both of them are the results after graph nodes reducing. And they also record the average processing time of each query before and after using cache strategy. It can be found from Fig. 10 that with the number of keyword nodes increasing, the process time also increases, but it is much more better than the effect on processing time brought by the increased nodes number in Bayesian Networks. Compared with Fig. 10 and 11, using cache strategy can low processing time remarkably. Because of query randomness, the ratio of results using cache strategy is uncertain, which makes the ratio of lowing time complexity uncertain and the effects of increasing keywords number on processing time less obvious.

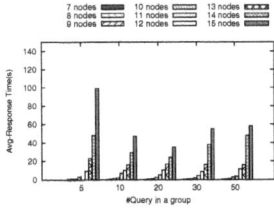
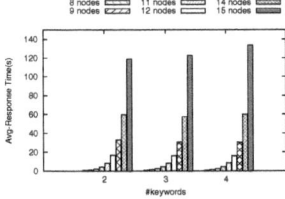
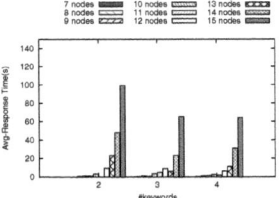

Fig. 9. #Query's Effect Using Cache Strategy

Fig. 10. #Keyword's Effect

Fig. 11. #Keyword's Effect Using Cache

6   Conclusions and Future Work

To represent the general dependency correlations between sibling nodes in probabilistic XML, this paper proposes Bayesian Networks based probabilistic XML model, PrXML-BN, and combines SLCA semantic meaning into Bayesian Networks. Besides, it also implements keywords filtering based on SLCA. And the optimizations including graph nodes reducing and cache strategy are also applied in the system to reduce algorithm time complexity. Experiment studies have shown that for SLCA keywords filtering based on Bayesian Networks, the main factor influencing time complexity is the size of Bayesian Networks. The query features, such as keywords number, has less effect on processing time. Cache strategy can reduce processing time remarkably. And it is easy to model data using Bayesian Networks and can represent more general dependency correlations, but the time complexity will increase exponentially. The future works are how to further reduce the time cost and apply the probabilistic XML data management in the distributed environment.

Acknowledgments. This work was supported by the National Major Projects on Science and Technology under grant number 2010ZX01042-002-003-004, NSFC grant (No. 61033007, 60903014 and 61170085), 973 project(No. 2010CB328106), Program for New Century Excellent Talents in China (No.NCET-10-0388).

References

1. Nierman, A., Jagadish, H.V.: Protdb: Probabilistic data in xml. In: VLDB, pp. 646–657 (2002)
2. Hung, E., Getoor, L., Subrahmanian, V.S.: Pxml: A probabilistic semistructured data model and algebra. In: ICDE, pp. 467–478 (2003)
3. van Keulen, M., de Keijzer, A., Alink, W.: A probabilistic xml approach to data integration. In: ICDE, pp. 459–470 (2005)
4. Abiteboul, S., Kimelfeld, B., Sagiv, Y., Senellart, P.: On the expressiveness of probabilistic xml models. VLDB J. 18(5), 1041–1064 (2009)
5. Kimelfeld, B., Sagiv, Y.: Matching twigs in probabilistic xml. In: VLDB, pp. 27–38 (2007)
6. Kimelfeld, B., Kosharovsky, Y., Sagiv, Y.: Query efficiency in probabilistic xml models. In: SIGMOD Conference, pp. 701–714 (2008)
7. Chang, L., Yu, J.X., Qin, L.: Query ranking in probabilistic xml data. In: EDBT, pp. 156–167 (2009)
8. Kimelfeld, B., Kosharovsky, Y., Sagiv, Y.: Query evaluation over probabilistic xml. VLDB J. 18(5), 1117–1140 (2009)
9. Li, J., Liu, C., Zhou, R., Wang, W.: Top-k keyword search over probabilistic xml data. In: ICDE, pp. 673–684 (2011)
10. Xu, Y., Papakonstantinou, Y.: Efficient keyword search for smallest lcas in xml databases. In: SIGMOD Conference, pp. 537–538 (2005)
11. Sun, C., Chan, C.Y., Goenka, A.K.: Multiway slca-based keyword search in xml data. In: WWW, pp. 1043–1052 (2007)
12. Wang, W., Wang, X., Zhou, A.: Hash-Search: An Efficient SLCA-Based Keyword Search Algorithm on XML Documents. In: Zhou, X., Yokota, H., Deng, K., Liu, Q. (eds.) DASFAA 2009. LNCS, vol. 5463, pp. 496–510. Springer, Heidelberg (2009)

Provenance Based Conflict Handling Strategies

Domenico Beneventano

Department of Computer Science
University of Modena and Reggio Emilia
Via Vignolese 905, 41125 Modena, Italy
domenico.beneventano@unimore.it

Abstract. A fundamental task in data integration is *data fusion*, the process of fusing multiple records representing the same real-world object into a consistent representation; data fusion involves the resolution of possible conflicts between data coming from different sources; several high level strategies to handle inconsistent data have been described and classified in [8].

The MOMIS Data Integration System [2] uses either *conflict avoiding* strategies (such as the *trust your friends* strategy which takes the value of a preferred source) and *resolution* strategies (such as the *meet in the middle* strategy which takes an average value).

In this paper we consider other strategies proposed in literature to handle inconsistent data and we discuss how they can be adopted and extended in the MOMIS Data Integration System. First of all, we consider the methods introduced by the *Trio* system [1,6] and based on the idea to tackle data conflicts by explicitly including information on provenance to represent uncertainty and use it to answer queries. Other possible strategies are to ignore conflicting values at the global level (i.e., only *consistent values* are considered) and to consider at the global level *all conflicting values*.

The original contribution of this paper is a provenance-based framework which includes all the above mentioned conflict handling strategies and use them as different *search strategies* for querying the integrated sources.

1 Introduction

A fundamental task in data integration is *data fusion*, the process of fusing multiple records representing the same real-world object into a consistent representation; data fusion involves the resolution of possible conflicts between data coming from different sources; several high level strategies to handle inconsistent data have been described and classified in [8].

MOMIS (Mediator envirOnment for Multiple Information Sources) is a framework to perform integration of structured and semi-structured data sources [3,2]. The MOMIS DAta Integration System is characterized by a classical wrapper/mediator architecture: the local data sources contain the real data, while a Global Schema (GS) provides a *reconciled, integrated, read-only view* of the underlying sources. The GS and the mapping between GS and the local sources have to be defined at design time by the Integration Designer; end-users can then pose queries over this GS. MOMIS has been

H. Yu et al. (Eds.): DASFAA Workshops 2012, LNCS 7240, pp. 286–297, 2012.

developed by the DBGROUP of the University of Modena and Reggio Emilia[1]. An open source version of the MOMIS system is delivered and maintained by the academic spin-off DataRiver[2].

The Data Fusion framework of the MOMIS System uses either *conflict avoiding* strategies (such as the *trust your friends* strategy which takes the value of a preferred source) and *resolution* strategies (such as the *meet in the middle* strategy which takes an average value); these strategies are implemented by means of the so-called *full outerjoin-merge operator* proposed in [16] and adapted to the MOMIS System in [2].

In this paper we consider other strategies proposed in literature to handle inconsistent data and we discuss how they can be adopted and extended in the MOMIS System. In particular, we consider the techniques introduced into the *Trio* system [1,6] and based on the idea to tackle data conflicts by explicitly including information on *lineage* to represent uncertainty and use it to answer queries.

Lineage, or *provenance*, in its most general definition, describes where data came from, how it was derived and how it was modified over time. Lineage provides valuable information that can be exploited for many purposes, ranging form simple statistical resumes presented to the end-user, to more complex applications such as managing data uncertainty or identifying and correcting data errors. For these reasons, in the last few years the research activity in the Information Management System area has been increasingly focused on this topic. In particular, lineage has been studied extensively in data warehouse systems [11,10]. However, in Data Integration systems, lineage is still considered as an open research problem [14,13].

In [4] we introduced the notion of provenance into the MOMIS framework, by defining the provenance for the *full outerjoin-merge operator*; this definition is based on the concept of *PI-CS*-provenance (Perm Influence Contribution Semantics) proposed in *Perm* [12] to produce more precise provenance information for outerjoins. Another important reason behind the choice to use the *PI-CS*-provenance is that it is implemented in an open-source provenance management system called *Perm* [12] (Provenance Extension of the Relational Model) that is capable of computing, storing and querying provenance for relational databases. We are using the *Perm* system as the SQL engine of the MOMIS system, so obtaining the provenance in our Data Integration System. On the other hand, from a theoretical point of view, in [4] we argued some differences between the *PI-CS*-provenance of the full outerjoin-merge operator and of the outerjoin operator and we are extending the *Perm* system to take into account these differences.

The original contribution of this paper is a provenance-based framework which includes several conflict handling strategies and use them as different *search strategies* for querying the Global Schema. Besides the techniques introduced into the *Trio* system [1,6], we will also take into account the strategy to ignore conflicting values at the global level, i.e., only *consistent values* are considered (in [8] this strategy is called *No Gossiping* and classified as a *conflict avoiding* strategy). Finally, another possible strategy, is to consider at the global level *all conflicting values* (in [8] this strategy is called *Considering all possibilities* and classified as a *conflict ignoring* strategy).

---

[1] http://www.dbgroup.unimore.it
[2] http://www.datariver.it

The remainder of the paper is organized as follows. In section 2 we will introduce the basic definitions of the MOMIS framework that will be used along the paper; section 2.1 briefly describes *conflict avoiding* and *resolution* strategies implemented in MOMIS by means of the *full outerjoin-merge operator*, whose provenance is introduced in section 2.2. The novel Provenance based Conflict Handling Strategies will be described in section 3 Finally, conclusions and future works are sketched in section 4.

2   The MOMIS Data Fusion System

In this section we will introduce the basic definition of the MOMIS framework [3,2] that will be used along this paper. A MOMIS Data Integration System is constituted by: a set of *local schemas* $\{LS_1, \ldots, LS_k\}$, a *global schema GS* and *Global-As-View* (*GAV*) mapping assertions [15] between GS and $\{LS_1, \ldots, LS_k\}$. A global schema GS is a set of *global classes*, denoted by G. A local schema LS is a set of *local classes*, denoted by L. Both the global and the local schemas are expressed in the ODL_{I^3} language [7]. However, for the scope of this paper, we consider both the GS and the LS_i as relational schemas, but we will refer to their elements respectively as global and local classes to comply with the MOMIS terminology.

For each global class G, a *Mapping Table (MT)* is defined, whose columns represent a set of local classes $\{L_1, \ldots, L_n\}$; an element $MT[GA][L]$ represents the local attribute of L which is mapped onto the global attribute GA, or $MT[GA][L]$ is empty (there is no local attribute of L mapped onto the global attribute GA)[3].

A small example, similar to the one used in [1], is shown in Figure 1; both local classes LA and LB contain data about crime vehicle sightings coming from two different data sources; Figure 1 also shows the Mapping Table of the global class SAW, with schema SAW(Viewer,Age,Car), obtained by integrating these two local classes LA and LB.

GAV mapping assertions are expressed by specifying for each global class G a query over $\mathcal{L}(G)$, called *mapping query* and denoted by $\mathcal{MQ}^G$, which defines the instance of G starting from the instances of its local classes. The mapping query $\mathcal{MQ}^G$ is defined to make a global class perform *Data Fusion* among its local class instances [8]: multiple *local tuples* coming from local classes and representing the same real-world object are fused into a single and consistent *global tuple* of the global class. To identify multiple local tuples coming from local classes and representing the same real-world object, we assume that error-free and shared object identifiers exist among different sources: two local tuples with the same object identifier ID indicate the same object in different sources; thus we can use L^{id} to denote the tuple t of a local class L with object identifier ID equal to id, i.e. $t[ID] = id$. In our example, we assume viewer as an object identifier; then the first tuple of the local class LA, i.e. the tuple with viewer $= Amy$, will be denoted with LA^{Amy}.

---

[3] In this paper, for the sake of simplicity, we consider a simplified version of the MOMIS framework proposed in [3,2], where $MT[A][L]$ is a set of local attributes and *Data Transformation Functions* specify how local attribute values have to be transformed into corresponding global attribute values. Moreover we assume $S(G) = \cup_i S(L_i)$, i.e. global and local attribute names are the same.

Local classes
LA and LB:

LA

viewer	age	car
Amy	21	Honda
Betty	NULL	Honda
Sam	NULL	Toyota
Joe	32	Toyota

LB

viewer	car
Amy	Toyota
Betty	Honda
Sam	Honda

Mapping Table of the global
class Saw:

SAW	LA	LB
Viewer	viewer	viewer
Age	age	
Car	car	car

Fig. 1. Example: two local classes with a conflicting attribute

In fusing data from different sources into one consistent representation, several high level strategies have been described and classified in [8]. With respect to this classification, in MOMIS we used either *conflict avoiding* strategies (such as the *trust your friends* strategy which takes the value of a preferred source) and *resolution* strategies (such as the *meet in the middle* strategy which takes an average value); these solutions will be briefly described in section 2.1.

In this paper we consider how other strategies, proposed in literature, to handle inconsistent data, can be adopted and extended in the MOMIS Data Integration system. First of all, we consider the methods introduced into the *Trio* system [1,6] based on the idea to tackle data conflicts by explicitly including information on lineage to represents uncertainty and use it to answer queries. Another strategy is to ignore conflicting values at the global level, i.e., only *consistent values* are considered (in [8] this strategy is called *No Gossiping* and classified as a *conflict avoiding* strategy). Finally, another possible strategy, is to consider at the global level *all conflicting values* (in [8] this strategy is called *Considering all possibilities* and classified as a *conflict ignoring* strategy).

2.1   Data Fusion Strategies in MOMIS

In the MOMIS system, data fusion is performed at the level of a global class G by defining, for each conflicting attribute of G, a *Resolution Function* and by defining its mapping query $\mathcal{MQ}^G$ by means of the *full outerjoin-merge operator*, proposed in [16] and adapted to the MOMIS framework in [2]. Intuitively, this corresponds to the following two operations: (1) Computation of the *Full Outer Join*, on the basis of the shared object identifier, of the *local classes* of G; (2) Application of the resolution functions. Thus $\mathcal{MQ}^G$ can be formulated by standard SQL; for example, for the global class SAW:

```
MQ^SAW : SELECT viewer AS Viewer, age AS Age,
                COALESCE(SAW_B.car,SAW_A.car) AS Car
           FROM LA FULL OUTER JOIN LB
                          USING (viewer)
```

where COALESCE is the standard SQL function which returns its first non-null parameter value. The obtained instance of the global class SAW is shown in Table 1. The global attribute Viewer, derived from the shared object identifier viewer, is an object identifier for the global class SAW; the global tuple with Viewer = Amy, is denoted with SAW^{Amy}.

Table 1. Instance of the global class SAW

Viewer	Age	Car
Amy	21	Toyota
Betty	NULL	Honda
Sam	NULL	Honda
Joe	32	Toyota

2.2   Provenance in the MOMIS System

In [4] we defined the provenance for the full outerjoin-merge operator by using the concept of *PI-CS*-Provenance. The *PI-CS*-Provenance of an output tuple is a set of *witness lists*, where each witness list represents one combination of input relation tuples that were used together to derive the output tuple; a witness list contains a tuple from each input of an operator or the special value ⊥ which indicates that no tuple from an input relation was used to derive the output tuple and, therefore, is useful in modeling outerjoins. For example, the *PI-CS*-Provenance of the global tuple SAW^{Amy} is the set $\{\langle LA^{Amy}, LB^{Amy} \rangle\}$ and the *PI-CS*-Provenance of SAW^{Joe} is the set $\{\langle LA^{Joe}, \bot \rangle\}$ which indicates that LA^{Joe} paired with no tuples from LB influences SAW^{Joe}.

Another important reason behind the choice to use *PI-CS*-Provenance is that it is implemented in a provenance management system called *Perm* [12] (Provenance Extension of the Relational Model) that is capable of computing, storing and querying provenance for relational databases. From an implementation point of view, we are using the *Perm* system as the SQL engine of the MOMIS system, so obtaining the provenance in our Data Integration System. On the other hand, from a theoretical point of view, in [4] we argued some differences between the *PI-CS*-provenance of the full outerjoin-merge operator and of the outer join operator; however, this aspect is not relevant for the scope of this paper.

As an example of provenance, let us consider the following query on the global class SAW (the user is searching for the suspected cars defined as the car sighted by viewers with age greater than 18):

```
SUSPECTED_CAR = SELECT DISTINCT CAR
                FROM SAW
                WHERE AGE > 18
```

The query returns the tuple $(Toyota)$; the *PI-CS*-Provenance for this tuple is a set with two witness lists (see Figure 2). In the *Perm* system witness lists are represented in a relational form, as shown in Figure 2: each witness list of a result tuple is represented by a single tuple. This is the data provenance information obtained by executing the query on the MOMIS system integrated with the *Perm* system.

Car	*PI-CS* Provenance as a set of witness lists
Toyota	$\{ \langle \mathrm{LA}^{Amy}, \mathrm{LB}^{Amy} \rangle, \langle \mathrm{LA}^{Joe}, \perp \rangle \}$

Relational Representation of the *PI-CS* Provenance

Car	LA.viewer	LA.age	LA.car	LB.viewer	LB.car
Toyota	Amy	21	Honda	Amy	Toyota
Toyota	Joe	32	Toyota		

Fig. 2. Example: *PI-CS* Provenance for the query SUSPECTED_CAR

3 Provenance Based Conflict Handling Strategies

The proposed solution to handle inconsistent data is inspired to the methods introduced into the *Trio* system [1,6], which is based on the *ULDB* model [5], where *uncertainty* and *lineage* of the data are considered as first-class concepts. The idea is to tackle data conflicts by explicitly including information on lineage (and accuracy) into its data model and thereby into the data itself; as stated in [5], any application that integrates information from multiple sources may be uncertain about which data is correct, and the original source and derivation of data may offer helpful additional information.

With the concept of *alternatives* introduced in *ULDB*, relations have a set of *certain* attributes and a set of *uncertain* attributes; each tuple in a *ULDB* relation has one value for each certain attribute, and a set of possible values for the uncertain attributes. The *ULDB* model is based on the idea that *provenance* enables simple and consistent representation of uncertain data; provenance in *ULDB* is recorded at the granularity of alternatives: provenance connects a tuple-alternative to those tuple-alternatives from which it was derived. Specifically, provenance is defined as a function λ over tuple-alternatives: $\lambda(t)$ is a boolean formula over the tuple-alternatives from which the alternative t was derived. In the following, we will refer to the provenance defined in *ULDB* as *Trio*-Lineage.

Intuitively, by applying this concept of *alternatives* to our context of data fusion, *non-conflicting* attributes are modelled as *certain* attributes and *conflicting* attributes are modelled as *uncertain* attributes where conflicting values are considered as *mutually exclusive* values for the global tuple; in this way, the global class SAW is considered as an *uncertain relation* which represents a set of possible relation instances. We will refer to this conflict handling strategy as the *TRIO*-strategy. In our example (see Table 2), with the *TRIO*-strategy we have *four possible instances* for the global class SAW: two choices for Amy's car times two choices for Sam's car. We use G^{id} to denote the global tuple of G with object identifier ID equal to id, then we will use (G^{id}, j) to denote the jth tuple-alternative of the global tuple G^{id}; as an example, for the global tuple SAW^{Amy} there are two alternatives, $(SAW^{Amy}, 1)$ and $(SAW^{Amy}, 2)$. Given a global class constituted by n local classes and with only a conflicting attribute GA (the general case with more than one conflicting attribute will be considered in section 3.2) mapped on k local classes $L_1, \ldots, L_k$, with $1 < k \leq n$, the number of tuple-alternatives for a global tuple is of course equal or less than k.

Table 2. Global class SAW as an *uncertain* relation: global tuple-alternatives

Viewer	Age	Car
Amy	21	Honda | | Toyota
Betty	NULL	Honda
Sam	NULL	Toyota | | Honda
Joe	32	Toyota

For a global class, the provenance connects a *global tuple* to local tuples from which it was derived; by using the *ULDB* to model a global class as an uncertain relation, *Trio*-Lineage connects a *global tuple-alternative* to those local tuples from which it was derived. More precisely, the *Trio*-Lineage of a global tuple-alternative is a boolean formula over the local tuples from which the alternative was derived; this is shown in Table 3 where each *global tuple-alternative* of SAW is represented by a row of the table with the related *TRIO*-Lineage.

Table 3. Global tuple-alternatives of SAW, with the related *TRIO*-Lineage

Viewer	Age	Car	*TRIO*-Lineage
Amy	21	Honda	$\lambda(\text{SAW}^{Amy}, 1) = \text{LA}^{Amy}$
Amy	21	Toyota	$\lambda(\text{SAW}^{Amy}, 2) = \text{LA}^{Amy} \wedge \text{LB}^{Amy}$
Betty	NULL	Honda	$\lambda(\text{SAW}^{Betty}, 1) = \text{LA}^{Betty} \vee \text{LB}^{Betty}$
Sam	NULL	Toyota	$\lambda(\text{SAW}^{Sam}, 1) = \text{LA}^{Sam}$
Sam	NULL	Honda	$\lambda(\text{SAW}^{Sam}, 2) = \text{LB}^{Sam}$
Joe	32	Toyota	$\lambda(\text{SAW}^{Joe}, 1) = \text{LB}^{Joe}$

With the *TRIO*-strategy a global class is considered as an *uncertain relation* which represents a set of possible instances. Another possible strategy is to consider *a unique instance* containing *all conflicting values*, i.e., an instance with all the global tuple-alternatives shown in Table 3; we will refer to this strategy as the *ALL* strategy. Finally, another possible strategy is to ignore conflicting values at the global level, i.e., only *consistent values* are considered; this strategy called *No Gossiping* and classified as an *avoiding* strategy in [8], is similar to the concept of *Consistent Query Answering* [9]. With this strategy, which we will call *CQA* strategy, there is a unique instance for the global class SAW : $\{\text{SAW}^{Betty}, \text{SAW}^{Joe}\}$.

The original contribution of this paper is a provenance-based framework which includes all the above conflict handling strategies. In the next section, we will discuss how these different conflict handling strategies correspond to different *search strategies* for querying the Global Schema. In the remaining of this section we show how to implement the proposed framework.

Since in the MOMIS System we are already using the *Perm* system as the basis to obtain the provenance of the integrated data, the proposal is to implement the framework using the *Perm* system. This choice is motivated by several reasons. Firstly, besides the *Trio*-strategy, we want to obtain a framework which also contain the other two kind of strategies, then extensions to the *Trio* framework are needed. Moreover, as shown in section 2.1, the computation of the global class is based on a outerjoin operation, then we need a provenance model where this operation is supported. In the *ULDB* model [5] and in the *Trio* system [1,6] outerjoins are not considered.

In the following we will illustrate how to compute in the MOMIS system the Global tuple-alternatives of a global class, with the related *TRIO*-Lineage (see the example of SAW in Table 3). First of all, a relational representation of the *Trio*-Lineage is intuitively obtained as follows; the *Trio*-Lineage of a global tuple-alternative is a boolean formula over the local tuples; we consider the disjunctive normal form of this formula and each conjunctive clause of this formula is represented by a row; for example, since $\lambda(\text{SAW}^{Betty}, 1) = \text{LA}^{Betty} \vee \text{LB}^{Betty}$ we will obtain two rows for the alternative $(\text{SAW}^{Betty}, 1)$, the first one representing LA^{Betty} and the second one representing LB^{Betty} (see Table 4). For a global class G, the instance G^{all} containing all the global

Table 4. Relational Representation of the *Trio*-Lineage for SAW

Viewer	Age	Car	LA.viewer	LA.age	LA.car	LB.viewer	LB.car
Amy	21	Honda	Amy	21	Honda		
Amy	21	Toyota	Amy	21	Honda	Amy	Toyota
Betty	NULL	Honda	Betty	NULL	Honda		
Betty	NULL	Honda				Betty	Honda
Joe	32	Toyota				Joe	Toyota
Sam	NULL	Toyota	Sam	NULL	Toyota		
Sam	NULL	Honda				Sam	Honda

tuple-alternatives of G, with their relational representation of the *Trio*-Lineage, is obtained as follows:

- non-conflicting attributes $GA_1, \ldots, GA_n$ are computed by the full outerjoin-merge operator defined in section 2.1 (see MQ^SAW); we denote by $G1(ID, GA_1, \ldots, GA_n)$ the obtained relation (where ID is the object identifier);
- the conflicting attribute GA_{n+1} is computed by the union of k queries on the local classes $L_1, \ldots, L_k$ where it is mapped; we denote by $G2(ID, GA_{1+1})$ the obtained relation; as an example, for the conflicting attribute Car of SAW:

```
G2  :  SELECT viewer, car from   LA
       UNION
       SELECT viewer, car from   LB
```

The relation G^{all} is then obtained as the natural full outerjoin between G1 and G2; the result for the global class SAW is shown in Table 4, where global tuple with more than one alternative are highlighted.

3.1   Strategies for Querying the Global Schema

In this section we discuss how the different conflict handling strategies introduced in the previous section can be used as different *search strategies*: the same query on the

Global Schema has a different interpretation on the basis of the chosen conflict handling strategy. In particular, with the *TRIO*-strategy, the *membership problems* of uncertain databases are considered: *Tuple Membership*: Is a given tuple in some possible instance of an uncertain relation?

Tuple Certainty: Is a given tuple in every instance of an uncertain relation?

In [17] is shown that the tuple membership and certainty problems can be solved in polynomial time and algorithms are given.

Given a query Q, in the following we will define (in an informal way) and compare the query answers obtained with the different conflict handling strategies (as an example we consider the SUSPECTED_CAR of page 2.2).

CQA-strategy: $Q^{consistent}$ is the answer obtained by considering only *consistent values*; in the example: SUSPECTED_CAR^{consistent} = $\{Toyota\}$.

ALL-strategy: Q^{all} is the answer obtained by considering *all values*; in the example: SUSPECTED_CAR^{all} = $\{Honda, Toyota\}$.

TRIO-strategy: $Q^{possible}$ is the set of all *possible* tuples of the uncertain relation Q, i.e., is the set of tuples which are in some possible instance of Q (*Tuple Membership*);

in the example: SUSPECTED_CAR^{possible} = $\{Honda, Toyota\}$.

TRIO-strategy: $Q^{certain}$ is the set of all *certain* tuples of the uncertain relation Q, i.e., is the set of tuples which are in all possible instances of Q (*Tuple Certainty*);

in the example: SUSPECTED_CAR^{certain} = $\{Toyota\}$.

It easy to verify that the following relationships hold among these query answers:

$$Q^{consistent} \subseteq Q^{certain} \subseteq Q^{possible} \subseteq Q^{all}$$

On the basis of these relationships, the *CQA* strategy is the *most restrictive* search strategy (i.e., brings back fewer results) and the *ALL* strategy is the *least restrictive* one.

Table 5 shows Q^{all} and, for each tuple of this result, the *most restrictive* search strategy which brings back the tuple: in this way the user knows that $Toyota$ can be obtained from *consistent values* while to obtain $Honda$ also *conflicting values* need to be considered. Then the user can ask for information about the provenance, by obtaining the result shown in Table 6: in this way the user knows that $Toyota$ has two derivations, the first one from *consistent values* and the second one from a possible instance, i.e., from *conflicting values*.

Table 5. All tuples in SUSPECTED_CAR with the related *most restrictive* search strategy

Car	Strategy
Toyota	consistent
Honda	possible

3.2   Dependent and Independent Conflicting Attributes

So far we considered only one conflicting attribute; with two ore more conflicting attributes, another dimension need to be take into account for the conflict handling problem; with reference to Figure 3 the question is whether to consider the tuple

Table 6. Relational Representation of the *Trio* Lineage for SUSPECTED_CAR

Car	Strategy	LA.viewer	LA.age	LA.car	LB.viewer	LB.car
Toyota	consistent	Joe	32	Toyota		
Toyota	possible	Amy	21	Honda	Amy	Toyota
Honda	possible	Amy	21	Honda		

$(Amy, 21, Honda, brown)$ as a possible instance of the global class SAW? The answer is affirmative if we consider the two conflicting attributes Color and Car as *independents*: Car is taken from local class LA and Color is taken from local class LB. The answer is negative if we consider the two conflicting attributes as *dependents*: Color and Car are taken from the same local class. As shown in Figure 3, if we consider the conflicting attributes as *independents*:

- for the global tuple SAW^{Amy} there are two further alternatives
- for the global tuple SAW^{Betty} there are no further alternatives but the alternative with $(Honda, red)$ has now two different derivation, i.e. its *TRIO*-Lineage is $\lambda(\text{SAW}^{Betty}, 1) = \text{LA}^{Betty} \vee (\text{LA}^{Betty} \wedge \text{LB}^{Betty})$.

LA

viewer	age	car	color
Amy	21	Honda	red
Betty	NULL	Honda	red
Sam	NULL	Toyota	red
Joe	32	Toyota	brown

LB

viewer	car	color
Amy	Toyota	brown
Betty	Honda	brown
Sam	Honda	red

Global class SAW as an *uncertain* relation: global tuple-alternatives

Viewer	Age	(Car,Color)
Amy	21	(Honda,red) || (Toyota, brown) || (Honda,brown) || (Toyota, red)
Betty	NULL	(Honda,red) || (Honda, brown)
Sam	NULL	(Toyota, red) || (Honda,red)
Joe	32	(Toyota, brown)

Fig. 3. Example: two local classes with two conflicting attributes

Thus, we can consider different instances for a global class with two (or more) conflicting attributes: an instance obtained by considering *dependent* attributes is, of course, a subset of the one obtained by considering the attributes as *independents*. As a consequence, given a query Q on a global class, for each conflict handling strategies S (except for the *CQA* strategy), we can consider two query answers : Q^S_{depend} and $Q^S_{independ}$, with $Q^S_{depend} \subseteq Q^S_{independ}$. The *CQA* strategy is the *most restrictive* search strategy and the *ALL* strategy with *independent* attributes is the *least restrictive* one. To give an example, let us refine the SUSPECTED_CAR query as follows:

```
SUSPECTED_RED_CAR = SELECT DISTINCT CAR
                    FROM G
                    WHERE AGE > 18
                    AND COLOR = 'RED'
```

For this query, we have the following search strategies:

- *most restrictive*: $\text{SUSPECTED\_RED\_CAR}^{consistent} = \emptyset$
- *least restrictive* with dependent attributes: $\text{SUSPECTED\_RED\_CAR}^{all}_{dep} = \{Toyota\}$
- *least restrictive* with independent attributes: $\text{SUSPECTED\_RED\_CAR}^{all}_{indep} = \{Honda, Toyota\}$

Table 7 shows $\text{SUSPECTED\_RED\_CAR}^{all}_{indep}$ and, for each tuple of this result, the *most restrictive* search strategy which brings back the tuple: in this way the user knows that to obtain *Honda* and *Toyota conflicting values* need to be considered. and that *Toyota* can be obtained only considering *independent* attributes. The provenance is shown in Table 8.

Table 7. SUSPECTED_RED_CAR

Car	Strategy
Honda	possible-dependent
Toyota	possible-independent

Table 8. Relational Representation of the *Trio* Lineage for SUSPECTED_RED_CAR

Car	Strategy	LA.viewer	LA.age	LA.car	LA.color	LB.viewer	LB.car	LB.color
Honda	possible dependent	Amy	21	Honda	red			
Toyota	possible independent	Amy	21	Honda	red	Amy	Toyota	blue

4   Conclusion and Future Work

In this paper we presented our work in progress to implement data provenance techniques in the MOMIS Data Integration System; currently, the provenance is obtained by using as SQL engine of our Data Integration System the open-source provenance management system *Perm* [12]. The original contribution of this paper are some idea to develop a provenance-based framework which enables different conflict handling strategies and use them as different search strategies for querying the Global Schema. In particular, we considered the provenance-based techniques to tackle data conflicts introduced into the *Trio* system [1,6] and we discussed how they can be adopted, extended and implemented in our system.

As future works, from a theoretical point of view, a formal account for the introduced ideas will be given. From an implementation point of view, since also the *Trio* source code is freely available and the system is based also on PostgreSQL, another possible implementation is to use the *Trio* as SQL engine of our Data Integration System and then evaluate which extensions are necessary (such as the outerjoins) to realize the proposed framework.

References

1. Agrawal, P., Benjelloun, O., Sarma, A.D., Hayworth, C., Nabar, S., Sugihara, T., Widom, J.: Trio: a system for data, uncertainty, and lineage. In: VLDB 2006: Proceedings of the 32nd International Conference on Very Large Data Bases, pp. 1151–1154. VLDB Endowment (2006)
2. Beneventano, D., Bergamaschi, S., Guerra, F., Orsini, M.: Data integration. In: Embley, D., Thalheim, B. (eds.) Handbook of Conceptual Modelling. Springer, Heidelberg (2010), http://dbgroup.unimo.it/SSE/SSE.pdf
3. Beneventano, D., Bergamaschi, S., Guerra, F., Vincini, M.: Synthesizing an integrated ontology. IEEE Internet Computing 7(5), 42–51 (2003)
4. Beneventano, D., Dannoui, A.R., Sala, A.: Data lineage in the momis data fusion system. In: ICDE Workshops of the 27th International Conference on Data Engineering, ICDE 2011, Hannover, Germany, April 11-16, pp. 53–58 (2011)
5. Benjelloun, O., Sarma, A.D., Halevy, A., Widom, J.: Uldbs: databases with uncertainty and lineage. In: VLDB 2006: Proceedings of the 32nd International Conference on Very Large Data Bases, pp. 953–964. VLDB Endowment (2006)
6. Benjelloun, O., Sarma, A.D., Hayworth, C., Widom, J.: An introduction to uldbs and the trio system. IEEE Data Eng. Bull. 29(1), 5–16 (2006)
7. Bergamaschi, S., Castano, S., Vincini, M., Beneventano, D.: Semantic integration of heterogeneous information sources. Data Knowl. Eng. 36(3), 215–249 (2001)
8. Bleiholder, J., Naumann, F.: Data fusion. ACM Comput. Surv. 41(1), 1–41 (2008)
9. Chomicki, J.: Consistent Query Answering: Five Easy Pieces. In: Schwentick, T., Suciu, D. (eds.) ICDT 2007. LNCS, vol. 4353, pp. 1–17. Springer, Heidelberg (2006)
10. Cui, Y., Widom, J.: Lineage tracing for general data warehouse transformations. The VLDB Journal 12(1), 41–58 (2003)
11. Cui, Y., Widom, J., Wiener, J.L.: Tracing the lineage of view data in a warehousing environment. ACM Trans. Database Syst. 25(2), 179–227 (2000)
12. Glavic, B., Alonso, G.: Perm: Processing provenance and data on the same data model through query rewriting. In: Proceedings of the 2009 IEEE International Conference on Data Engineering, pp. 174–185. IEEE Computer Society, Washington, DC (2009)
13. Halevy, A., Li, C.: Information integration research: Summary of nsf idm workshop breakout session. In: NSF IDM Workshop (2003)
14. Halevy, A., Rajaraman, A., Ordille, J.: Data integration: the teenage years. In: VLDB 2006: Proceedings of the 32nd International Conference on Very Large Data Bases, pp. 9–16. VLDB Endowment (2006)
15. Lenzerini, M.: Data integration: A theoretical perspective. In: PODS, pp. 233–246 (2002)
16. Naumann, F., Freytag, J.C., Leser, U.: Completeness of integrated information sources. Inf. Syst. 29(7), 583–615 (2004)
17. Sarma, A.D., Benjelloun, O., Halevy, A., Nabar, S., Widom, J.: Representing uncertain data: models, properties, and algorithms. The VLDB Journal 18(5), 989–1019 (2009)

Incomplete Databases:
Missing Records and Missing Values

Werner Nutt, Simon Razniewski, and Gil Vegliach

Free University of Bozen-Bolzano, Dominikanerplatz 3, 39100 Bozen, Italy
{nutt,razniewski}@inf.unibz.it, gil.vegliach@gmail.com

Abstract. Data completeness is an essential aspect of data quality as in many scenarios it is crucial to guarantee the completeness of query answers. Data might be incomplete in two ways: records may be missing as a whole, or attribute values of a record may be absent, indicated by a null. We extend previous work by two of the authors [10] that dealt only with the first aspect, to cover both missing records and missing attribute values. To this end, we refine the formalization of incomplete databases and identify the important special case where values of key attributes are always known. We show that in the presence of nulls, completeness of queries can be defined in several ways. We also generalize a previous approach stating completeness of parts of a database, using so-called table completeness statements. With this formalization in place, we define the main inferences for completeness reasoning over incomplete databases and present first results.

1 Introduction

Data quality deals with the question how well data serves its purpose. Aspects of data quality concern accuracy, currency, correctness, and similar issues. In settings such as manual data insertion or data integration, completeness of data plays a key role. A question whether data is complete enough for its aims and what might be potentially missing. In particular, in relational databases incompleteness comes in two flavors: records (rows) in tables might be missing entirely or attribute values might be *null*.

Consider as a driving example the management of school data in the province of Bolzano, Italy, which motivated the technical work reported here. The schools in the province are largely autonomous in their administration: although the provincial IT department runs a central database for administering data about pupils, teachers and the like, the schools might choose to what extent to use this system. This freedom leads to many kinds of incompleteness of the underlying database, especially when data submission is optional: for example, when statistics about schools are collected, missing records and *null* values of attributes might have two meanings, either facts that do not hold in reality, or not submitted data.

The IT department already has some information about how the schools use the central database, but not yet a systematic approach to use it. They would like to have a generic technique to tell whether their database is complete enough to answer a certain query, and furthermore what data is needed for the query to be answered completely.

We believe similar problems also occur in other application domains and identify the following research questions:

H. Yu et al. (Eds.): DASFAA Workshops 2012, LNCS 7240, pp. 298–310, 2012.
© Springer-Verlag Berlin Heidelberg 2012

1. How can one describe completeness of parts of a possibly incomplete database?
2. How can one characterize the completeness of query answers?
3. How can one infer completeness of query answers from such completeness descriptions?

There has been some previous work by two of the authors [9,10] on these questions, which only considered incompleteness in the form of missing records. In practice however, incompleteness in the form of *null* values is at least as important.

A problem with SQL *null*s is their ambiguity, as they may mean both that an attribute value exists but is unknown, and that no value applies to that attribute. The established models of null values in database research, such as Codd, v-, and c-tables [6], usually avoid this ambiguity by concentrating on the aspect of unknown values. In this work, we consider the ambiguous standard SQL *null* values [2], because those are the ones commonly used in practice. We will show later that some ambiguity can be resolved, when meta information about database completeness is present.

In this paper, we define a formal framework to study problems that arise when reasoning about the completeness of query answers over databases with *null* values and potentially missing records. We proceed as follows: in section 2 we introduce two example scenarios, in section 3 we formalize incomplete databases, query completeness, and table completeness, in section 4 we show how canonical table completeness statements link table and query completeness, in section 5 we introduce the reasoning problems and we present preliminary results for some of them, in section 6 we discuss related work and with section 7 we conclude this paper.

2   Example Scenarios

We define two scenarios that display incompleteness in the form of missing records and of *null* values. The scenarios differ in that in the first the values of the key attributes are known while in the second some key values are unknown. As we will see later on, in the first case completeness of queries can be detected more easily than in the second.

A School Database. We consider a school database that contains, inter alia, the following two tables, where key attributes are underlined:
- student (<u>sid</u>, name, level, code, hometown)
- class (<u>level</u>, <u>code</u>, formTeacher, viceFormTeacher, profile).

The student table stores for each student their unique student ID, their name, the level and the code of the student's class, such as '3' 'A', and the student's hometown. The class table stores for each class its level and code, the ID of its form teacher and its vice form teacher, and the profile of the class, such as 'science' or 'commerce'. The attributes level and code uniquely determine a class while the student ID uniquely determines a student.

Student tuples could be missing because students enroll by submitting a paper form, while the data are only later entered into the electronic database. Similarly, the formation of classes is decided during an administrative meeting, yet this information is not always immediately recorded in the database.

Values for the attributes `level` and `code` of the `student` table may be *null* because (i) students are assigned to classes only some time after their enrollment, and (ii) the decision may not be recorded immediately. For similar reasons, the value of the attributes `formTeacher, viceFormTeacher,` or `profile` of the `class` table could be *null*.

In contrast, the keys of tuples can never be *null*. To insert a student record into the database, it is necessary to assign a student ID. Similarly, to insert a class it is necessary to specify its level and code.

Over this school database, some queries will return the same answer over an instance with *null*s and where records are missing as they would over an ideal complete instance. For instance, if all classes have been entered, we can tell the number of classes, even if the form teachers have not yet been entered. Similarly, if each type of profile has been entered once, we know the complete spectrum of profiles, even if the `profile` of some specific class record is *null* or if the class record is missing altogether.

We will sketch later on how to formalize such facts and how to reason about them.

An Integrated Business Database. We consider a company that wants to integrate data about business contacts from different departments into one database. We assume that so far the sales, purchase and research department maintained their own databases with different schemas and custom IDs.

In the integrated database, contacts are identified by their name, address, and city. Even if the original databases were complete, records can be missing because a business contact is kept by another department than those three and values can be missing because the integrated schema contains attributes not present in one of the original databases. Also key values can be missing, because information that makes up the key was missing in one of the original databases.

3 Formalization

3.1 Standard Definitions

In the following we summarize the standard formalization of relational databases and conjunctive queries (cf.[1]). The latter model the widely-used single-block SQL queries. We extend this formalization to take into account SQL-style null values and the semantics of queries over databases with nulls, following the approach in [5].

We assume a set of relation symbols Σ, the *schema*, and an infinite set of constants *dom*, including the rational numbers. A *term* is a constant or a variable.

A *relational atom* is an expression $R(\bar{t})$, where R is a relation symbol and $\bar{t}$ is a vector of terms. A *comparison* is an atom with one of the predicates $<, \leq, =$, or $\neq$. A *condition* G is a set of relational and comparison atoms. We write a condition as a sequence of atoms, separated by commas. A condition is *safe* if each of its variables occurs in a relational atom. We generically use the symbol L for the subcondition of G containing the relational atoms and M for the subcondition containing the comparisons. A *conjunctive query* is written in the form $Q(\bar{s}) :- B$, where B is a safe condition, $\bar{s}$ is a vector of terms, and every variable in $\bar{s}$ occurs in B. As usual, we call $Q(\bar{s})$ the *head* and B the *body* of Q. A variable in B is called a *join variable* if it occurs at least twice in B. We often refer to the entire query by the symbol Q.

A *database instance* D is a finite set of relational ground atoms, that is, atoms without variables, which may contain the special constant *null*. For a relation symbol $R \in \Sigma$ we write $R(D)$ to denote the set of R-records in D, that is, the set of atoms in D with relation symbol R.

An *assignment* α for the query Q is a mapping that maps the variables of Q to elements of *dom* $\cup \{null\}$. If A is an atom, then αA is the (ground) atom obtained by replacing every variable x in A with the constant $\alpha(x)$. We say that α *satisfies* the condition $G = L, M$ over the instance D if (i) α maps all join variables of G to constants $\neq null$, (ii) $\alpha A \in D$ for every relational atom $A \in L$, and (iii) α satisfies all comparisons in M. Note that part (i) captures the semantics of *nulls* in SQL, where an equality or comparison involving *null* has the truth value "unknown" and thus does not contribute to a query result.

In SQL, a query $Q(\bar{s}) :- B$ can be evaluated under two semantics. Under *bag semantics*, which is the default, applying Q to D results in the multiset $Q^b(D) = \{\!| \alpha(\bar{s}) \mid \alpha$ satisfies $B |\!\}$, which contains as many tuples, including duplicates, as there are satisfying assignments for the body of Q. Under *set semantics*, which is enforced by adding the keyword DISTINCT, Q returns $Q^s(D)$, the set version of $Q^b(D)$, obtained by dropping duplicates.

As usual, we say that a query Q is *set contained* (*bag contained*) in a query Q', if $Q^s(D) \subseteq Q'^s(D)$ $(Q^b(D) \subseteq Q'^b(D))$ for all database instances D. For both semantics, there is a rich body of literature on algorithms and complexity of containment for conjunctive queries (cd. [1]). The problem has also been studied over databases with SQL-style null values [5].

3.2 Incomplete Databases

A database can be incomplete only with respect to a comparison database, considered to be complete. Consequently, we model an incomplete database in the style of Levy [7] as a pair $\mathcal{D} = (D^i, D^a)$ of database instances: D^i, the *ideal* state, with complete information, and D^a, the *available* state, with possibly incomplete information. We require that D^a contains no more information than D^i and formalize this by the concept of dominance. An atom $R(\bar{s})$ is *dominated* by an atom $R(\bar{t})$, written $R(\bar{s}) \preceq R(\bar{t})$, if $R(\bar{s})$ is the same as $R(\bar{t})$, except that it may have more *nulls*. Now, for $\mathcal{D}$ to be an incomplete database, the instance D^a must be *dominated* by the the instance D^i, written $D^a \preceq D^i$, in the sense that every atom in D^a is dominated by some atom in D^i.

We say that $\mathcal{D} = (D^i, D^a)$ satisfies the principle of *unique dominance* when the dominance can be established without using any tuple in the ideal database twice, that is, if $R(\bar{s}_1) \preceq R(\bar{t})$ and $R(\bar{s}_2) \preceq R(\bar{t})$ implies $R(\bar{s}_1) = R(\bar{s}_2)$ for all $R(\bar{s}_1), R(\bar{s}_2) \in D^a$ and $R(\bar{t}) \in D^i$. For instance, unique dominance is satisfied by an incomplete database where a key is defined for each relation and no key attribute is *null*. Then every record in the available database is dominated by the record with the same unique identifier in the ideal database.

In data integration, however, key values may uniquely identify records in each of the source databases, but may fail to identify the entities in the integrated database. In such a case, two records from two distinct sources may represent the same entity, but may erroneously be mapped to two records in the integrated database, which are then dominated by a single record in the ideal database. In such a scenario, unique dominance does not hold.

Example 1. Table 1 shows an incomplete database in the Bolzano school scenario. Information present in the ideal but missing in the available database is written in *italics*. For example, the student Diego is missing entirely and for class 1B we do not know the form teacher and the vice form teacher. Note that we also have *null* values in the ideal database, which express, first, that class 2A has no vice form teacher and, second, that Andrea is registered as an external student not belonging to any class.

Table 1. An incomplete school database

D^i		D^a	
class	student	class	student
(1, A, 101, *103*, science)	(702, Paul, 1, A, *Bolzano*)	(1, A, 101, *null*, science)	(702, Paul, 1, A, *null*)
(1, B, *104*, *109*, commerce)	(781, Maria, 1, A, *Merano*)	(1, B, *null*, *null*, commerce)	(781, Maria, 1, B, *null*)
(2, A, *102*, null, *science*)	(739, Andrea, null, null, Brunico)	(2, A, *null*, *null*, *null*)	(739, Andrea, null, null, Brunico)
	(*754, Diego, 2, A, Bolzano*)		—

In the school example, unique dominance did hold. We now give an example from the business scenario where unique dominance is not satisfied.

Example 2. The sales, purchase, and research department of a company maintain databases with business contacts, each with a different schema, as shown in Table 2.

Table 2. Contact databases of the three departments

Sales	Purchase	Research
customer(name, street, city)	supplier(name, city)	partner(name, city, long_term)
(Johnson Corp., North St., Boston)	(Smith Inc., Detroit)	(Johnson Corp., Boston, no)

In addition to Johnson Corp. and Smith Inc., the company is also in contact with Miller & Co., through its human resources department. Thus, each of the three should appear once in the integrated database, shown in Table 3. However, due to a failure in entity resolution, Johnson Corp. shows up twice, while Miller Inc. is missing completely. Moreover, some attribute values are *null*, as the corresponding information was missing in the original sources (information missing in D^a is in *italics*).

Table 3. Incomplete integrated contact database

D^i	D^a
contact(name, street, city, long_term)	contact(name, street, city, long_term)
(Johnson Corp., North St., Boston, *no*)	(Johnson Corp., North St., Boston, *null*)
	(Johnson Corp., *null*, Boston, no)
(Smith Inc., *Main St.*, Detroit, *yes*)	(Smith Inc., *null*, Detroit, *null*)
(*Miller & Co., Central Rd., New York, no*)	—

Observe that Johnson Corp. appears twice in the available database, coming from the sales department and the research department, but only once in the ideal database. Hence, unique dominance is not satisfied in this example.

3.3 Query Completeness

We now want to define formally when a query Q is complete over an incomplete database $\mathcal{D} = (D^i, D^a)$. Considering bag and set semantics and the concept of dominance, there are three meaningful definitions:

Bag Completeness: The query returns the same *bag* of answers over the available and over the ideal database, that is, $Q^b(D^a) = Q^b(D^i)$;

Set Completeness: The query returns the same *set* of answers over the available and over the ideal database, that is, $Q^s(D^a) = Q^s(D^i)$;

Set Completeness Modulo Redundancy: The query returns all the answers from the ideal database also over the available one, and every (additional) tuple over the available database is dominated by some tuple over the ideal database, thus being redundant, that is, $Q^s(D^i) \subseteq Q^s(D^a)$ and $Q^s(D^a) \preceq Q^s(D^i)$.

If for a query Q, an incomplete database $\mathcal{D}$ satisfies query completeness in one of those cases, we write $\mathcal{D} \models Compl(Q)$. When the meaning is not clear from the context, we add $\cdot^s$, $\cdot^b$, or $\cdot^{sred}$ as superscript to the statement. We call these expressions *query completeness* statements or for short *QC* statements.

Example 3. In the school scenario, consider the query $Q_{lev1}() :- student(s, n, 1, c, h)$. Note that under bag semantics, Q_{lev1} returns a copy of the empty tuple for each student at class level 1 and thus reports the number of such students. Since Q_{lev1} returns the empty tuple twice over both the ideal and the available database, it is bag complete.

Consider also the query $Q_{sci}(n) :- student(s, n, l, c, h), class(l, c, fT, vFT, science)$, which asks for the names of all students in science classes. As 'Diego' is returned over the ideal, but not over the available database, Q_{sci} is neither bag nor set complete.

Clearly, $Compl^b(Q)$ entails $Compl^s(Q)$. Also, $Compl^s(Q)$ entails $Compl^{sred}(Q)$. However, in general, the converse does not hold as we show next.

Example 4. In the business scenario, the query $Q_{Bo}(n, lt) :- contact(n, s, Boston, lt)$ asks for the name and long term status of contacts from Boston. Over the ideal database it returns $Q_{Bo}^s(D^i) = \{(Johnson\ Corp., no)\}$, while over the available database it returns $Q_{Bo}^s(D^a) = \{(Johnson\ Corp., no), (Johnson\ Corp., null)\}$. Thus, Q_{Bo} is not set complete over this incomplete database. However, as the additionally returned tuple is dominated by the record returned over the ideal database, Q_{Bo} is set complete modulo redundancy.

3.4 Table Completeness

In addition to the completeness of queries, which can be expressed by QC statements, we want to state the completenes of parts of an incomplete database. To this end, we generalize the table completeness (TC) statements introduced in [10]. A TC statement says that a specific fragment of a relation is complete without requiring other parts of the database to be complete. The challenge is to come up with a formalization that is both intuitive and allows one to infer query completeness from table completeness.

A TC statement, written $Compl(R(\bar{s}); P; G)$, has three components: (i) a relational atom $R(\bar{s})$, (ii) a set of numbers $P \subseteq \{1, \ldots, arity(R)\}$, and (iii) a condition G such that $R(\bar{s}), G$ is safe. The numbers in P are interpreted as attribute positions of R. For instance, if R is the table student, then $\{2, 5\}$ would refer to the attributes name and hometown. Intuitively, such a TC statement says that if we take the ideal records $R(\bar{t}) \in R(D^i)$, satisfying the conditions $\bar{t} = \bar{s}$ and G over D^i, and project these records onto P, then these projections are also present in $\pi_P(R(D^a))$, the projection of R onto P over the available database. Clearly, we obtain two different semantics, depending on whether the projection returns a bag or a set of records.

Example 5. Taking into account that profile is the 5th attribute of the table class in the school database, the TC statement $C_{prf} = Compl(\text{class}(l, c, fT, vFT, p); \{5\}; true)$ says, under set semantics, that all class profiles are present in the available database. In our example, C_{prf} holds as both 'science' and 'commerce' are in the available database. The TC statement $C_{outBZ} = Compl(\text{student}(id, n, l, c, h); \{\}; h \neq \text{Bolzano})$ has an empty set of positions and thus talks about empty tuples. Interpreted under bag semantics, it says that there are as many students from outside Bolzano in the available database as there are in the ideal database. In our example, C_{outBZ} does not hold under bag semantics, as two such students can be found in the ideal, but only one in the available database. It holds under set semantics, though, as both ideal and available database contain at least one student from outside Bolzano.

To make this formal, we associate to the TC statement $C = Compl(R(\bar{s}); P; G)$ the query $Q_C(\bar{s}) :- R(\bar{s}), G$, which returns the R-records $\bar{t}$ satisfying $\bar{t} = \bar{s}$ and G. This query will be evaluated under set semantics over D^i, as we are only interested in the R-records as such, not how many times they can be derived using G. Recall that evaluation of Q_C under set semantics is indicated by the superscript $\cdot^s$. Similarly, we use the operators π_P^b and π_P^s to distinguish between projection on P under bag and set semantics. We now define that $\mathcal{D} = (D^i, D^a)$ *satisfies* C under bag or set semantics, respectively, if

$$\pi_P^\star(Q_C^s(D^i)) \subseteq \pi_P^\star(R(D^a)),$$

where $\star \in \{b, s\}$. The inclusion $\subseteq$ is bag inclusion if $\star = b$ and set inclusion if $\star = s$. To indicate whether a TC statement is to be interpreted under bag or set semantics, we write $Compl_b$ and $Compl_s$, when necessary. In the special case that P comprises all attributes of R, which was the case studied in [10], we can drop the projection and need not distinguish between bag and set semantics, as C is satisfied by $\mathcal{D}$ if $Q_C^s(D^i) \subseteq R(D^a)$.

The next example shows that table completeness statements can also resolve the inherent ambiguity of *null* values found in an available database.

Example 6. Consider the record student(739, Andrea, *null*, *null*, Brunico) in our available school database, with *null* values for class level and code. Without further information, we do not know whether level and code are missing or whether the student is an external student not assigned to any class. If we knew, however, that our partial database satisfied $Compl(\text{student}(id, n, l, c, h); \{1, 2, 3, 4, 5\}; true)$, that is, the student table is complete, we could conclude that the *null*s can only have the meaning that no level and code apply to the student, and hence he is an external.

4 Canonical Table Completeness Statements

The overall goal of reasoning about completeness is to infer QC statements from information about the content of a database, expressed by TC statements.

In previous work on completeness reasoning for databases without *nulls* [10], a powerful approach consisted in translating a completeness statement about a query Q into a set C_Q of so-called *canonical* TC statements for Q that, intuitively, express which parts of which tables should be complete to guarantee completeness of Q.

Canonical TC statements were then used to reduce the problem of deciding whether an arbitrary set of TC statements C entails the QC statement $Compl(Q)$ (called *TC-QC reasoning*) to checking whether C entails the canonical TC statements C_Q, which is a special case of deciding entailment between sets of TC statements (called *TC-TC reasoning*). TC-TC reasoning was then reduced to the well-studied query containment.

In this section we report on approach to generalize this work to the richer setting accommodating *nulls*.

4.1 Definition of Canonical TC Statements

We first want to single out those attributes of relations that must be complete in the available database so that we can answer a query Q completely. These should be the attributes that occur in selections, in joins, and that are output by Q.

Let $Q :- A_1, \ldots, A_n, M$ be a query with relational atoms A_j and a set of comparisons M. A term t occurring in Q is *essential* if (i) t is a constant or (ii) t is a variable occurring more than once in Q. Intuitively, essential terms are those that express a selection condition, a join condition, or that appear in the head of Q. A position p in the relational atom A_i is *essential* if the term occurring at position p in A_i is essential in Q. The set of essential positions of A_i in Q is denoted as $EPos(A_i, Q)$.

The *canonical completeness statement* C_{A_i} for A_i has the form

$$Compl(A_i; EPos(A_i, Q); A_1, \ldots, A_{i-1}, A_{i+1}, \ldots, A_n, M).$$

Intuitively, a canonical statement $C_{R(\bar{s})}$ states that the projection on the essential positions of $R(\bar{s})$ is complete for those tuples $\bar{t}$ in R that satisfy $\bar{t} = \bar{s}$ and the condition composed by all the other atoms in the query. The *set of all canonical completeness statements* for a query Q is denoted as C_Q. As other TC statements, a canonical statement can be interpreted with respect to set and bag projection, which is indicated by the superscripts $\cdot^s$ and $\cdot^b$ as in $C_{A_i}^s$ and C_Q^b.

4.2 Properties of Canonical TC Statements

Canonical statements are a link between QC and TC statements.

Theorem 1. *Let Q be a conjunctive query, $\mathcal{D}$ an incomplete database. Then we have:*

1. *If $\mathcal{D} \models C_Q^s$, then $\mathcal{D} \models Compl^{sred}(Q)$;*
2. *If $\mathcal{D}$ satisfies unique dominance, then $\mathcal{D} \models C_Q^b$ if and only if $\mathcal{D} \models Compl^b(Q)$.*

The theorem says that canonical statements under set semantics are a sufficient condition for set completeness modulo redundancy. Moreover, in the presence of unique dominance, canonical statements under bag semantics completely characterize bag completeness of a query. The proof is omitted due to space constraints.

As a corollary we note that we can find sufficient conditions for TC-QC entailment in terms of TC-TC entailment. We do not know whether the converse holds, too.

Corollary 1. *Let C be a set of TC statements and Q be a conjunctive query. Then*
1. $C^s \models C^s_Q \Rightarrow C^s \models Compl^{sred}(Q)$;
2. $C^b \models C^b_Q \Leftrightarrow C^b \models Compl^b(Q)$.

The next corollary is a trivial consequence of Theorem 1.1 and is stated explicitly as a contrast to Theorem 2 below.

Corollary 2. *Let $\mathcal{D}$ be a partial database that satisfies all possible TC-statements and let Q be a conjunctive query. Then $\mathcal{D} \models Compl^{sred}(Q)$.*

We next show that the assumption about unique dominance above cannot be dropped.

Theorem 2. *There exist an incomplete database $\mathcal{D}$ without unique dominance and a conjunctive query Q such that*
1. *$\mathcal{D}$ is complete for all possible TC statements, both under bag and set semantics;*
2. *Q is neither bag nor set complete over $\mathcal{D}$.*

Proof. Let $\mathcal{D}$ consist of $D^i = \{R(a,b)\}$ and $D^a = \{R(a,b), R(a, null)\}$. Unique dominance does not hold due to the record $R(a, null)$. Clearly, $\mathcal{D}$ satisfies all possible TC statements under any semantics, since all records from the ideal database are also in the available database. Consider the query $Q(y) :- R(x, y)$ that projects R on the second argument. Then $Q^b(D^i) = \{b\}$, while $Q^b(D^a) = \{b, null\}$, which implies that Q is neither bag nor set complete over $\mathcal{D}$.

Theorem 2 shows that without unique dominance the unexpected situation can arise that all tables of a database are complete, according to the TC statements, yet some query is bag and set incomplete. Intuitively, a reason for this is that TC statements assert that a query result over the ideal database is *included* in a projection over the available database, while bag and set completeness require *equalities* to hold. Such equalities, however, may fail to hold because records in the available database may contain nulls where there are constants in the corresponding records in the ideal database.

The interplay of bag semantics and unique dominance can prevent this. Several copies of the same record in the result of a conjunctive query can be obtained from several combinations of records in the database. If the canonical TC statements hold under bag semantics, then for each such combination of records in the ideal database, there must be a corresponding combination in the available database. Morover, unique dominance ensures that two different combinations in the available database correspond to different combinations in the ideal database. This can be seen as the intuition behind Theorem 1.2.

The next theorem shows that the situation is different for TC statements under set semantics and set completeness of a query.

Theorem 3. *There exist a conjunctive query Q and an incomplete database $\mathcal{D}$ with unique dominance such that*

1. *$\mathcal{D}$ satisfies C_Q^s, the canonical TC statements for Q under set semantics;*
2. *$\mathcal{D}$ does not satisfy $Compl^s(Q)$, that is, set completeness of Q.*

Proof. Consider the query $Q(y) :\!-\, R(x, y)$, which projects R onto the second position. As Q has only the atom $A = R(x, y)$, there is a single canonical TC statement for Q, namely, $C_A = Compl(R(x, y); \{2\}; true)$.

Next, consider the partial database $\mathcal{D}$ consisting of $D^i = \{R(1, a), R(2, a)\}$ and $D^a = \{R(1, a), R(2, null)\}$. Clearly, $\mathcal{D}$ satisfies the principle of unique dominance. We easily check that $\mathcal{D}$ satisfies C_A, as $\pi_{\{2\}}^s(Q_{C_A}(D^i)) = \{a\} \subseteq \{a, null\} = \pi_{\{2\}}^s(R(D^a))$. However, $Q^s(D^i) = \{a\} \neq \{a, null\} = Q^s(D^a)$. Hence, Q is not set complete over $\mathcal{D}$.

5 Reasoning Problems and Preliminary Results

In this section we present the four central reasoning problems involving query completeness (QC) and table completeness (TC) and some preliminary results on them.

Problem 1: QC Characterization. Given a conjunctive query Q and a set of TC statements C, is C characterizing $Compl^\star(Q)$, that is, do we have $\mathcal{D} \models C$ if and only if $\mathcal{D} \models Compl^\star(Q)$ for all incomplete databases $\mathcal{D}$?

Preliminary Results. For bag completeness, the canonical TC statements under bag semantics are characterizing according to Theorem 1.2. An arbitrary set C is therefore characterizing for $Compl^b(Q)$ if it is equivalent to C_Q^b. We have shown that in general for set completeness and set completeness modulo redundancy, query completeness cannot be characterized by a set of TC statements. An intuition is that for a tuple in the result of a such query there can be several derivations and for the two set semantics, just one of the many possible derivations is needed, which cannot be expressed by our TC statements, since a special kind of existential quantification would be needed.

The next problem is to find whether some canonical completeness statements can ensure query completeness.

Problem 2: TC-QC Entailment. Given a query Q, when and under which semantics do the canonical TC statements imply query completeness?

Preliminary Results. From Theorem 1.2 we know that the canonical statements under bag semantics entail bag completeness if we allow only incomplete databases satisfying unique dominance. It can be shown that under set semantics, they do not. Since query completeness under bag semantics entails query completeness under set semantics, the canonical statements under bag semantics entail QC under the two set semantics, provided we have unique dominance. According to Theorem 1.1, in the general case, canonical statements under set semantics entail set completeness modulo redundancy but, according to Theorem 3, may not entail set completeness proper. What holds for other combinations is an open question

If there are some TC statements that entail completeness of Q, a follow-up question is whether there exists a most general set of TC statements that entail completeness of Q, meaning a set that requires as little database completeness as possible.

Problem 3: Weakest Preconditions for TC-QC Entailment. Given a query Q, does there exist a set of TC statements C_0 such that $C_0 \models Compl^\star(Q)$, $\star \in \{s, b, sred\}$, such that $C \models C_0$ for any other set C with this property?

Preliminary Results. For queries under bag semantics, it follows again from Theorem 1.2 that the canonical statements under bag semantics fulfil this requirement in the presence of unique dominance. For the two set semantics the problem is still open.

Finally, the most important problem is to decide whether a query can be answered completely, given knowledge about the completeness of parts of an incomplete database.

Problem 4: Deciding TC-QC Entailment. Given a query Q and a set of TC statements C, how can once check that whenever an incomplete database satisfies C it also satisfies $Compl^\star(Q)$, $\star \in \{s, b, sred\}$.

Preliminary Results. For queries under bag semantics, TC-QC can be reduced to TC-TC entailement by Corollary 2. However, we do not know yet how to decide this. There are indications that it can be reduced to query containment under a combination of bag and set semantics over databases with null values.

For queries under set semantics modulo redundancy, the entailment $C \models C_Q^s$ is a sufficient condition. This problem, again, can be mapped to a problem of query containment under set semantics over databases with null values as a sufficient condition. It is open whether these sufficient conditions are also necessary.

6   Related Work

Motro [8] investigated query completeness as an aspect of query integrity. He introduced the notion of partially incomplete and incorrect databases as databases that can both miss facts that hold in the real world and contain facts that do not hold there. He described partial completeness in terms of *query completeness* (QC) statements under set semantics. To infer completeness of a given query from a set of queries known to be complete, he would search for a conjunctive rewriting of the given query in terms of the complete queries. This solution is correct, but not complete, as later results on query determinacy show [11].

Halevy [7] suggested *local completeness* statements, which we call table completeness (TC) statements, as an alternate formalism for asserting partial completeness of an incomplete database. The main problem he addressed was how to derive query completeness from table completeness (TC-QC reasoning). However, his approach led only to a decision procedure applicable to trivial cases.

Fan and Geerts [3] discussed the problem of query completeness in the presence of master data. Their work is not directly comparable to the one presented here because in addition to the different setting it always considers a database instance. In follow-up work, they considered incomplete data also in the form of missing but constrained attribute values [4], which they represented by *c-tables* [6].

Recently, Razniewski and Nutt picked up Levy's problem of TC-QC entailment over databases that can miss records [9,10]. They showed that TC-QC entailment is decidable for all languages of positive conjunctive queries used for formulating TC and QC statements and analysed the complexity of the problem in detail, finding combined complexities ranging from PTIME to Π_2^P.

7 Conclusion

We have introduced the concept of incomplete databases with missing tuples and missing values, represented by SQL-style *nulls* and identified as an important special case the one where unique record identifiers are always known, which leads to a property called *unique dominance*.

We introduced three different ways to define query completeness (QC) over an incomplete database, which are based on bag and set semantics of queries and take into account partiality of information in records with *nulls*.

We generalised Levy's approach to describing complete parts of tables by table completeness (TC) statements to TC statements that describe completeness of projections of parts of tables. Depending on whether projections are performed under set or bag semantics, we defined two diffent semantics for these generalized TC statements.

We also generalized the canonical TC statements for queries from our previous work in such a way that they capture those projections of tables that are needed to answer a query. First results show how such generalized canonical TC statements can be used infer query completeness from other TC statements.

Finally, we have defined and discussed four reasoning problems: (1) finding a set of TC statements that characterize query completeness, (2) checking whether canonical TC statements under some semantic entail query completeness, (3) finding TC statements that are weakest preconditions query completeness, and (4) checking TC-QC entailment. For some of the problems we presented results while for others we sketched possible approaches.

In future work we aim to answer the open questions. Furthermore, we want to investigate the impact of schema constraints (keys, foreign keys, finite domains) on completeness reasoning.

References

1. Abiteboul, S., Hull, R., Vianu, V.: Foundations of databases. Addison-Wesley (1995)
2. Codd, E.F.: Understanding relations (installment #7). FDT – Bulletin of ACM SIGMOD 7(3), 23–28 (1975)
3. Fan, W., Geerts, F.: Relative information completeness. In: PODS, pp. 97–106 (2009)
4. Fan, W., Geerts, F.: Capturing missing tuples and missing values. In: PODS, pp. 169–178 (2010)
5. Farré, C., Nutt, W., Teniente, E., Urpí, T.: Containment of Conjunctive Queries over Databases with Null Values. In: Schwentick, T., Suciu, D. (eds.) ICDT 2007. LNCS, vol. 4353, pp. 389–403. Springer, Heidelberg (2006)
6. Imieliński, T., Lipski Jr., W.: Incomplete information in relational databases. J. ACM 31, 761–791 (1984)

7. Levy, A.: Obtaining complete answers from incomplete databases. In: Proc. VLDB, pp. 402–412 (1996)
8. Motro, A.: Integrity = Validity + Completeness. ACM TODS 14(4), 480–502 (1989)
9. Razniewski, S., Nutt, W.: Checking query completeness over incomplete data. In: LID (2011)
10. Razniewski, S., Nutt, W.: Completeness of queries over incomplete databases. In: VLDB (2011)
11. Segoufin, L., Vianu, V.: Views and queries: Determinacy and rewriting. In: Proc. PODS, pp. 49–60 (2005)

User Interface Design Guidelines Arrangement in a Recommender System with Frame Ontology

Maxim Bakaev and Tatiana Avdeenko

Department of Economic Informatics, Novosibirsk State Technical University, Russia
maxis81@gmail.com, tavdeenko@mail.ru

Abstract. Design guidelines, which come from the extensive body of knowledge currently formed in HCI and usability engineering domains, remain poorly integrated. Guidelines and design patterns from various sources may contradict or duplicate each other, lack links to origins and justification, as well as contextual associations to concrete problems. The paper describes how the recommender system, developed to support interface design, resolves the issues of data integration and credibility via employing frame-based ontology model and guidelines "efficiency" evaluation algorithm based on fuzzy relations. Also, experimental investigation was carried out with 24 subjects of different age groups to assess the quality of the system work. The results suggests reasonable applicability of the proposed approaches, as the success rate for the website created with the system nearly doubled the one for the control group, and guidelines efficiencies were significantly higher for relevant target user groups.

Keywords: Data integration, frame ontology, metadata, fuzzy relations model.

1 Introduction

As the amount of data worldwide continues to grow exponentially, and the diversity of applications and devices is increasing, ensuring high quality of human-computer interaction (HCI) remains a vital problem. The practical adoption of usability engineering methods continues to be far from universal [1], while positive results of their employment are highly dependent on usability specialists' skills. One possible allocation of layers for interface design-related expert knowledge is the following:

1. Laws – high-level theoretical constructs or statements describing significant aspects of interaction or users.
2. Principles – more or less universal canons concerning design decisions or design process in general. Heuristics and design standards go into this layer as well.
3. Guidelines – pieces of practical advice on how to implement features for a successful interface or reminders about common pitfalls.

A recent strong trend in HCI is promotion of design patterns – approved model solutions for common design problems – as the next layer of knowledge, even more concrete than guidelines (see, e.g. [2, p.360]). However, it is still recognized that

H. Yu et al. (Eds.): DASFAA Workshops 2012, LNCS 7240, pp. 311–322, 2012.

"guidelines remain the most widely accepted form of presenting experience and knowledge" [3, p.160], so we in our paper chose to focus on guidelines and consider design patterns a sufficiently similar notion.

Among the layers of knowledge marked out above, guidelines constitute the most extensive and most used in interface design practice one. Both guidelines and design patterns [2, pp.365-367] should be applied on all stages of HCI engineering process, however certain problems concerned with their practical use are noted [4, pp.82-98]:

1. Average time spent to find and implement one guideline is 15 minutes.
2. Difficulties with correct interpretation of found guidelines by interface designers (in 30% of the cases).
3. Decreased guidelines utility due to obscurity of implied application context.

Indeed, nowadays design guidelines originate from a multitude of sources, so they routinely contradict or duplicate each other, lack theoretical justification as well as indication of appropriate application context, etc. [5, p.52]. Thus we undertook a review of HCI knowledge organization tools and the approaches they use to integrate guidelines or design patterns collected from various sources.

1.1   Design Guidelines Organization in Knowledge-Based Systems (KBS)

Our review of knowledge-based interface design support systems allows to suggest that the majority of them are indeed tools for working with guidelines. A Special Interest Group "Tools for Working with Guidelines" was established in the 1990s, when the works of J. Vanderdonckt and his co-authors set the basis for development of such systems [4] and subsequently proposed concrete implementations [6]. The result was the MetroWeb system, the detailed description of which is provided in [7], together with experimental data showing positive effect of its application in interface design. The guidelines in the system did include relations to model(s) and context, however the latter was understood as a set consisting of development stages, evaluation methods, etc., rather than as design context that would consider target user group, the product features and qualities, and so on [7 p.53]. On the whole, the researchers noted that designers were not engaged with the system and used it unenthusiastically, but even then it led to more user-centered approaches.

Another example is ontology-based system for organizing design patterns, whose core component was named BORE and which was in development since 1997, but currently seemingly disappeared from publications. The ontology implements a metamodel for design patterns, with attributes and formal relations such as "contains", "alternate to", "disjoint with", etc. [5, p.50]. Obviously, the problem of the patterns duplications and contradictions was recognized and attended to, but it remained unclear whether such strict associations are sufficient to adequately represent the domain, and if an expert supporting a working system would be able to manually maintain such relations.

Our further enquiry, supplemented by a detailed review of the tools existing in HCI domain provided in [8], led us to the conclusion that the issue of design guidelines organization is yet to be fully resolved, and this is the main problem we address in our paper. To ensure proper guidelines arrangement, a recommender system in particular must:

1. Address the issue of guidelines integration from multitude of sources, providing means for resolving contradictions and identification of duplicates. This may be done via introduction of data "significance" determination mechanism and guidelines classification, both preferably maintained by the system.
2. Ensure data provenance in terms of both tracing the source of guideline and its justification, i.e. relation to higher-level knowledge, such as laws or principles.
3. Include guideline application context and be able to match it to the context in which recommendation is made. Thus the system must contain metadata on interface design domain that could be both assigned to guidelines and used to describe design context.

1.2   The Recommender System to Support Interface Design

In our project dedicated to the development of recommender system to support interface design, we came to the conclusion that the design context is primarily defined by target users attributes and requirements, mostly non-functional ones. The outcome of the system should be the set of human-readable guidelines relevant to the project context, with the addition of interface wireframe that together should be used by designer to construct the interface prototype (see Fig. 1).

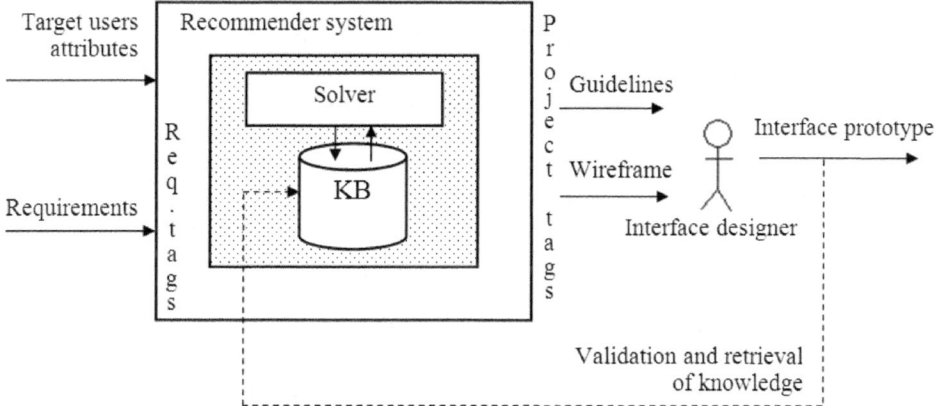

Fig. 1. General structure of the recommender system, input and output information

We propose to establish the recommender system on the knowledge base (KB) built upon ontology which model is described in the *Method* chapter of the paper. Among various possible ontology models, we developed the one based on such data structures as frames, which were first described by M. Minsky in the 1970s, as the ontology needs to combine domain specification and linguistic features. The former permits the classification of guidelines by assigning "tags" (domain concepts), while the latter would allow design context extraction directly from requirements specifications in natural language. In addition, the relationships peculiar to ontologies permit natural data provenance, as guidelines may be linked to both their sources and justifying knowledge. To evaluate the relative "significance" of guidelines, which is essential given their current multitude and poor organization, we also propose the model, which is based on fuzzy relations.

Although detailed description of the recommender system creation and application is beyond the scope of the current paper, which is dedicated to the proposed solutions for guidelines organization in the system, we briefly outline the development process in the chapter *The Recommender System Implementation*. The chapter *Experiment* describes the experimental investigation carried out to assess the feasibility of the proposed approaches to design guidelines arrangement.

2   Method

2.1   Knowledge Representation Model

Numerous researchers have been criticizing T. Gruber's initial definition of ontology as the "explicit specification of a conceptualization" – conceptualization being <D,R>, where D is domain and R are relations on the domain, – and working on its development (among them N. Guarino [9, p.5], one of the founders of ontology application in information science). Currently, the specification of ontology made a transition to a following form [10, p.180]:

$$O = <C, R, A, TDL, A_X>, \tag{1}$$

where C is a set of classes corresponding to certain concepts of domain;

$R = \{R_I\} \cup \{R_P\} \cup \{R_A\}$ – a set of binary relations defined on classes C:

R_I – binary transitive relation of inheritance, giving child classes both attributes and relations of the parent class (and the corresponding inverse relation);

R_P – binary transitive relations of inclusion (e.g. "has-part" – "is-part-of");

R_A – a finite set of other possible relations;

$A \subseteq C \times TDL \cup R \times TDL$ – a set of attributes for classes or relations;

$TDL = T \cup D \cup L$, correspond to set of attributes types ($T = \{integer, string, \ldots\}$), data domains ($D = \{D_1, \ldots, D_n\}$) or constraints for the attributes values ($L = \{L_1, \ldots, L_m\}$);

A_X – a set of "axioms" that represent certain intrinsic knowledge of the modeled domain. As simple constraints for attributes types, values, etc. may be set in TDL, subsequently axioms should be some logical rules or state-independent information (S_I) of the domain (see [9, p.10] on state-independent and state-dependent information).

If an information system or a KBS must process textual resources, then it has to contain a thesaurus T_H – a structured vocabulary of terms corresponding to the domain, in one or several languages [10, pp.188-189]:

$$T_H = <T_R, R_T, A_T, A_{TX}>, \tag{2}$$

where T_R is a set of terms that generally contain a specially defined subset T_{DEF}, with preferred (default) terms for a concept or relation;

$R_T = \{R_{TI}\} \cup \{R_{TS}\} \cup \{R_{TE}\} \cup \{R_{TA}\} \cup R_{TO}$ – a set of binary relations defined for T_H:

R_{TI} – relations connecting a term with more general term (and inverse relations);

R_{TS} – relations connecting terms from T_{DEF} and their less preferred synonyms;

R_{TE} – symmetrical relations of lexical equivalence for terms in different languages;

R_{TA} – a finite set of other possible relations between the terms;

R_{TO} – a relation connecting a thesaurus term to a class or relation of ontology;

A_T – a set of attributes describing the semantics of terms from T_R;

A_{TX} – a set of axioms describing the semantics of thesaurus T_H.

It is obvious that the specification of thesaurus complies with the one of ontology, and indeed their integration is a feasible method in information systems development, although approaches based on separation of vocabulary and domain ontology also exist [11]. There are also somehow intermediate solutions: for example in WordNet (http://wordnet.princeton.edu), the global lexical database, the nodes in thesaurus are not individual terms, but "synsets" – datasets that contain several synonymous terms and represent a distinct concept. However, WordNet is not concerned with relations between concepts except for linguistic ones, so it may hardly be considered a domain ontology (but possibly a language ontology).

We propose a specification of ontology based on frame model, as the following:

$$O_F = <F_C; F_R; F_A; F_E; I_F>, \tag{3}$$

where $F_C = <N_C; T_R; a_C; r_C>$ is a set of frames-concepts (correspond to Minsky's frames-prototypes). The frame name $N_C \in T_{DEF}$, i.e. it is the preferred term from thesaurus T_H, while the set T_R contains other terms, in different languages. There are also a set of frame-concept attributes $a_C \subseteq F_A$ and relations with other frames-concepts, $r_C \subseteq F_R$ (since a frame slot's value may be another frame);

$F_R = <N_R; R; a_R>$ – a set of frames-relations linking concepts: N_R is the frame name, $a_R \subseteq F_A$;

$F_A = <N_A; A_F; TDL>$ – a set of frames-attributes for concepts or relations: N_A is the frame name, $A_F = A \cup A_T$;

F_E – a set of frames-instances created based on frames-concepts (prototypes) and representing state-dependent knowledge of the domain, thus partially covering A_X;

I_F – a set of logical rules establishing semantic correctness of the domain (thus covering another fraction of A_X) or implementing procedural component in a KBS.

It should be noted that this approach removes the necessity of relations R_{TS}, R_{TE}, and R_{TO}, while T_{DEF} is formed automatically from the names of frames-concepts: $T_{DEF} = \{N_C\}$. Yet, $\{N_R\}$ and $\{N_A\}$ are not included in the thesaurus, so if these terms are deemed necessary, the corresponding frame-concept must be created. The purpose and substance of axioms, whose definition in literature is generally ambiguous, becomes more clear, and the set I_F may be also extended with rules, e.g. ones of production rule system that is a potent supplement to the naturally declarative frame model.

Production system supplementing the frame model may be presented in the following simplified form:

$$I_F = <N_I; L_{HS}; R_{HS}>, \tag{4}$$

where N_I is name assigned to the rule, to ensure at least some order in naturally non-structured production system;

$L_{HS} = U(F_C; F_R; F_A; F_E; S_D)$ – "left-hand side", a logical expression U. To define whether it is true, inference engine assesses certain values in the current frames condition or in state-dependent data (S_D);

$R_{HS} = V(F_C; F_R; F_A; F_E; S_D)$ – "right-hand side", a set V of certain operations (actions) performed on frames or S_D if U in L_{HS} turned out to be true.

2.2   Guidelines Efficiency Evaluation Model

Due to considerable amount of existing design guidelines, we suggest they must not only be classified with tags and associated with design context, but also ranked by relative significance. Such an "efficiency" index may be evaluated based on the information obtained via explicit data collection, common for recommender systems, and in accordance with the following algorithm.

Let the number of guidelines (G) in the system equals N (thus $G = \{g_1, g_2, \ldots g_N\}$), and each is assigned efficiency index ($a_n \in [0;1]$, default is 0.5), while the number of stored interface designs (I) equals to R (thus $I = \{i_1, i_2, \ldots, i_R\}$). We can define a binary fuzzy relation GI:

$$GI = \{< g_n, i_r >, \mu_{GI}(< g_n, i_r >)\}, \tag{5}$$

where $<g_n, i_r>$ is a tuple of two elements, each of which is taken from the corresponding set: $g_n \in G$, $i_r \in I$. The membership function $\mu_{GI}(<g_n, i_r>)$ is continuous in the range [0;1] and semantically corresponds to the extent (generally evaluated by an expert) to which interface design i_r conforms to guideline g_n.

The system also stores several possible interface quality metrics (Q), whose total number is M (thus $Q = \{q_1, q_2, \ldots, q_M\}$). Let us define another binary fuzzy relation, IQ:

$$IQ = \{< i_r, q_m >, \mu_{IQ}(< i_r, q_m >)\}, \tag{6}$$

where $<i_r, q_m>$ is a tuple of two elements, each of which is taken from the corresponding set: $i_r \in I$, $q_m \in Q$. The membership function $\mu_{IQ}(<i_r, q_m>)$ is continuous in the range [0;1] and corresponds to interface design i_r quality rating on metric q_m. The quality ratings may be obtained via various usability engineering methods: usability testing (e.g. success rate), survey (e.g. subjective satisfaction), etc. As $\mu_{IQ}(<i_r, q_m>) \in [0;1]$, in some cases normalization of quality ratings would be required.

Then, based on the above, we need to define a fuzzy relation GQ meaning guideline g_n effect on enhancing quality metric q_m. We can do that via GI and IQ fuzzy relations composition with product-averaging, so that membership function will be the following:

$$\mu_{GQ}(< g_n, q_m >) = \underset{i \in I}{average}\{\mu_{GI}(< g_n, i_r >) * \mu_{IQ}(< i_r, q_m >)\}. \tag{7}$$

Considering so far the relative importance of all M quality metrics being equal, we can finally calculate efficiency index for each of N guidelines:

$$a_n = \underset{q \in Q}{average}\{\mu_{GQ}(< g_n, q_m >)\}, n = \overline{1, N}. \tag{8}$$

The values a_n may be then normalized so that their mean $\overline{a} = 0.5$.

3   The Recommender System Implementation

3.1   Development Tools

Among existing ontology editors, we chose Protégé-Frames created within Stanford University (http://protege.stanford.edu), as it fully supports frame-based ontologies in the form that we proposed in (3). The developed ontology was imported into CLIPS (C Language Integrated Production System, http://clipsrules.sourceforge.net) that allows efficient construction of expert systems and incorporates capabilities for both logic and procedural programming, supplemented by object-oriented paradigm (OOP). The prototype of the system was made available for online access at http://clips.vgroup.su.

In our case, concepts (F_C) of the imported frame-based ontology corresponded to OOP classes and frames-instances (F_E) – to objects, while the slots (F_A) and relationships (F_R) were reflected as data fields and their types. The I_F component is based on CLIPS rules, whereas state-dependent information, such as context of a particular project, is represented with unordered facts.

3.2   The Frame Ontology and Knowledge Base

The ontology design process combined top-down and bottom-up approaches and was carried out in several iterations. The current version of the system incorporated more than 150 frames-concepts, with such classes as *Design property, HCI engineering task, HCI knowledge, Interface design, Interface element, Requirement, Target user, User attribute, Website, Website element*, etc. at the top level.

Fig. 2 outlines how the HCI knowledge structure was represented in the ontology, showing related classes, their attributes and relations. So, *Guideline* may be related to its *Source*, justification (*Law, Principle, Finding*) and assigned classification tags.

To extend the ontology into KB, we added more than 300 instances representing domain knowledge, them mostly being design guidelines that were collected from various sources ([12], [13], [14], etc.) by human experts. Assigning domain concepts (tags) to guidelines was also carried out by experts, although in the future versions of the system this process may be supplemented by automated information extraction as well.

3.3   The Application of the Recommender System

One of the practical tasks the system was applied for was the design of the official website for the People's Faculty of Novosibirsk State Technical University (NSTU) – a department that provides "computer literacy" courses for senior citizens. The specified target user attributes were reflected in the system, which then mined the project requirements for ontology terms ($T_{DEF} \cup T_R$) and produced a set of concepts describing the design context that was matched against web design-related guidelines stored in the KB. The output of the system was the ordered set of about 100 guidelines and the interface wireframe, which were used by the recommender system user to create the interface design prototype. The final version of the People's Faculty official website, improved based on results of the experiment described below, was open at http://nf.assoc.nstu.ru.

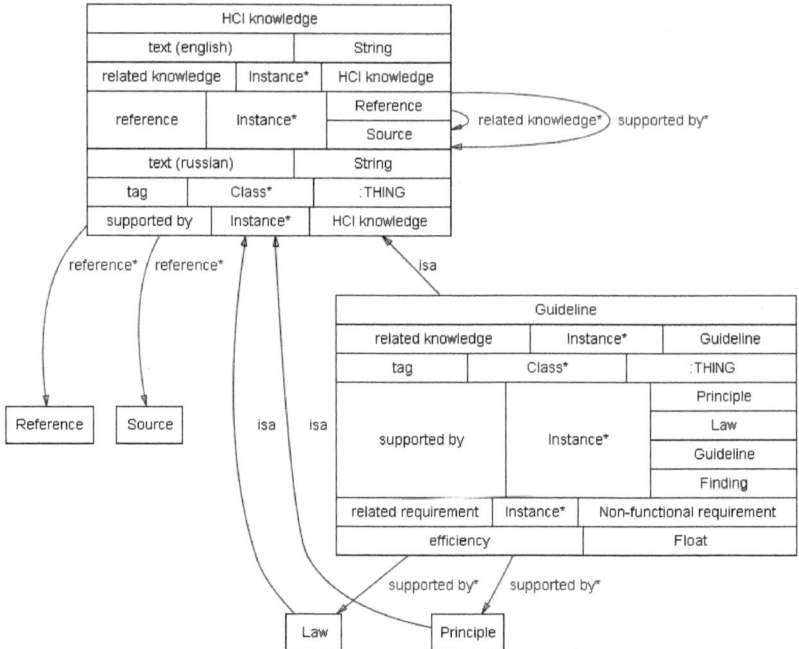

Fig. 2. HCI knowledge representation in the frame-based ontology

4 Experiment

4.1 The Experiment Description

Based on the output provided by the recommender system, designer developed the first version of the website interface, which was used in the experimental investigation together with 5 other existing websites. The goal of the experiment was twofold: a) to assess the quality of the recommender system-aided design for the target users and b) to explore if the proposed guideline efficiency evaluation model produces eloquent results. As the target user group for the People's Faculty website was senior citizens, the control group websites (#1, #2, #3, and #4) were selected from similar third-party websites also having seniors as one of target user groups. Website #5 was also a functioning one, created about 6 months before the experiment with the participation of a usability specialist affiliated with the authors of the current paper.

As for guidelines, 32 of them were elected for the experiment, on the basis of being relevant for all the 6 websites. They were divided into two types: 15 of them, which had associated tag *Elder user* in the KB, were assigned "for seniors" type (*Sen*), while all the others became of "general" type (*Yng*). Accordingly, younger participants were also employed in the experiment, so that the guidelines efficiency could be evaluated for the two subject groups independently, based on their corresponding quality metrics. The hypothesis is that guidelines of *Sen* type should have higher efficiency for senior group, while *Yng* type guidelines will have higher evaluations for younger one.

Subjects. Senior subjects were recruited among the recent graduates of computer literacy courses for seniors provided by the People's Faculty. The sampling was not random, as higher priority was given to graduates with more intense online experience, which was deemed necessary to better simulate the current and future senior population online, or even the alumni themselves in a few months from the graduation date. In total, there were 11 senior subjects (2 male, 9 female) in the experiment, whose age ranged from 58 to 71 years (mean = 62.5, SD = 4.1).

Also, there were 13 younger subjects (4 male, 9 female), recruited among undergraduate students of NSTU's Business Faculty, aged from 17 to 19 years (mean = 17.8, SD = .69). All the participants took part in the experiment voluntary and provided informed consent after reading through the tasks and learning the instructions.

Procedure. The experimental design and settings were quite typical for a user testing session, and there were in total 12 specially developed tasks, for a planned experimental session duration of 90 minutes. For each of the attempted tasks, success rate was measured by the instructor, and 0 was assigned for completely failed tasks, 0.3 – for tasks involving major errors possibly requiring support from the instructor, 0.7 – for tasks involving minor errors possibly requiring encouragement from the instructor, and 1 – for successfully completed tasks (similar approach was proposed by J. Nielsen in [15]). After completing all the tasks with all the 6 websites, the subjects were also asked to evaluate their overall impression of the websites by ranking them on a scale from 1 (worst) to 5 (best).

To evaluate guidelines efficiency separately for the two subject groups, their respective subjective impressions (normalized) were used as quality ratings $\mu_{IQ}(<i_r, q_m>)$. Success rate quality metric was not used in guideline efficiency calculation, as, expectedly, its values obtained in the experiment were stably higher for younger participants. The values for the membership function $\mu_{GI}(<g_n, i_r>)$, i.e. the degrees of interfaces correspondence to guidelines, were provided by human experts.

Thus, independent variables in the experiment were website group (third-party website, expert designer's website or website developed with recommender system), subject group, and guideline type. Dependent variables were task success rates, user subjective impressions of the websites, and evaluated guidelines efficiencies.

4.2   The Experiment Results

In total, 262 tasks were attempted by the participants and the overall mean success rate was 63.4% for the target group of senior participants. Their mean success rate for the control group of websites (#1, #2, #3 and #4) was 40.8%, while for the website developed with the system (#6) it ran up to 85.9%, however for the website developed with a usability specialist (#5) the success rate was even higher, 86.4%. Table 1 provides information on the experimental websites, together with respective success rates and subjective impressions (SI).

Table 1. Experimental websites descriptions and obtained quality metrics values

Website ID	Website description	Success rate seniors	SI seniors, Mean (SD)	SI younger, Mean (SD)
#1	http://pensionerki.ru A forum for pensioners.	23.2%	3.86 (.69)	3.18 (1.25)
#2	http://npfraiffeisen.ru A non-state pension fund.	20.0%	3.29 (1.25)	4.00 (.77)
#3	http://euro-kurses.ru A business education center.	82.2%	4.29 (.76)	4.30 (.82)
#4	http://moscow.apteka.ru An online medical shop.	52.0%	4.20 (.45)	3.64 (.81)
#5	http://vgroup.ru A web development company.	86.4%	4.33 (.52)	3.64 (1.43)
#6	http://nf.assoc.nstu.ru The People's Faculty.	85.9%	4.50 (.84)	4.50 (.71)

4.3   The Guidelines Efficiency Evaluation

Efficiencies for the 32 guidelines employed in the experiment were evaluated according to the proposed model, for the two groups of subjects independently (Table 2).

Table 2. Normalized guidelines efficiencies for younger (*Yng*) and senior (*Sen*) subjects

	Guideline		Efficiency	
N	Text	Type	*Yng*	*Sen*
1	Adequately large font size, possibly with option to increase size	*Sen*	**.278**	.276
2	Uncondensed text and without long fragments in italic style	*Sen*	.517	**.519**
3	High-contrast for reading text (dark on white or beige background w/out pattern)	*Yng*	**.604**	.585
4	Senior-friendly colors (avoiding blue, green and violet tones)	*Sen*	**.500**	.492
5	Increased size for interface elements	*Sen*	.269	**.272**
6	Increased or non-existent time-outs (e.g. when filling web-forms)	*Sen*	.737	**.744**
7	Avoiding dynamic/moving elements in interface, especially in navigation	*Sen*	.722	**.730**
8	Most of webpage text visible without scrolling	*Sen*	.433	**.449**
9	Pop-up windows avoided or at least not implying scrolling	*Yng*	**.595**	.585
10	Search of adequate quality and with errors correction	*Yng*	.341	**.346**
11	Minimum amount of data input and strict formats, max use of default values	*Yng*	**.551**	.544
12	Concise web-pages with a small number of topics covered on one page	*Sen*	.513	**.515**
13	"Flat", rather than "deep", website hierarchy	*Sen*	.638	**.649**
14	Avoiding ads, especially animated or interface-like banners	*Yng*	**.599**	.597
15	Highlighted hyperlinks and consistency of their style throughout the website	*Yng*	**.419**	.408
16	Visited hyperlinks change color, preferably to more "worn-out"	*Yng*	.217	**.232**
17	Explained hyperlinks (in text or with titles), rather than "click here"	*Yng*	.247	**.251**
18	Dedicated explanation on how to use the website (e.g. Help chapter)	*Sen*	.353	**.368**
19	Consistent placement (layout) of interface elements on web pages	*Yng*	**.558**	.553
20	Simpler and familiar navigation interface (e.g. with tabs)	*Sen*	**.452**	.451
21	Avoiding multimedia materials or at least providing alternate text	*Sen*	.699	**.716**
22	Providing alternate text for all images used as interface elements	*Sen*	.613	**.615**
23	All the website content in HTML (no .pdf, .doc or .zip files)	*Sen*	.589	**.610**
24	Avoiding hyperlinks leading to the same page (anchors)	*Yng*	**.605**	.583
25	More "personal" address to website visitor in texts	*Yng*	**.421**	.409
26	Classical design (clear, orderly, pleasant, etc.) rather than "expressive"	*Sen*	.656	**.668**
27	Saturated (with less grayish) primary and secondary colors in design	*Yng*	**.473**	.459
28	Non-radical difference in primary and secondary colors hues	*Yng*	**.379**	.376
29	Trust-enhancing details about company, certifications, awards, etc.	*Yng*	**.449**	.439
30	Avoiding "marketese" text style ("new", "best", "unique", etc.)	*Yng*	.545	**.552**
31	Adequate quality of text typography, no misspellings or misprints	*Yng*	.704	**.705**
32	High-quality and high-resolution photos	*Yng*	**.324**	.301
	Mean:		.500	.500

Further, for each guideline we calculated E_d, the difference between *Yng* and *Sen* efficiencies, – it is negative if guideline is more significant in interface design for senior users rather than younger ones and positive otherwise. Employed ANOVA test showed highly significant effect of guideline type on E_d ($F_{1,30}$=15.2; p=.001), with overall means of -.007 and .0062 for senior and younger subject groups respectively.

5   Conclusion

The recommender system was developed to support interface design for modern diverse applications and websites via providing ordered set of context-dependent guidelines and interface wireframe. The system is founded upon knowledge base storing human-readable guidelines, which is built from ontology combining domain model and thesaurus used for information extraction from software project requirements.

One of the major tasks when developing the system was organization of design guidelines, which currently come from multitude of sources and are poorly integrated. The employment of knowledge representation model incorporating frame ontology with metadata allowed classification of guidelines via assigning domain concepts to them and their subsequent intersection with similar tags describing the design context. The frame ontology model also could naturally support data provenance, in particularly guidelines relations to their origin and justification in other layers of HCI knowledge (laws, principles or findings). Further, fuzzy relations-based algorithm for guideline efficiency index evaluation was proposed, which can also contribute to automated identification of duplicating guidelines in the knowledge base or providing credible recommendation in case of contradictions.

The system was built as a web application, with the core programmed in CLIPS language and the ontology developed in Protégé-Frames editor, and the intermediate online version was made available at http://clips.vgroup.su. About 150 guidelines were extracted from various sources and saved into the knowledge base of the system, which allowed its application in a practical project – the development of the official website for the People's Faculty of NSTU (http://nf.assoc.nstu.ru).

The results of the recommender system use suggest its applicability to support interface design activities, mostly to save specialists' time and avoid serious drawbacks in design. Such objective interface quality metric as success rate was 85.9% for the developed website, compared to only 40.8% for the control group. Website created with a usability specialist, however, predictably achieved even higher value of 86.4%.

Further, subjective impressions of the websites from the two subject groups, senior and younger users, were used to evaluate the guidelines efficiencies according to the proposed algorithm. The results of these values statistic analysis suggest that guidelines efficiencies were significantly higher for their relevant target user groups. The findings confirm the applicability of the recommender system and of the proposed solutions for guidelines arrangement, which could be probably also considered for complex data integration in other domains.

References

1. Bygstad, B., Ghinea, G., Brevik, E.: Software development methods and usability: Perspectives from a survey in the software industry in Norway. Interacting with Computers 20(3), 375–385 (2008)
2. Borchers, J.O.: A pattern approach to interaction design. AI & Soc. 15(4), 359–376 (2001)
3. Koukouletsos, K., Khazaei, B., Dearden, A., Ozcan, M.: Creativity and HCI: From Experience to Design in Education. IFIP, vol. 289, pp. 159–174. Springer, Boston (2009)
4. Vanderdonckt, J.: Development milestones towards a tool for working with guidelines. Interacting with Computers 12, 81–118 (1999)
5. Henninger, S., Ashokkumar, P.: An Ontology-Based Infrastructure for Usability Design Patterns. In: Proc. Semantic Web Enabled Software Engineering (SWESE), Galway, Ireland, pp. 41–55 (2005)
6. Furtado, E., Furtado, V., Soares, S.K., Vanderdonckt, J., Limbourg, Q.: KnowiXML: A Knowledge-Based System Generating Multiple Abstract User Interfaces in UsiXML. In: Proc. of 3rd Int. Workshop on Task Models and Diagrams for User Interface Design, TAMODIA 2004, pp. 121–128. ACM Press, New York (2004)
7. Chevalier, A., Fouquereau, N., Vanderdonckt, J.: The Influence of a Knowledge-Based System on Designers' Cognitive Activities: a study involving Professional Web Designers. Behaviour & Information Technology 28(1), 45–62 (2009)
8. Dearden, A., Finlay, J.: Pattern languages in HCI: A critical review. Human-Computer Interaction 21(1), 49–102 (2006)
9. Guarino, N.: Formal Ontology in Information Systems. In: Proceedings of FOIS 1998, Trento, Italy, pp. 3–15. IOS Press, Amsterdam (1998)
10. Zagorulko, Y.A., Borovikova, O.I.: Models and methods for building information systems based on ontologies (in Russian). In: Marchuk, A.G. (ed.) Sistemnaya Informatika 2011, pp. 175–207. SO RAN, Novosibirsk (2011)
11. Zagorulko, Y.A., Sidorova, E.A.: Document analysis technology in information systems for supporting research and commercial activities. Journal Optoelectronics, Instrumentation and Data Processing 45(6), 520–525 (2009)
12. Coyne, K.P., Nielsen, J.: Web usability for senior citizens. The Nielsen Norman Group, Fremont (2002)
13. Nielsen, J., Snyder, C., Molich, R., Farrell, S.: E-Commerce User Experience. Nielsen Norman Group (2001)
14. U.S. Department of Health and Human Services. Research-based web design & usability guidelines, http://www.usability.gov/pdfs/chap.html
15. Nielsen, J.: Success Rate: The Simplest Usability Metric. Alertbox, February 18 (2001), http://www.useit.com/alertbox/20010218.html

Assessing Information Quality by Six Sigma Method

Sang Hyun Lee and Abrar Haider

School of Computer and Information Science,
University of South Australia,
Australia
leesy116@mymail.unisa.edu.au
abrar.haider@unisa.edu.au

Abstract. Information is the most critical resource of any organization. However, its value to the organization is dependent on its quality. In order to manage information quality, the foremost requirement is the ability of the organization to evaluate the quality of information held in their systems and the measures in place to ensure that only quality information is captured, processed, and contained with their information systems. However, obtaining accurate measurement and cost-effective assessments of information quality has been prevented by the complexities of information systems and the subjective nature of information quality. This research introduces a new approach to information quality measurement and employs six-sigma approach to information quality assessment. The fundamental contribution of this research is the focuses on continuous improvement of information quality by a systematic assessment of multiple information quality dimensions.

Keywords: Information quality, Information quality assessment, six-sigma.

1   Introduction

Contemporary business paradigm is information intensive. With the increase in information, associated issues like quality, security, and usefulness of information are also gaining attention. Information quality, however, has become the most strategic item on the agenda of quality management, decision support, and business intelligence. As a result, research and industry practice has seen scores of different information and data quality management frameworks. Despite of the large number, these frameworks only work for specific application context [1]. Additionally, most of these frameworks are ad-hoc, less intuitive and incomplete, and cannot produce robust and systematic measurement models [2]. Nevertheless, in order to manage information quality, the first requirement is to evaluate the quality of information held in their systems. However, obtaining accurate measurement and cost-effective assessments of information quality has been extremely difficult due to the complexities of information systems and the subjective nature of information quality [3]. To redress these challenges, it is important to take a product perspective of information, where information systems are regarded as information manufacturers [4]. Taking a product

H. Yu et al. (Eds.): DASFAA Workshops 2012, LNCS 7240, pp. 323–334, 2012.

perspective helps in understanding information stakeholder's information needs; managing information as the product of a well-defined production process (information processing in an information system); managing information as a product with a lifecycle; and monitoring information on a regular basis for continuous improvement in information quality.

Six-sigma is a defect reduction mechanism that helps in reducing variation in the output and thus leading to yield improvement. It has evolved from Total Quality Management and integrates tools and methodologies, such as Cause-and-Effect Diagram, Poka-Yoke device (mistake-proof mechanism to prevent obvious errors), Kaizen (continuous improvement), Lean Manufacturing (reducing and preventing wastes), and ISO 9000 (process system approach) [5]. Six-sigma when applied to the product perspective of information provides a comprehensive view of existing information quality as well as indicators the areas for improvement in the quality of information. The significant contributions of using this approach are the correlation of information quality dimensions (i.e. impact of a quality dimension on other dimensions), objective definitions of safe limits for quality dimensions, monitoring of information quality at various levels, and most importantly the ability to relate the 'fitness of use' perspective to each stakeholder and system.

Fitness for use for an information product will be different for different users at different levels. Even though information users and top management in organizations use same information, their required information quality specifications will be different because of the different purposes for which they use information. For example, sigma level specifications for an information system controlling the finished products on an assembly or production line will certainly not fit the sigma level for an information system (using the same information) at management level. As an example timeliness for the information user at production line may have an interval between 1 and 15 minutes, whereas for the production manager timelines may have an accepted interval of 0 to 0.5 day. Now both the production management system (being used by the person working at the production line) and the finished product inventory system (being used by his/her manager) are taking inputs from production scheduling system, but they are using that information for different purposes. One is interested in how much needs to be manufactured and the other one is interested in knowing how much has been manufactured. In this case if the manager applies same specification as that of the production line user, the sigma level will decline. This is because quality is defined differently at different levels and different viewpoints. In other words, sigma level of IQ dimensions' specification will be different for different systems.

This paper starts with an extension of plan, do check, act cycle to information quality, which leads into an illustration of six-sigma based information quality management framework. The next section provides a detailed discussion of the framework and highlights how it could be applied to assess, monitor, and improve information quality. The paper concludes with recommendations for future work.

2   The PDCA Cycle and EIIA Cycle

For this research we apply the concept of Plan, DO, Check, Act (PDCA) Cycle to propose an IQ assessment that leads to continuous improvement of information resources. The original concept of the PDCA Cycle was developed by Walter Shewhart. This cycle is often taken up in quality management area for continuous improvement. The PDCA cycle emphasizes that improvements should be initialized from the planning of quality, be carried out on a small-scale first, (if successful) the results should lead to effective action, and further action be taken on the basis of the learning from the study. The PDCA, thus, embodies

- Plan stage, i.e. to establish objectives and processes necessary to deliver results in accordance with the expected output.
- Do stage, i.e. a small scale change or experimental test. In this way there is minimal disruption to routine activity while the new initiatives are being analyzed to see if they will work or not.
- Check stage, i.e. to measure the new processes and compare the results against the expected results. Ascertain and differences to determine if the plan is producing the desired outcomes. If the outcome is not successful it is necessary to return to the plan stage and start the cycle again. It is also worth considering in the check phase that the quality of outcomes is maintained in other key areas of self-management support.
- Act stage, i.e. to implement changes. If the results are positive, then it is time to Act and implement changes on a larger scale. These initiatives will now become part of the daily routine. Once the first problem has been solved, return to the plan stage to identify a new issue to be improved on. It may be necessary to involve other personnel who may be influenced by the change if they were not part of the initial planning stage.

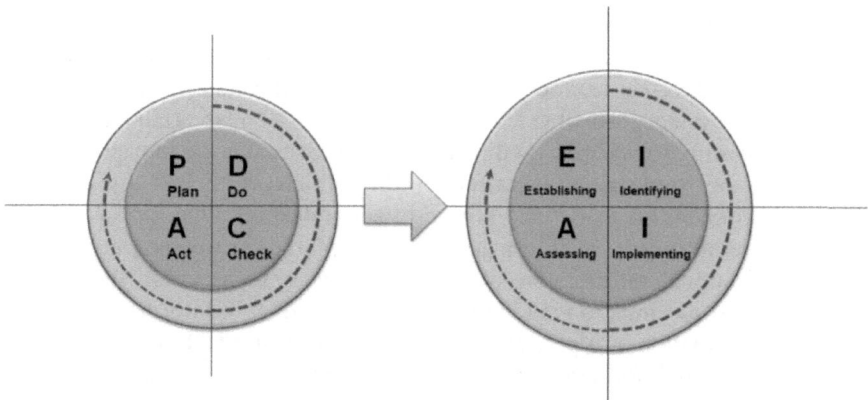

Fig. 1. PDCA Cycle and proposed EIIA Cycle

When the concept of PDCA is applied to IQ, the picture as illustrated in figure 1 emerges EIIA Cycle. This includes,

- Establishing stage, i.e. to establish IQ objectives and requirements from customers of information to ascertain their information related issues/problems/complaints, and how to transform the complaints in accordance with the business rules.
- Identifying stage, i.e. the pre-processing of the IQ measurement and analysis. As IQ cannot be 'objectively' measured due to the complexities of IQ dimensions; unlike the Do stage in PDCA cycle that aims at small scale change or experimental test, the 'identifying stage' reveals details, correlation, and impact of IQ dimensions on each other as well as overall IQ.
- Implementing stage, i.e. the measurement of IQ. In this stage, the identified IQ dimensions in identifying stage are applied. IQ dimensions are measured and analyzed by revealing critical factors and root causes of poor information. Just like the PDCA Cycle, if the outcome is not successful it is necessary to return to the establish stage and start the cycle again.
- Assessing stage, i.e. to compare current IQ measurement results with improved IQ measurement results though continuous IQ monitoring.

3   Six-Sigma Methodology for Information Quality Management

In recent years, numerous studies have attempted to find and explore ways of improving IQ. However, most of the research effort has focused on IQ management than IQ assessment, whereby IQ studies have endeavored to improve IQ by re-designing an existing IQ framework or correcting poor information under IQ management scope. Wang *et al.* [4] argue that IQ should be controlled first before an attempt is made to manage it. In other words, IQ control is prerequisite clause of IQ management. Without controlling IQ, problems relating to poor information cannot be resolved completely and sophisticated IQ management framework cannot be derived. In order to control IQ, it is important to assess quality if information residing in the organizational information systems. However, it is relatively easy to ascertain the quality of information relating to specific IQ dimensions, but assessing the impact of an IQ dimension on other dimensions is extremely difficult. Nevertheless, by understating how each IQ dimension works and how it affects other IQ dimension is critical for controlling overall quality of information being captured, processed, and maintained in the organizational information systems.   However, to ascertain the IQ control requirements it is essential to view information as a product of information systems.

Treating information as a product is to provide a well-defined product process and produce high quality information product rather than treating information solely as the by-product of business process execution. This is the key idea that this paper uses to apply six-sigma methodology to IQ area for quality improvement and IQ assessment. Six-sigma is an organized and systematic method for manufacturing process improvement that relies on statistical methods and the scientific method to make reductions in defect rates [6]. Applying six-sigma methodology to IQ assessment, therefore, provides benefits, such as defining critical factors to quality, measuring current quality (sigma) level, analyzing deficiencies in information and identifying the root causes of poor information, improving quality of information products, and controlling standardized IQ assessment framework. Table 1 illustrates the analogy between product manufacturing and information manufacturing to manage information as a product.

Table 1. Product vs. information manufacturing [7]

	Product Manufacturing	**Information Manufacturing**
Input	Raw Material	Raw Data
Process	Assembly Line	Information System
Output	Physical Products	Information Products

Using a product perspective of information, figure 2 presents a six-sigma based IQ assessment framework. Although this framework appears as assessing IQ, its fundamental core is based on continuous IQ improvement.

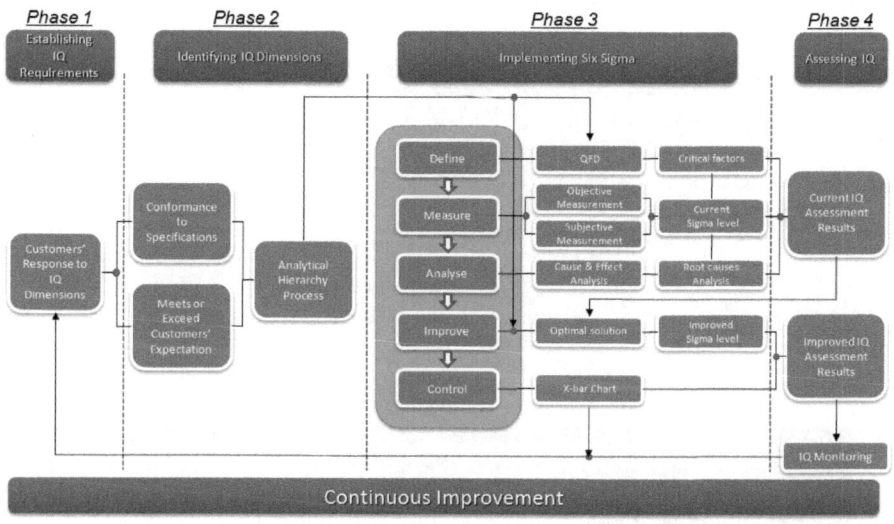

Fig. 2. IQ Assessment Framework Using EIIA Approach

3.1   Establishing IQ Requirements (Phase 1)

At the initial stage, the proposed framework seeks information stakeholders' requirements in terms of IQ. These requirements are then translated or mapped to the various IQ dimensions, such as accuracy, timeliness, completeness, accessibility, and security. The authors propose to survey information stakeholders to collect their responses to IQ.

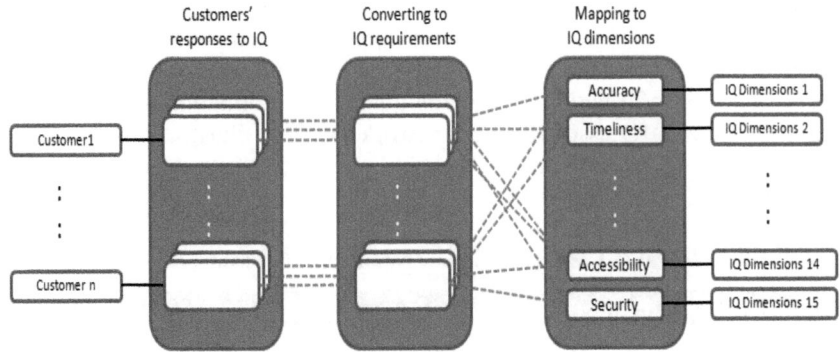

Fig. 3. Converting and Mapping customer's response to IQ dimensions

As shown in figure 3, the customers are the IQ stakeholders who use information for a variety of different purposes. Considering the fact that these stakeholders represent a variety of job functions, their interpretation of IQ and related dimensions is not standard. It is therefore important to view customer's response with regards to business rules, in order to reduce the level of abstraction.

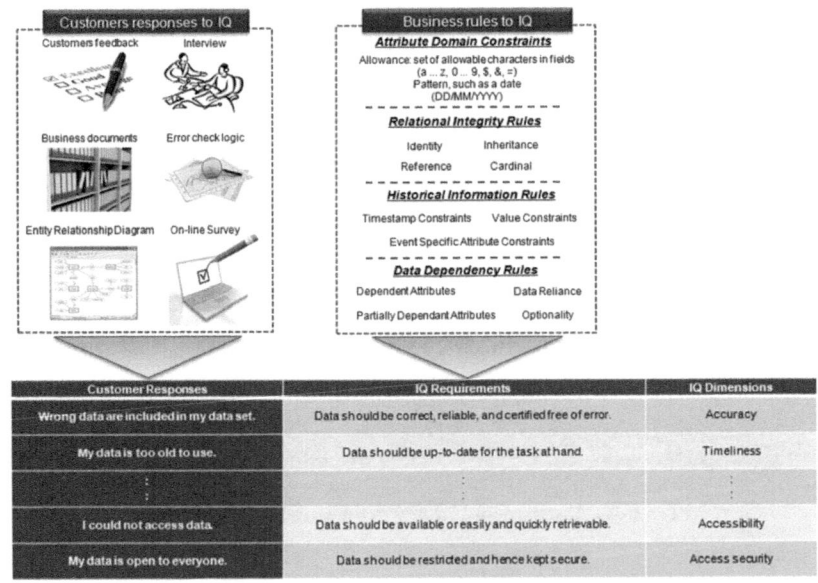

Fig. 4. How to represent customers' responses to IQ dimensions

Here, the business rules are categorized by attribute domain constraints, relational integrity, historical information, and data dependency rules to ensure quality of data. The IQ requirements of customers are, thus, linked to IQ dimensions. Figure 4 describes how customer responses are mapped to IQ dimensions.

3.2   Identifying IQ Dimensions (Phase 2)

At this stage, IQ dimensions from phase 1 are corresponded to the IQ hierarchy. The IQ dimensions are categorized by the 'conformance to specifications' i.e. product quality; and the 'meets or exceeds customers' expectation', i.e. service quality, respectively. Here, the product quality or information quality implies quality dimensions associated with information, and the service quality include dimensions that are related to the process of delivering right information at right time to right stakeholders. The end result of this exercise is a set of hierarchy or IQ dimensions as shown in table 2. Analytical Hierarchy Process (AHP) is utilized to correlate these dimensions and assign weights of relative importance. AHP is a hierarchical representation of a system assigning weights to a group of elements by a pair-wise comparison [8-9]. The pair-wise comparisons operate by comparing two elements at one time regarding their relative importance throughout the whole hierarchy. Therefore, it helps to capture the importance of desired measurement objects in comparison to other objects in the same hierarchy. The assigned weights to IQ dimensions are applied to quality function deployment (QFD) by providing the weights of importance as a scale of importance.

Table 2. Hierarchy of IQ dimensions (Adopted from [10])

	Categories	Quality Perspective	Assessments Items
Information Quality Dimensions for Assessments	Conformance to Specifications	Product Quality	Free of Error
			Concise
			Completeness
			Consistent Representation
		Service Quality	Timeliness
			Security
	Meets or Exceeds Customer's Expectations	Product Quality	Appropriate Amount
			Relevancy
			Ease of Understanding
			Interpretability
			Objectivity
		Service Quality	Believability
			Accessibility
			Easy of Operation
			Reputation

3.3   Implementing Six-Sigma (Phase 3)

This phase applies six-sigma methodology to IQ dimensions, based on DMAIC (Define, Measure, Analyze, Improve, and Control). Table 3 shows each perspective, procedure, and expected output.

Table 3. Six Sigma perspectives to IQ perspectives

Six-sigma Phases	Six-sigma Perspective	IQ Perspective	Procedure	Expected Output
Define	Defining the process and customers' satisfactions.	Representing customers' requirements (Specification) to IQ and Identifying IQ problems related IQ assessment.	Mapping and representing customer's satisfactions to IQ dimensions and requirements.	Failure Mode & Effects Analysis QFD (Quality Function Deployment).
Measure	Collecting and comparing data to determine problems.	Identifying criteria, systems, scales and scope for IQ measurement. Identifying data sampling method.	Implementing measurement of each IQ dimensions and calculating sigma level.	Determining current sigma level (Performance).
Analyze	Analyze the causes of defects.	Identifying poor information and root cause of poor Information.	Critical factor analysis based on measurement results. Applying correlations matrix.	Root causes of poor information Fish-born diagram/Logic tree.
Improve	Eliminating variations and creating alternative process.	IQ improvement by eliminating root cause of poor information.	Identify and define specific process improvements for information system.	Assessment results IQ assessment framework.
Control	Monitoring and controlling the improved process.	Standardize IQ.	Interpret and report information quality. Document improvement.	X-bar Chart for IQ monitoring.

- Define Phase

The define phase consists of three stages: Process, Scope, and Requirements. At process stage, overall structure of information flow is drawn to provide a top down view of IQ from a business perspective. In the scope stage, the scope of IQ from an information system perspective is defined to profile IQ dimensions and to identify IQ problems related IQ assessment. In the requirements stage, the specifications of each IQ dimension and IQ rule are defined to meet the customers' requirements utilizing the results of the phase 2 by creating the QFD to identify the correlation of each IQ dimension. In order to implement six-sigma into IQ assessment, the specifications will be different according to different users at different levels. This is because quality is defined differently at different levels and with different viewpoints. Therefore, customized IQ specifications based on IQ dimensions as assessment criteria must be established in this phase. Each IQ dimension criteria can be utilized in the assessing IQ phase as inspection list. Simultaneously, definition of CTQ (Critical to Quality) must be conducted at the beginning of this phase. CTQs must be interpreted from qualitative customers' requirements and be measured in the measure phase.

- Measure Phase

The measure phase consists of two stages, i.e. information collection and information measurement. In the information collecting stage, identifying criteria, measurement systems, scales and sampling methods are considered. Once the information collecting stage is complete, the information measurement calculates current sigma level with the specification of each IQ dimension and IQ rules. In the information measurements stage, the method of IQ measurement can be categorized to objective and subjective measurement.

- Analyze Phase

The analyze phase consists of two stages: Identifying poor information and Analysing root cause. In the identifying poor information stage, the deficiencies of information products are revealed according to the results of the measure phase and tracing the mapping between the CTQs and IQ rules. In the analysing root cause stage, the cause and effect analysis is utilized by generating comprehensive lists of possible causes to discover the reason for a particular effect and understand of how information products may become deficient in information systems. In this stage, a cause and effect diagram is designed based on the results of define and the measure phases to perform root causes analysis. This root cause analysis focuses on specific problems by resolving into basic elements of problems.

- Improve Phase

The improve phase consists of two stages, i.e. eliminating root causes and improving sigma level. The main objective of this phase is to identify an improvement of information systems by increasing quality of information products. In the eliminating root causes stage, determining of an optimal solution and finding the optimal trade-off values of IQ dimensions for IQ improvement is designed by eliminating the root causes which are discovered in the analyse phase. In the improving sigma level stage, all the results from define, measure, and analyse phases are integrated to lead the improved sigma level for IQ improvement.

• Control Phase

The control phase consists of two stages, i.e. controlling and monitoring. The main objective of this phase is to maintain high quality of information. In the controlling stage, representing IQ assessment results for information system, standardizing the IQ assessment framework are conducted, and documents are generated. In the monitoring stage, an X-bar chart representing each IQ dimensions with upper and lower control level and inspection lists of each IQ dimension are designed. The X-bar chart is a control chart used for monitoring information by collecting sample at regular intervals [7]. Each IQ dimension of sampled information in information system is monitored by using the X-bar chart at regular intervals to prevent production of poor information and to ensure the high quality of information.

3.4   Assessing IQ (Phase 4)

Based on phase 3 (Implementing Six Sigma), current IQ and improved IQ assessment results are compared to evaluate IQ assessment results. In order to ensure improved IQ on continuous basis, the X-bar chart derived in the control phase of the phase 3 is applied. By using the X-bar chart, if a certain dimension exceeds the specified limits, then the X-bar chart would raise alarm about that dimension.

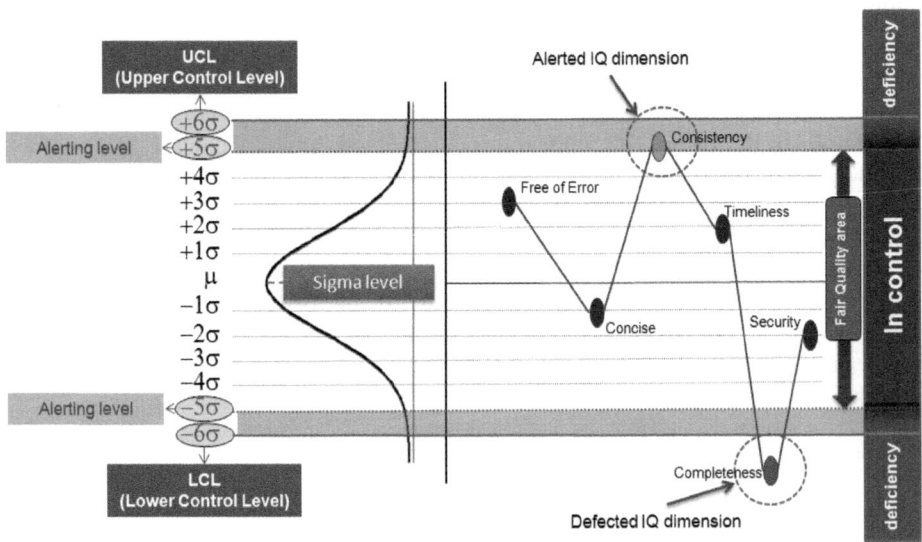

Fig. 5. X-bar chart for IQ monitoring

Figure 5 shows an example of IQ dimensions' monitoring. In this case, "Consistency" dimension is placed at the alert level. The consistency dimension can be described as that the information is always presented in the same format and is compatible with already existing information. Therefore, a consistent format of information based on business rules is the critical index for its assessment and it can be assessed by an objective method, i.e. software tools or a subjective method, i.e. survey from information users. In this example we took the subjective method for straightforward explanation. Suppose the surveyed information users (units) using a 10-question (average opportunities per unit) survey as to the quality level of consistency dimension, and 20% information users answered a total of 30 questions regarding consistency of information as unacceptable (defects) (i.e., less than 3 in the 9 scale). In this case, the DPMO (defects per million opportunities) is 30,000 and its corresponding sigma value is roughly 3.3 (see for reference of six-sigma conversion table and formula, [11]). According to the different tolerance level of each IQ dimension in an organization, the signal value can be converted based on the upper or lower control (specification level).

This is because quality is defined differently at different levels and different viewpoints. Hence, the tolerance of sigma level needs to be carefully designed while identifying IQ dimensions.

4   Conclusion

This paper has attempted to sketch out a framework for IQ assessment based on six-sigma methodology. The purpose of this paper has been to apply EIIA cycle for information quality improvement. This paper has applied the product perspective to information and has thus used six-sigma to develop a proactive IQ assessment, monitoring, and management framework. By applying this framework, business organization can define critical factors for quality of information; measure current levels of the quality of information held in their information systems; assess deficiencies in information quality; identify the root causes of poor information; and control and manage IQ through continuous assessment of information acquired, captured, and processed by their information systems. There are, however, a number of problems that remain to be explored. Since the proposed framework is still at conceptual level, the next step is to evaluate it in a real practice to verify its effectiveness. In addition, as the tolerance level of each IQ dimensions varies, research about determining IQ dimensions' specification level also must be carried out for more specific IQ assessments.

References

1. Knight, S., Burn, J.: Developing a framework for assessing information quality on the World Wide Web. Informing Science: International Journal of an Emerging Transdiscipline 8, 159–172 (2005)

2. Stvilia, B., Gasser, L., Michael, B., Linda, C.: A framework for information quality assessment. Journal of the American Society for Information Science and Technology 58(12), 1720–1729 (2007)
3. Madnick, S.E., Wang, R.Y., Lee, Y.W., Zhu, H.: Overview and Framework for Data and Information Quality Research. Journal of Data and Information Quality (JDIQ) 1(1), 1–22 (2009)
4. Wang, R.Y., Lee, Y.W., Pipino, L.L., Strong, D.M.: Manage your information as a product. Sloan Management Review 39(4), 95–105 (1998)
5. Dedhia, N.S.: Six sigma basics. Total Quality Management & Business Excellence 16(5), 567–574 (2005)
6. Linderman, K., Schroeder, R.G., Zaheer, S., Choo, A.S.: Six-sigma: a goal-theoretic perspective. Journal of Operations Management 21(2), 193–203 (2003)
7. Wang, R.Y.: A product perspective on total data quality management. Communications of the ACM 41(2), 58–65 (1998)
8. Cheng, E.W.L., Li, H.: Analytic hierarchy process: an approach to determine measures for business performance. Measuring Business Excellence 5(3), 30–37 (2001)
9. Saaty, T.L.: Time dependent decision-making; dynamic priorities in the AHP/ANP: Generalizing from points to functions and from real to complex variables. Mathematical and Computer Modelling 46(7-8), 860–891 (2007)
10. Kahn, B.K., Strong, D.M.: Product and service performance model for information quality: an update. In: Proceedings of the Conference on Information Quality, pp. 102–115 (1998)
11. Gupta, P.: Six Sigma Performance Handbook: A Statistical Guide to Optimizing Results. McGraw-Hill, New York (2004)

Author Index